INTEGRATED PLANT DISEASE MANAGEMENT

INTEGRATED PLANT DISEASE MANAGEMENT

Editors

R.C. Sharma

and

J.N. Sharma

Department of Mycology and Plant Pathology

Dr. Y.S. Parmar University of Horticulture and Forestry

Nauni, Solan – 173 230 (HP)

SCIENTIFIC® PUBLISHERS

Published by:

SCIENTIFIC PUBLISHERS (INDIA)
5A, New Pali Road, P.O. Box 91
Jodhpur 342 001, India
E-mail: info@scientificpub.com
Website: www.scientificpubonline.com

Print: 2026

ISBN: 978-93-86102-56-0 (P/B
eISBN: 978-93-87307-82-7

Printed in India

FOREWORD

After India's attaining the self-sufficiency in food production, the emphasis is now on horticultural and forest crops keeping in view the perspective of the growers and consumers' preference to exploit the opportunities in the global market today. One of the most important considerations in the world today is, increasing the production of nutritious food and providing fibre, feed, fuel, timber, etc. Consequently, in recent years, public interest in horticultural and forest crops and naturally, occurring diseases has greatly increased. Emergence of recent epiphytotics, such as : Marssonina blotch of apple and Ganoderma root rot of shisham in north India, has become a matter of concern for all of us.

Accordingly, the Indian Phytopathological Society, North Zone Chapter had organized a symposium, "Challenging Problems in Horticultural and Forest Pathology" on November 14 - 15, 2003 at Dr. Y.S. Parmar University of Horticulture and Forestry, Solan (H.P.). The research papers presented therein have been compiled in this book covering diseases of important fruits, vegetables, ornamentals, cereals, pulses, oilseeds, forest, medicinal plants etc. It has useful and valuable informations pertaining to integrated disease management practices of important horticultural, agricultural and forestry crops.

The editors, Dr. R.C. Sharma and Dr. J.N. Sharma, deserve sincere appreciation for their dedicated efforts in bringing out this excellent publication. This book, I am sure, would benefit not only the researchers and students but also others interested in managing plant diseases.

S.M. Paul Khurana

Vice-Chancellor,
Rani Durgavati University
Jabalpur (M.P.) – 482 001

PREFACE

Man has been aptly recognized as the guest of plants. The losses caused by plant diseases have affected the mankind all through different eras of civilization. The advancement of mankind depends upon the developments in the science and technology. But the human greed for better food, feed, fibre, fuel, timber etc. has resulted in the over-exploitation of natural resources – leading to imbalance in the ecosystem. Excessive monoculture and faulty practices have resulted in the emergence of various plant diseases. Their destructive impact has led to the development of various strategies to protect the plants from them.

The concept of disease management does not aim at the complete control of diseases but, to manage them at the economic threshold level. Integrated disease management requires detailed knowledge, understanding and integration of all aspects of crop production including, information on economics, environment, pathogen biology, genetics, epidemiology etc. All these parameters and other input data direct the management decisions. Minimal use of pesticides is desirable from economic view point, resistance development, environment and health hazard considerations. It is in this context that, different inputs of integrated disease management, such as : cultural practices, sanitation, eradication, organic amendments, plant growth promoting microorganisms, resistant varieties, botanicals and biocontrol agents become important.

The book, "Integrated Plant Disease Management" is based on the research papers presented in the symposium, "Challenging Problems in Horticultural and Forest Pathology" held on November 14 - 15, 2003 at Dr. Y.S. Parmar University of Horticulture and Forestry, Nauni, Solan. The symposium was organised by North Zone Chapter of Indian Phytopathological Society in its sixteenth annual meeting. About 140 delegates attended this symposium and discussed various issues related to integrated disease management. This volume contains 55 chapters under different sections viz., fruit crops; vegetable crops; ornamental crops; cereals, pulses and oilseed crops; forest and medicinal plants; and mushrooms. This book provides an insight into the major pathological problems and integrated disease management in different crops, and horticultural crops in particular. The book will be useful to plant pathologists, horticulturists, microbiologists, breeders, nematologists, mushroom scientists and students.

On behalf of Indian Phytopathological Society (North Zone Chapter), we thank all the authors for their scientific contributions. We are thankful to Dr. S.S. Negi, Vice-Chancellor and other authorities of the university for allowing the symposium to be held in the campus and publication of the proceedings. Financial assistance received from Directorate of Horticulture, Shimla; HPMC, Shimla; State Bank of India, Zonal Office, Shimla; Indofil Chemicals Co., Mumbai and other organizations for the symposium is thankfully acknowledged. We express our sincere thanks to all the staff members of Department of Mycology and Plant Pathology for their encouragement and cooperation during the symposium. The editors wish to express their sincere gratitude to Dr. Anil Handa for his help and Sh. K.K. Sharma for his untiring and invaluable assistance in bringing out this book.

R.C. Sharma
J.N. Sharma

CONTENTS

FRUIT CROPS

VEGETABLE CROPS

ORNAMENTAL CROPS

CEREAL, PULSE AND OILSEED CROPS

FOREST AND MEDICINAL PLANTS

MUSHROOMS

FRUIT CROPS

Integrated Plant Disease Management (2005)
Eds. Sharma, R.C. and Sharma, J.N.
Scientific Publishers (India), Jodhpur, pp. 1-7

1

Studies on Epidemiology and Management of Marssonina Blotch, the cause of Premature Leaf fall in Apple

J.N. SHARMA, P.K. SHARMA and ANITA SHARMA

Department of Mycology and Plant Pathology, Dr. Y.S. Parmar University of Horticulture and Forestry, Nauni, Solan 173 230 (H.P.)

ABSTRACT

Marssonina blotch caused by *Marssonina coronaria* (Ell. and J.J. Davis) J.J.Davis (Syn. *Marssonina mali* (P. Henn.) Ito) is responsible for widespread defoliation in apple. First recorded in 1992 in some of the unsprayed orchards in Kotkhai and it assumed epiphytotic proportions by 1995, thereby affecting the apple plantations in all the major apple growing districts of Himachal Pradesh. It caused up to 50 per cent defoliation of apple trees from July to September, when fruits were still hanging on the defoliated branches and rendered the trees barren in the following years. This disease was found to be favoured by moderate temperature of 20-22°C and high relative humidity following rains in the summer months. The perfect state of the pathogen *Diplocarpon mali* was intercepted in over-wintering leaves, though less frequently. The pathogenic fungus was found to perennate in the overwintering leaf litter in the form of conidia, which showed up to 20 per cent viability in the next season. New crops of conidia were also produced in the existing fruiting bodies on the infected leaves left on the orchard floor and were able to cause disease under controlled conditions. Protective sprays of copper oxychloride (0.3%), zineb (0.3%), dodine (0.075%), carbendazim (0.05%), dithianon (0.05%), mancozeb flowable (0.35%), Companion (0.25%) and ziram (0.3%) exhibited complete protection of apple trees in the field.

INTRODUCTION

Marssonina blotch, caused by *Marssonina coronaria* (Ell. and J.J. Davis) J.J. Davis (Syn. *Marssonina mali* (P. Henn.) Ito) has become one of the foremost diseases of apple in the last few years. It is responsible for causing widespread defoliation of the apple trees in midseason thereby affecting the tree vigour and fruit bearing capacity of the trees adversely (Sharma and Gautam, 1997). The spots on the fruits also reduce their quality and marketability. First detected in some of the isolated orchards in Kotkhai in 1992, it became epidemic in 1995 affecting the apple orchards in the entire apple growing area of the state. Its incidence has also been reported from Uttaranchal and Jammu and Kashmir. First reported from Japanese orchards as early as 1907, there is a wide gap in the information regarding its survival and management. In the present paper, some light has been thrown on some of the epidemiological and management aspects.

MATERIALS AND METHODS

Disease prevalence

To record the prevalence and severity of the disease, periodical surveys of the apple growing areas located in Shimla district of Himachal Pradesh were conducted during the growing season. The incidence of apple leaf blotch under natural epiphytotic conditions was recorded as under:

$$\text{Disease incidence (\%)} = \frac{\text{Number of orchards infected}}{\text{Total number of orchards survyed}} \times 100$$

To calculate the per cent disease severity of Marssonina blotch on apple leaves, the plants were subjected to 0–8 scale according to degree of infection and each plant was given a numeral rating as follows:

Numerical rating	Description
0	No evidence of disease on leaves
1	Up to 10 per cent leaves showing spotting/yellowing due to Marssonina disease and no defoliation is apparent
2	Leaf fall initiated but less than 10 per cent, and 25 per cent of the remaining leaves showing disease symptoms
3	Up to 25 per cent of the leaves have fallen, and remaining 25 per cent showing leaf spotting and/or yellowing symptoms
4	Up to 33 per cent of leaves dropped, and more than 50 per cent showing disease symptoms
5	Up to 50 per cent leaf fall has taken place, and up to 25 per cent of the rest showing infection
6	Up to 75 per cent leaves have fallen, and the remaining leaves showing infection
7	Up to 90 per cent leaf fall has occurred, and the remaining leaves showing infection
8	More than 90 per cent leaves have fallen, and remaining leaves also showing disease symptoms

Per cent disease severity was calculated according to the following formula, given by McKinney (1923) as follows:

$$\text{Disease severity (\%)} = \frac{\text{Sum of all disease ratings}}{\text{Total number of ratings x Maximum grade}} \times 100$$

Survival of *Marssonina coronaria*

To study the survival of the fungus *M. coronaria* in the winter, severely infected apple leaves were collected from the orchards in autumn in the month of October. They were kept in three different conditions viz., (i) in a layer of sand (5 cm), (ii) in nylon net bags on a layer of sand (5 cm) and (iii) in nylon net bags on the orchard floor under natural conditions. Observations were recorded on the formation of perfect state and/or the production of new conidia after over-wintering from March onwards.

Studies were also conducted in order to ascertain the role of fallen leaves in the over-wintering of *M. coronaria*, the cause of leaf blotch of apple on the orchard floor. Infected leaves of three types viz., very old surface leaves, old decayed leaves and comparatively intact leaves were cut into 1-cm diameter discs containing fruiting bodies of fungus. Number of acervuli per leaf disc were examined and counted with the help of stereoscopic binocular microscope. The discs were then incubated at 20°C. Seven leaf discs of each type of leaves were kept in sterilized Petri plates containing sterilized moist blotting paper so as to maintain proper humidity. Observations were recorded for any increase in number of acervuli per leaf disc at 24 h intervals up to 168 h.

Disease management

A field trial was conducted to ascertain the bio-efficacy of different fungicides against Marssonina blotch and other associated diseases of apple at Sarahan during 2002 season with 11 fungicides viz., mancozeb (0.3%), copper oxychloride (0.3%), zineb (0.3%), dodine (0.075%), carbendazim (0.05%), dithianon (0.05%) mancozeb flowable (0.35%), Companion (0.25%), ziram (0.3%), hexaconazole (0.05%) and captan (0.3%). All of these fungicides were sprayed on three trees each starting from fruit set stage till harvest. Data were recorded on disease severity and per cent disease control was calculated. The data were subjected to the analysis of variance in a randomized block design.

RESULTS AND DISCUSSION

Disease prevalence

Incidence of Marssonina blotch was first observed in the orchards located at lower elevations in Kotkhai during the first fortnight of June in 2001 and 2003 seasons, whereas it appeared in the first week of August during 2002 probably due to the severe drought experienced in the latter case. On an average, 55.3 per cent orchards were found infected with 36.2 per cent disease severity in Shimla district of Himachal Pradesh, Rohroo being the most severely infected fruit belt followed by Jubbal and Sarahan during this period (Table 1). After its outbreak in 1995, this disease has become endemic in major apple growing areas of Himachal Pradesh and is being reported consistently from all the districts of this state (Sharma, 1999, 2001, 2003).

Table 1. Incidence of Marssonina blotch, the cause of premature leaf fall in apple in Shimla district of Himachal Pradesh

Location	Per cent disease							
	2001		2002		2003		Mean	
	1	2	1	2	1	2	1	2
Kotkhai	45.0	59.0	41.0	22.6	40.0	19.0	41.6	33.5
Jubbal	52.0	63.0	60.0	19.5	60.0	28.0	57.3	36.8
Rohroo	—	—	70.0	31.4	—	—	70.0	31.4
Sarahan	62.0	69.0	45.0	30.0	50.0	30.0	52.3	43.0
Mean	53.0	63.6	57.7	53.8	50.0	25.7	55.3	36.2

1 = Per Cent Orchards Infected; 2 = Per Cent Disease Severity

Symptomatology

Symptoms of Marssonina blotch occurred both on leaves as well as fruits. The disease started with the development of dark green circular patches on the upper leaf surface in June end giving rise to 5 to 10 mm diameter brown leaf spots, which became dark brown in the due course. The leaf spots occurred initially on older leaves in the lower portion of the tree canopy. As the disease progressed, black pinhead-like fruiting bodies (acervuli) developed on the upper leaf surface and when the lesions were numerous, they coalesced to form large dark brown patches (blotches) and the surrounding areas turned yellow. Numerous leaf infections also resulted in the loss of chlorophyll giving rise to mosaic type appearance on the leaves. The infected leaves dropped off the trees, producing the characteristic symptoms of premature leaf fall, the name by which the disease is commonly known in the state. The disease symptoms were also recorded on fruits. The lesions were initially circular and 4 to 5 mm, which later became oval, depressed and dark brown with age and almost black at harvest time. Small black acervuli were also visible on the fruit lesions. Similar disease symptoms due to *Marssonina coronaria* were also observed on apple leaves and fruits by earlier workers (Sharma and Verma, 1999; Sharma and Kaul, 2000; Sharma, 2001).

Isolation and identification

During the surveys, the infected leaves were collected from different locations for the isolation and identification of the pathogen. The asexual fruiting bodies (acervuli) produced by the causal fungus *Marssonina coronaria* on the infected leaves and fruits were subcuticular and measured 100 to 200 μm in diameter. The conidia were one-septate, constricted at the septum, hyaline, guttulate and measured 20-24 x 6.5-8.5 μm in size and were borne on short clavate conidiophores. The fungus was isolated from diseased leaves as well as fruits in pure culture on potato dextrose agar medium. Observations recorded after one month revealed that the fungal colonies were sub-globoid, 5-8 mm in diameter and 2-3 mm raised with somewhat wrinkled surface and were dark brown in colour. These fungal characteristics are in accordance with the description given by Sharma and Kaul (2000) who identified it as *Marssonina coronaria* (Ell. and J.J. Davis) J.J. Davis (IMI 375985 and 377423).

Survival of the pathogen

At the start of the experiment few conidia were observed on the leaves, showing scant germination at 20°C up to 72 h. The leaves were examined for the formation of perfect state and/or the production of new conidia after over-wintering from March onwards. There was formation of new crops of conidia in the acervuli of infected over-wintered leaves in all the treatments in March and following months. Conidia production was observed right from the first observation i.e. March and 25-35 x 10^3 conidia were produced per gram of over-wintered leaves on the orchard floor by the month of May in nylon net bags kept on sand and orchard floor (Table 2).

Table 2. Production and viability of conidia of *Marssonina coronaria* in over-wintered apple leaves

Treatment (Over-wintering of leaves)	Rotting of leaves (%)			Number of conidia produced (x10^3/g leaves)			Viability (% germination) of conidia		
	March	April	May	March	April	May	March	April	May
Orchard floor (in nylon net bags)	24	40	75	10	22	35	60	65	70
Sand (lose)	100	—	—	4	—	—	45	—	—
Sand (in nylon bags)	50	75	90	13	17	25	50	60	65

The newly formed conidia on the leaves kept in orchard floor showed significant germination up to 70 per cent in one per cent sucrose solution and conidia extracted from such over-wintered leaves when inoculated on potted apple plants were able to produce typical symptoms of Marssonina blotch in the months of April and May.

Data (Table 3) revealed that number of acervuli were significantly higher in comparatively intact old leaves (395.8) as compared to very old leaves lying on the orchard floor (91.08) and old decayed leaves (60.67) and both were statistically at par with each other. Time interval showed no significant difference in the formation of acervuli per leaf disc.

Table 3. Number of acervuli per leaf disc after different time intervals in infected leaves

Leaf condition	Number of acervuli/leaf disc after (hours)							
	24	48	72	96	120	144	168	Mean
Old leaves on orchard floor	90.0	90.0	91.43	91.43	91.57	91.57	91.57	91.08
Old decayed leaves	59.71	59.71	60.71	61.14	61.14	61.14	61.14	60.67
Comparatively intact leaf discs	395.1	395.1	395.6	396.0	396.0	396.4	396.4	395.8
Mean	181.6	181.6	182.6	182.9	182.9	183.0	183.0	

$CD_{0.05}$

Type of leaves = 81.14
Interval = 123.94
Interaction = 214.72

The over-wintered leaves were also collected from Kotkhai, Jubbal and Sarahan orchards from May to July months. Microscopic examination also revealed formation of fresh conidia in the orchards at the lower elevations particularly Anu (Jubbal) in May. The conidial formation in the over-wintered leaf litter was also observed in other locations of Kotkhai and Sarahan in later months. Although, ascospores produced in the over-wintered leaves have been reported as the major source of primary infection in the orchards (Harada *et al.,* 1974; Takahashi and Sawamura, 1990); yet the anamorphic state seems to play the prime role under our conditions. Cook (1979) reported the development of new acervuli in the original lesions of black spot disease on rose, which released the spore to cause primary infection.

Management

Out of 11 test fungicides, mancozeb (0.3%), copper oxychloride (0.3%), zineb (0.3%), dodine (0.075%), mancozeb flowable (0.35%), Companion (0.25%) and ziram provided 100 per cent disease control in a protective spray programme followed by dithianon (0.05%). Hexaconazole (0.05%) and captan (0.3%) failed to control the disease in the field (Table 4).

Hexaconazole, Companion and carbendazim also provided good control of powdery mildew, which has gained some importance in the last few years. All the test fungicides exhibited excellent control of sooty blotch and flyspeck diseases, which are responsible for lowering the fruit quality. Hence, the effective fungicides viz., mancozeb, carbendazim, dodine, dithianon etc. should be suitably exploited for integrated management of diseases in apple orchards. Chemical control of Marssonina blotch has also been suggested earlier (Takahashi and Sawamura, 1990; Sharma, 1999; 2001; 2003).

Table 4. Efficacy of different fungicides in a protective spray programme at Sarahan in 2002

Fungicide	Per cent disease incidence					
	Marssonina blotch		Powdery mildew	Sooty blotch	Fly speck	Fruit finish
	August	September				
Mancozeb (0.3%)	0.00	0.00	15.00	0.00	0.00	Very good
Copper oxychloride (0.3%)	0.00	0.00	18.00	0.00	0.00	Good
Zineb (0.3%)	0.00	0.00	23.00	0.00	0.00	Very good
Dodine (0.075%)	0.00	0.00	16.00	0.00	0.00	Good
Carbendazim (0.05%)	0.00	0.00	10.00	0.00	0.00	Very good
Dithianon (0.05%)	0.00	3.00	11.00	0.00	0.00	Very good
Mancozeb flowable (0.35%)	0.00	0.00	13.00	0.00	0.00	Very good
Companion (0.25%)	0.00	0.00	0.00	0.00	0.00	Very good
Ziram (0.3%)	0.00	0.00	18.00	0.00	0.00	Very good
Hexaconazole (0.05%)	3.00	43.00	0.00	23.00	19.00	Good
Captan (0.3%)	5.00	48.00	14.00	0.00	0.00	Very good
Control	17.00	75.00	23.00	40.00	37.00	Poor
$CD_{0.05}$	1.62	7.61	4.32	3.82	3.57	—

REFERENCES

Cook, R.T.A. (1979). Production of *Diplocarpon rosae* conidia on over-wintering rose leaves. *Trans. Br. Mycol. Soc.* 73: 354-357.

Harada, Y., Sawamura, K. and Konno, K. (1974). *Diplocarpon mali* sp. nov., the perfect state of apple blotch fungus *Marssonina coronaria*. *Ann. Phytopathol. Soc. Jpn.* 40: 412-418.

McKinney, H.H. (1923). Influence of soil moisture and temperature on infection of wheat seedlings by *Helminthosporium sativum*. *J. Agric. Res.* 26: 195-217.

Sharma, I.M. (2003). Influence of environmental factors on the development of premature defoliation caused by *Marssonina coronaria* in apple and its management. *J. Mycol. Plant Pathol.* 33: 89-95.

Sharma, J.N. (1999). Marssonina blotch causes premature leaf fall in apple - A new disease. *Indian Phytopathol.* 52: 101-102.

Sharma, J.N. (2001). Diagnosis and control of premature leaf fall in apple. *J. Mycol. Plant Pathol.* 31: 305-310.

Sharma, J.N. and Gautam, D.R. (1997). Studies on premature leaf fall in apple - A new problem. *Indian J. Plant Prot.* 25: 8-12.

Sharma, J.N. and Kaul, J.L. (2000). Marssonina blotch - A new disease hits apple crop in Himachal Pradesh. In: *Proceedings of International Conference on Integrated Plant Disease Management for Sustainable Agriculture.* New Delhi, pp. 775-776.

Sharma, J.N. and Verma, L.R. (1999). Premature leaf fall problem in apple - Diagnosis and control. In: *Diseases of Horticultural Crops: Fruits* (Eds. Verma L.R. and Sharma, R.C.). Indus Publishing Co., New Delhi, pp. 80-88.

Takahashi, S. and Sawamura, K. (1990). Marssonina blotch. In: *Compendium of Apple and Pear Diseases.* (Eds. Jones, A.L. and Aldwinckle, H.S.). The American Phytopathological Society, St. Paul (USA), pp. 33.

Integrated Plant Disease Management (2005)
Eds. Sharma, R.C. and Sharma, J.N.
Scientific Publishers (India), Jodhpur, pp. 9-17.

2

Studies on morphology and physiology of *Marssonina coronaria* causing premature leaf fall in apple

P. K. SHARMA and J.N. SHARMA

Department of Mycology and Plant Pathology, Dr.Y.S. Parmar University of Horticulture and Forestry, Nauni, Solan- 173 230 (H.P.)

ABSTRACT

Apple, which has become the most important fruit crop in Himachal Pradesh is attacked by a number of fungal diseases. After the scourge of scab, apple plantations in Himachal Pradesh were ravaged by a relatively less known fungal pathogen *Marssonina coronaria* (Ell. & J.J. Davis) J.J. Davis (Syn. *Marssonina mali* (P. Henn) Ito) by causing midseason defoliation. This fungus, a deuteromycete produced typical hyaline, bi-celled guttulate conidia on short clavate conidiophores in acervuli with average size of 20-24 x 6.5- 8.5 μm. There is a wide variability in its morphology with respect to the size and shape of the conidia collected from different geographical locations of the state. The fungus produced small brown to dark brown colonies, which were round, sub-globoid and somewhat wrinkled. As far as physiological attributes are concerned, this pathogen showed maximum growth on minimal salts agar medium after 35 days. Also, methionine and maltose supported its best growth as nitrogen and carbon sources, respectively. This pathogen grew best at 20°C temperature and pH 5.0.

INTRODUCTION

Apple (*Malus x domestica* Borkh.) is the most important fruit crop of the temperate regions of the world and has become the mainstay of the economy of hilly farmers in Himachal Pradesh. It is attacked by a number of fungal diseases. During the last 10 years, apple cultivation in Himachal Pradesh has suffered heavy losses due to sudden outbreak of Marssonina blotch caused by *Marssonina coronaria* (Ell. & J.J. Davis) J.J. Davis (Syn. *Marssonina mali* (P. Henn) Ito). It has threatened the apple cultivation by causing premature leaf fall (Sharma and Gautam, 1997; Sharma and Kaul, 2000; Sharma, 2001). Keeping in view the huge losses caused by the pathogen, studies on its morphology, pathogenicity and physiology were conducted and are discussed in this paper.

MATERIALS AND METHODS

Morphological studies

To study the variability in morphology of *M. coronaria,* infected leaves were collected from four different locations viz., Kotkhai, Kullu and Oachghat in Himachal Pradesh and Nogaon area of Uttaranchal. The leaves were soaked for 15 min in sterilized distilled water and conidia were scrapped with the help of camel hairbrush. One drop of the conidial suspension (50-60 conidia per microscopic field) was put on glass slide and length and breadth of the conidia was measured with the help of standardized ocular micrometer. This value was multiplied with calibration factor (3.64 μm) of the microscope to get the actual length and breadth of the conidia. The homogeneity of the variability was statistically tested by using Bartlett's test (Gupta and Kapoor, 1970).

Pathogenicity

Inoculum containing approximately 3 x 10^4 conidia/ml in sterilized distilled water was harvested from the fruiting bodies of the leaf blotch pathogen *Marssonina coronaria* and was applied with a hand atomizer on the upper surface of leaves of one year old potted plants. After inoculation, the plants were immediately covered with the polythene bags individually for 48 h. The seedlings sprayed with distilled water served as control. Observations were recorded on disease appearance, incidence and severity after inoculation.

Physiological studies

A series of experiments were conducted to record the effect of different solid and liquid substrates, temperature and pH on the vegetative growth of *Marssonina coronaria.* Different carbon and nitrogen sources replaced sucrose and yeast extract in the basal medium in a way to provide equal amount of carbon (except for cellulose and starch, used @ 2%) and nitrogen. Petri plates and flasks containing 25 ml media were inoculated aseptically with uniform size bits (2 mm) taken from the margins of an actively growing culture and incubated at 20 ± 1°C. Each treatment was replicated four times.

Data pertaining to the mean radial growth and dry mycelial weight were recorded after 35 days of incubation. Dry mycelial weight of the pathogen was recorded by filtering the fungal mycelium through previously dried and weighed Whatman No 1 filter papers after drying to a constant weight at 60°C for 3 days. The weight of the filter papers was deducted from the total weight for obtaining the actual weight of the mycelium.

RESULTS AND DISSCUSION

Isolation and identification

The asexual fruiting bodies (acervuli) produced by the causal fungus *Marssonina coronaria* (Ell. & J.J. Davis) J.J. Davis (Syn. *Marssonina mali* (P. Henn.) Ito) on the infected leaves and fruits were sub-cuticular and measured 100 to 200 μm in diameter. The conidia were one-septate, constricted at the septum, hyaline, guttulate and measured 20-24 x 6.5-8.5 μm in size and were borne on short clavate conidiophores. The fungus was isolated from diseased leaves as well as fruits in pure culture on potato dextrose agar medium. Observations recorded after one month revealed that the fungal colonies were sub-globoid, 5-8 mm in diameter and 2-3 mm raised with somewhat wrinkled surface and were dark brown in colour. These fungal characteristics are in accordance with the description given by Sharma and Kaul (2000) who identified it as *Marssonina coronaria* (Ell. & J.J. Davis) J.J. Davis (IMI 375985 and 377423). First reported from Japanese orchards in 1907, it was also studied in detail by Parmelee (1971) who has merged two species of *Marssonina* i.e. *M. coronaria* and *M. mali* and preferred to designate it as *Marssonina coronaria* (Syn. *M. mali*). However, Takahashi and Sawamura (1990) still preferred to name it as *M. mali*. The scanning electron micrographs of this pathogen were also prepared, which showed typical *Marssonina*-type macro and micro-conidia being produced on small clavate conidiophores.

Table 1. Pathogenicity test of *Marssonina coronaria* on Starking Delicious apple plants

Incubation period (Days)	Disease progress after appearance of initial symptoms	
	Disease incidence (%)	Disease severity (%)
12	7.10	6.67
22	34.84	25.24
32	75.82	54.48
42	92.00	87.75
Mean	52.44	43.54

Pathogenicity

Typical symptoms of Marssonina blotch were observed 12 days after pathogen inoculation on the older leaves of the plants and younger ones were the last to get infected. Maximum disease incidence (92%) and severity (87.75%) were recorded 42 days after inoculation (Table 1). Defoliation of leaves started after 20 days of disease appearance, however, complete defoliation was recorded after 45 days of inoculation. These observations are in line with the findings of Bala (1999), who also reported appearance of symptoms after 10-12 days while, Sharma and Kaul (2000) got the disease symptoms after 12-14 days.

Morphological variability

The perusal of data (Table 2) revealed that calculated value of chi-square for conidia length (4.95) was less than the tabulated value (7.82) at 5 per cent level of significance and 3 degrees of freedom. Hence, the results were non-significant proving thereby that the variance in conidia length was homogeneous.

Table 2. Variability in morphology of conidia of *Marssonina coronaria*

	Length and breadth (µm) of conidia from different locations							
	Kotkhai		Kullu		Oachghat		Uttaranchal	
	Length	Breadth	Length	Breadth	Length	Breadth	Length	Breadth
Maximum	25.48	9.10	29.12	8.19	25.48	9.10	25.48	8.19
Minimum	18.75	6.37	18.20	4.55	14.56	3.64	17.29	5.46
S.E.(±)	0.35	0.11	0.43	0.16	0.54	0.20	0.49	0.15
Variance	3.03	0.30	4.63	0.67	7.38	1.04	6.05	0.53
Mean	21.23	7.17	22.24	6.99	19.84	6.92	20.31	6.99

Calculated chi-square value for conidia length = 4.95
Calculated chi-square value for conidia breadth = 8.83
Tabulated chi-square value at 5 per cent = 7.82

It was inferred that there was not much difference in the length of conidia of different locations. Conidial length was maximum at Kullu (22.24 ± 0.43) and minimum at Oachghat (19.84 ± 0.54). Data on conidia breadth showed that there was a difference in the breadth of conidia from different locations. Calculated value of chi-square for conidia breadth (8.83) was more than tabulated value (7.82). *Marssonina coronaria* conidia harvested from Kotkhai apple leaf samples were broader (7.17 ± 0.11) than the conidia from other locations. Conidial breadth was minimum (6.92 ± 0.20) in case of Oachghat samples.

Effect of media

All the test media supported vegetative growth of the fungus (Table 3). However, maximum growth of *Marssonina coronaria* was found on minimal salts agar medium (11.95 mm) after 35 days of incubation, which was followed by malt extract agar medium (8.81 mm), King's B medium (6.47 mm) and potato dextrose agar medium (6.27 mm). Maximum dry mycelial weight was recorded on the minimal salts solution (159.98 mg) followed by glucose asparagine (141.2 mg) and sucrose asparagine (140.0 mg).

Table 3. Effect of different solid and liquid media on the vegetative growth of *Marssonina coronaria*

Medium	Mean radial growth (mm)	Dry mycelial weight (mg)
Tomato extract	4.17[g]	58.40[g]
Czapek-Dox	2.98[ij]	27.88[j]
Sucrose asparagine	3.42[h]	140.00[b]
Coon's	3.12[i]	34.10[i]
Minimal salts	11.95[a]	159.98[a]
Potato dextrose	6.27[d]	97.75[f]
Bean extract	2.62[kl]	3.60[k]
Richards'	3.0[i]	5.60[k]
Carrot extract	2.47[e]	35.63[i]
Apple leaf extract	3.35[h]	120.20[d]
Glucose asparagine	4.75[f]	141.20[b]
Sucrose	5.28[e]	103.60[e]
King's B	6.47[c]	34.70[i]
Malt extract	8.80[b]	124.00[c]
PDA +Glycerol	4.34[g]	53.00[h]
Water	2.77[jk]	2.60[k]
$CD_{0.05}$	0.20	3.05

Figures followed by same letter in each column do not differ significantly

The findings of the present investigations are in accordance with those of Sharma (1999), who reported sucrose asparagine agar medium to be the best medium for mycelial growth of *M. coronaria*. However, Bala (1999) reported maximum germination and germtube length of conidia of *M. coronaria* on rose leaf extract followed by apple leaf extract, walnut leaf extract and sucrose asparagine medium. Simpson and Hayes (1978) reported best growth of *M. brunnea*, the causal agent of poplar leaf blight on minimal salts medium, intermediate on malt extract agar and least on Czapek-Dox agar. Galea *et al.* (1986) reported best linear growth of *M. panattoniana* causing ring spot of lettuce on malt extract agar followed by potato dextrose agar and least on water agar medium.

Effect of temperature

Studies on the effect of temperature on the fungal growth indicated that the fungus was capable of growing on a wide range of temperature from 5 to 30°C (Table 4). Maximum mean radial growth (14.79 mm) and dry mycelial weight (177.0 mg) was recorded at 20°C followed by 15 and 25°C. The least growth (2.79 mm) and dry mycelial weight was observed at 30°C.

Table 4. Effect of different temperatures on the vegetative growth of *Marssonina coronaria*

Temperature (±1°C)	Mean radial growth (mm)	Dry mycelial weight (mg)
5	4.97 [e]	55.00 [f]
10	8.22 [d]	83.00 [d]
15	13.95 [b]	153.00 [b]
20	14.79 [a]	177.00 [a]
25	9.97 [c]	106.00 [c]
30	2.79 [f]	23.40 [e]
$CD_{0.05}$	0.18	3.91

Figures followed by same letter in each column do not differ significantly

Almost similar results were reported by Harada *et al.* (1974) who reported 20°C as the optimum temperature for the growth of *M. coronaria.* Bala (1999) reported maximum conidial germination and germtube length of *M. coronaria* conidia at 20°C followed by 25°C. Dolan (1947) reported increased growth rate of two *Marssonina* species isolates with rising temperature from 9 to 24°C, whereas, Simpson and Hayes found 18 to 20°C temperature as the optimum for the growth of *Marssonina brunnea*. Later Galea *et al.* (1986) reported 15 to 20°C as the cardinal temperature for the growth of *M. panattoniana.*

Effect of hydrogen ion concentration (pH)

Radial growth and dry mycelial weight of *Marssonina coronaria* recorded at different pH levels revealed that with the increase in pH from 3.0-5.0, the radial growth of the fungus also increased (Table 5). Maximum radial growth (12.12 mm) and maximum dry mycelial weight (148.2 mg) of the fungus was recorded at pH 5.0 followed by 6.0 (10.01 mm and 128.1 mg) and pH 4.0 (9.98 mm and 118.3 mg). This study concluded that highly acidic (below 4.0) and alkaline pH ranges were not suitable for the growth of this pathogen.

Table 5. Effect of different pH levels on the vegetative growth of *Marssonina coronaria*

Hydrogen ion conc. (pH)	Mean radial growth (mm)	Dry mycelial weight (mg)
3.0	4.04 [d]	36.32 [f]
4.0	9.98 [b]	118.30 [c]
5.0	12.12 [a]	148.20 [a]
6.0	10.01 [b]	128.10 [b]
7.0	9.90 [b]	116.45 [d]
8.0	7.61 [c]	97.87 [e]
$CD_{0.05}$	0.40	4.93

Figures followed by same letter in the column do not differ significantly

Galea *et al.* (1986) reported optimum germination of conidia of *M. panattoniana* between pH 4.25-5.2 whereas, germination decreased rapidly above pH 6.0 and ceased at pH 8.0. While, Simpson and Hayes (1978)

reported pH range of 7.0 to 7.5 as optimum for the growth of *M. brunnea* on solid medium.

Effect of nitrogen sources

The data in the Table 6 indicated that all the eight different nitrogen sources tested influenced the mycelial growth of the test fungus, however, maximum radial growth was registered in case of methionine (11.57 mm). It was closely followed by potassium nitrate (10.72 mm), ammonium nitrate (10.69 mm) and yeast extract (10.50 mm).

Table 6. Effect of different nitrogen sources on the vegetative growth of *Marssonina coronaria*

Nitrogen source	Mean radial growth (mm)	Dry mycelial weight (mg)
Potassium nitrate	10.72 [b]	142.00 [a]
Ammonium chloride	6.59[d]	36.00[d]
Ammonium nitrate	10.69 [b]	121.00[c]
Sodium nitrate	4.56[e]	2.80[e]
Ammonium sulphate	9.99 [c]	128.00[bc]
Yeast extract	10.50 [b]	124.00[bc]
Methionine	11.57 [a]	134.00[ab]
Leucine	9.90[c]	126.00[bc]
$CD_{0.05}$	0.39	11.73

Figures followed by same letter in the column do not differ significantly

While recording the dry mycelial weight of the test fungus, potassium nitrate was found the best nitrogen source yielding 142.0 mg dry mycelial weight followed by methionine (134.0 mg) and ammonium sulphate (128.0 mg). These findings also corroborate those of Bala (1999) who reported asparagine as the best nitrogen source followed by potassium nitrate for the germination and germ tube formation of conidia of *M. coronaria*. Simpson and Hayes (1978) reported best growth of *Marssonina brunnea* on medium containing potassium nitriate and 4- amino n-butyric acid. Good growth of *M. brunnea* was also recorded when ammonium nitrate and ammonium sulphate replaced potassium nitrate; and leucine, alanine, glycine replaced 4-amino n-butyric acid in the minimal salt medium.

Effect of carbon sources

Among different carbon sources, maximum radial growth of the fungus (Table 7) was supported by maltose (12.24 mm) followed by lactose (12.08 mm) and fructose (11.95 mm). It is evident that when sucrose was replaced with maltose as sole carbon source in the liquid medium, it exhibited the maximum dry mycelial weight (299.0 mg). This was followed by sucrose (181.0 mg) and lactose (166.0 mg). Bala (1999) reported maximum germination and germ tube length of conidia of *M. coronaria* on sucrose followed by glucose.

Table 7. Effect of different carbon sources on the vegetative growth of *Marssonina coronaria*

Carbon source	Mean radial growth (mm)	Dry mycelial weight (mg)
Galactose	8.84 f	126.00 e
Fructose	11.95bc	149.00 d
Sucrose	11.42 d	181.00 b
Glucose	8.99f	148.00 d
Maltose	12.24a	299.00 a
Mannitol	8.38g	151.00 d
Cellulose	9.63e	9.70 g
Lactose	12.08ab	166.00 c
Starch	11.80c	68.00 f
$CD_{0.05}$	0.22	6.74

Figures followed by same letter in the column do not differ significantly

Dhanvantri (1967) also reported best growth of *Marssonina (Diplocarpon) earlianum* on glucose and sucrose. Simpson and Hayes (1978) reported glucose, sucrose and cellulose as the suitable carbon sources for the growth of *M. brunnea*. Recently, Hong *et al.* (1998) have reported glucose as the best carbon source for the growth of *Diplocarpon mali*.

REFERENCES

Bala, R. (1999). *Studies on Marssonina Leaf Blotch of Apple*. M.Sc. Thesis. Dr. Y.S. Parmar University of Horticulture and Forestry, Nauni, Solan. 74 p.

Dhanvantri, B.N. (1967). The leaf scorch disease of strawberry (*Diplocarpon earlianum)* and the nature of resistance to it. *Can. J. Bot.* 45 : 1525-1543.

Dolan, D. (1947). New anthracnose of melons. *Phytopathology* 37 : 583-596.

Galea, V.J., Price, T.V. and Sutton, B.C. (1986). Taxonomy and biology of the lettuce anthracnose fungus. *Trans. Br. Mycol. Soc.* 84 : 619-628.

Gupta, S.C. and Kapoor, V.K. (1970). *Fundamentals of Mathematical Statistics*. Sultan Chand and Sons. Educational Publishers, New Delhi, pp. 15.1-15.56.

Harada, Y., Sawamura, K. and Konno, K. (1974). *Diplocarpon mali* sp. nov., the perfect state of apple blotch fungus *Marssonina coronaria. Ann. Phytopathol. Soc. Jpn.* 40:412-418.

Hong, K.K., Moon, C.Y. and Soon, Y.M. (1998). Cultural characteristics of *Diplocarpon mali. RDA J. Crop Prot.* 40: 120-123.

Parmelee, J.A. (1971). Marssonina leaf spot of apple. *Can. Plant Dis. Surv.* 51: 91-92.

Sharma, J.N. (1999). Marssonina blotch- A new disease causing premature leaf fall in apple. *Indian Phytopathol.* 52: 101-102.

Sharma, J.N. (2001). Diagnosis and control of premature leaf fall problem in apple. *J. Mycol. Plant Pathol.* 31: 305-310.

Sharma, J.N. and Gautam D.R. (1997). Studies on premature leaf fall in apple - A new problem. *Indian J. Plant Prot.* 25: 8-12.

Sharma, J.N. and Kaul, J.L. (2000). Marssonina blotch – A new disease hits apple crop in Himachal Pradesh. In: *Proceedings of International Conference on Integrated Plant Disease Management for Sustainable Agriculture.* Indian Phytopathologial Society, New Delhi, pp. 775-777.

Simpson, B. and Hayes, A.J. (1978). Growth of *Marssonina brunnea. Trans. Br. Mycol. Soc.* 70: 249-255.

Takahashi, S. and Sawamura, K. (1990). Marssonina blotch. In: *Compendium of Apple and Pear Diseases.* (Eds. Jones, A.L. and Aldwinckle, H.S.). APS Press. St. Paul, Minnesota, pp. 23.

Integrated Plant Disease Management (2005)
Eds. Sharma, R.C. and Sharma, J.N.
Scientific Publishers (India), Jodhpur, pp. 19-23.

3

Effect of weather parameters on the epidemic of apple blotch in Himachal Pradesh

V.S. THAKUR, NIRUPMA SHARMA and SONALI VAIDYA

Regional Horticultural Research Station, Dr. Y.S. Parmar University of Horticulture and Forestry, Mashobra, Shimla-171 007 (H.P.)

ABSTRACT

Apple blotch caused by the fungus *Marssonina coronaria* is an important disease manifested recently in all apple growing states of India. The disease appears in the form of dark brown blotch on the upper side of the leaf in the second week of June irrespective of the weather conditions on all the commercial cultivars grown in India. Correlation between meteorological data and pre-mature defoliation disease index for the years 2000 and 2001 was studied. The correlation analysis revealed that the prevalence of blotch disease was influenced by leaf ageing followed by average temperature above 20°C and 70 per cent relative humidity at least continuously for 5-6 days which induce yellowing and defoliation process. Prolong hot and dry weather affect the primary infection and disease conditions adversely. Under favourable weather conditions, the defoliation begins within 3-5 days whereas under interrupted/unfavourable conditions the infected leaves are retained up to 3 months.

INTRODUCTION

Apple (*Malus x domestica* Borkh.) is a major temperate fruit crop grown in Himachal Pradesh. The fruit production of apple is declining every year in Himachal Pradesh ever since the crop has been hit by the premature leaf fall. The problem starts in the months of June-July and manifested on all commercial cultivars of apple grown in India. It has been found that a fungal disease Marssonina blotch is solely responsible for causing the premature leaf abscission in apple orchards caused by *Marssonina coronaria* (Ell. & J.J. Davis) J.J. Davis (Sharma and Kaul, 2000). Disease symptoms appear both on leaves and fruits. Dark green circular patches develop on upper surface of leaves giving rise to 5-10 mm diameter brown leaf spots, which become dark brown in due course. Small lesions coalesce to form large dark brown blotches and surrounding areas turn yellow. These symptoms are followed by severe leaf shedding in mid-summer, which may be responsible for poor growth and development of fruiting buds that result into low productivity in ensuing season. Disease severity index increases with leaf maturity stage when plant starts producing more ethylene, the ripening/ageing hormone. Once the infection had taken place the rate of blotch development depends on the prevailing weather conditions. The purpose of this study was to ascertain the influence of meteorological parameters in the development of this disease under field conditions.

MATERIALS AND METHODS

The field experiments were conducted at Regional Horticultural Research Station, Mashobra, Shimla (31°N, 77°E; 2286 m amsl) during 2000 and 2001. For studying the influence of weather parameters (temperature, relative humidity and rainfall) on the development of apple blotch, infected leaves were examined at weekly intervals. While recording the leaf blotch intensity, tree was divided into upper and lower segments and four branches on four directions in each segment were selected and visual observations on 0-5 scale (James, 1974) were recorded as below:

0- No leaf fall and no symptoms (No yellowing leaf spots) on leaves.

1- No leaf fall but 2-3 leaves showing 1-2 leaf spots per branch.

2- Leaf fall initiated but less than 25 per cent and/or 3-4 leaves with spots/ yellowing per branch.

3- One third of leaves defoliated and/or 25 per cent of leaves with spots/ yellowing.

4- Leaf defoliation up to 50 per cent and/or several (about 50%) leaves with yellowing and spots.

5- Leaf defoliation more than 50 per cent.

Per cent disease index was calculated by the following formula given by Mckinney (1923).

$$\text{Disease index (\%)} = \frac{\text{Sum of all ratings}}{\text{Total number of ratings x Maximum grade}} \times 100$$

Weekly data of meteorological parameters for the years 2000 and 2001 were recorded by Metos weather station. Studies were conducted from second week of June up to first week of October. Meteorological data and premature defoliation disease severity obtained for the years 2000 and 2001 were subjected to analysis of variance as described by Gomez and Gomez (1985).

RESULTS AND DISCUSSION

The data (Figs. 1 & 2) on weather parameters and disease development revealed that initiation of symptoms started in the second week of July in 2000, but the disease progress was slow (Fig. 1) while it started appearing in the last week of June during 2001 (Fig. 2). Rapid spread of the disease was recorded during the months of July, August and September in both the years. The disease index varied from 1.3 to 80.2 per cent during 2000 and 2001. Since the disease progress was dependent on temperature, relative humidity and rainfall; simple, partial and multiple correlations were worked out separately for both the years to establish the relative contribution of these factors in the spread of the disease.

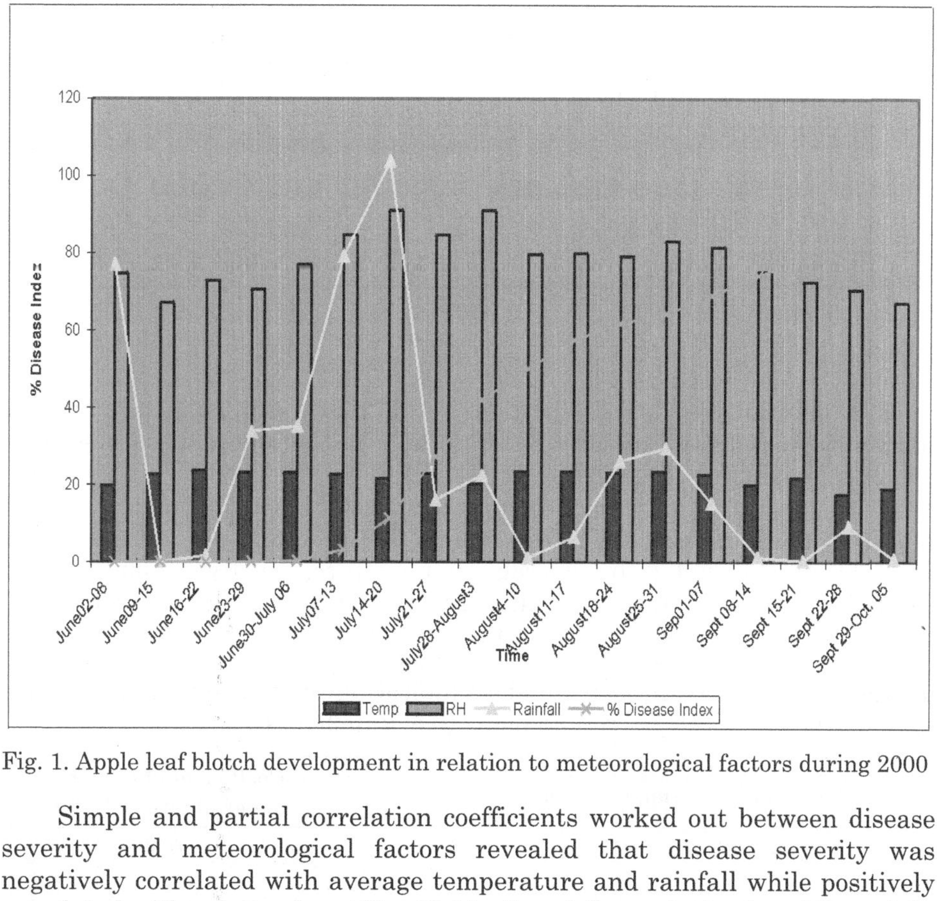

Fig. 1. Apple leaf blotch development in relation to meteorological factors during 2000

Simple and partial correlation coefficients worked out between disease severity and meteorological factors revealed that disease severity was negatively correlated with average temperature and rainfall while positively correlated with relative humidity (Table 1) and thus relative humidity might change the severity of the blotch disease every year.

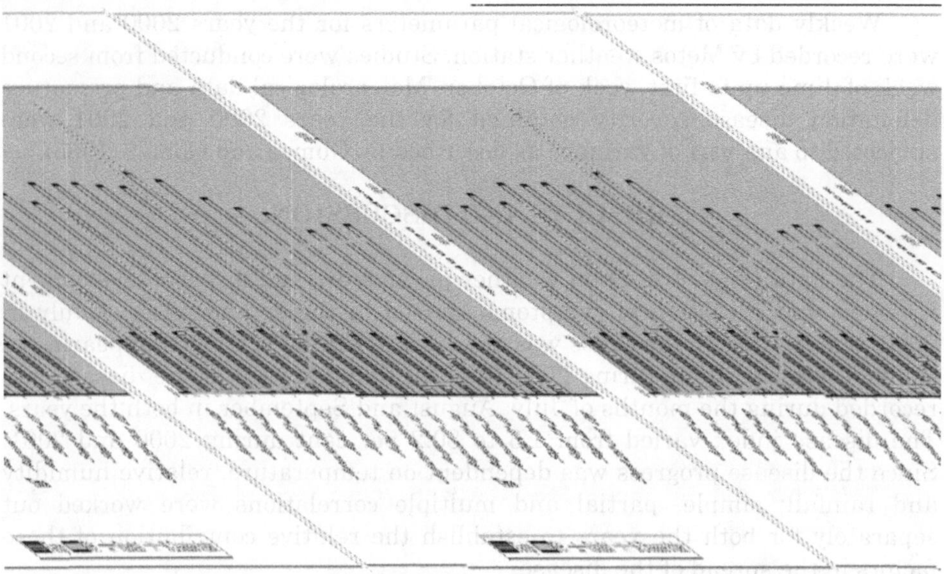

Fig. 2. Apple leaf blotch development in relation to meteorological factors during 2001

Table 1. Correlation coefficients between apple leaf blotch disease and meteorological factors

Factor	Correlation coefficient			
	Simple		Partial	
	2000	2001	2000	2001
Temperature	-0.28	-0.37	-0.17	-0.56
Relative humidity	0.41	-0.07	0.57	0.45
Rainfall	-0.05	-0.53	-0.37	-0.68

The multiple regression equations reveal that a unit change in average temperature, relative humidity and rainfall could influence disease severity up to an extent of 4.49, 6.69 and 0.04 units, respectively during 2000 and 8.69, 1.91 and 0.79 units, respectively during 2001. From the coefficient of multiple determination, it is evident that 70.38 and 53.47 per cent variations have been contributed by average temperature, relative humidity and rainfall, whereas the rest of variations were due to unexplained factors and the factors not included in the investigations (Table 2).

Table 2. Multiple regression equation indicating relationship between apple blotch development and meteorological factors

Year	Multiple regression equation	Coefficient of multiple determination R^2 (%)
2000	951.43-4.49T+6.69RH-0.04RF (6.78) (1.94) (0.03)	70.38
2001	101.21-8.69T+1.91RH-0.79RF (3.46) (1.00) (0.23)	53.47

Epidemic was noticed only when the monsoon rains receded and relative humidity ranged between 65 to 98 per cent and temperature around 20°C. Two years correlations between disease severity and meteorological parameters indicated that temperature and rainfall had negative and varied effect on the disease however, the cumulative dependence of the disease on weather parameters was 70.38 and 53.47 per cent for 2000 and 2001, respectively. These results are in accordance with the findings of Sharma (2003). The study thus reveals that the blotch symptoms appear only after June and during the investigations the per cent dependence of meteorological factors had varied results.

REFERENCES

Gomez, K.A. and Gomez, A.A. (1985). *Statistical Procedures for Agricultural Research* (3rd ed.). John Willey and Sons, Singapore, 680 p.

James, W.C. (1974). Assessment of plant diseases and losses. *Annu. Rev. Phytopathol.* 12: 27-48.

McKinney, H.H. (1923). Influence of soil temperature and moisture on infection of wheat seedlings by *Helminthosporium sativum. J. Agric. Res.* 26: 195-217.

Sharma, I.M. (2003). Influence of environmental factors on the development of premature defoliation disease caused by *Marssonina coronaria* in apple and its management. *J. Mycol. Plant Pathol.* 33: 89-95.

Sharma, J.N. and Kaul. J. L. (2000). Marssonina blotch- A new disease hits apple crop in Himachal Pradesh. In: *Proceedings of International Conference on Integrated Plant Disease Management for Sustainable Agriculture.* Indian Phytopathological Society, New Delhi, pp. 775-777.

Integrated Plant Disease Management (2005)
Eds. Sharma, R.C. and Sharma, J.N.
Scientific Publishers (India), Jodhpur, pp. 25-31.

4

Occurrence of Alternaria leaf spot (*Alternaria alternata*) on apple and its management

J.N. SHARMA, DHARMESH GUPTA, L.N. BHARDWAJ and RAVINDER KUMAR

Department of Mycology and Plant Pathology, Dr.Y.S. Parmar University of Horticulture and Forestry, Nauni, Solan- 173 230 (H.P.)

ABSTRACT

After the onslaught of scab and Marssonina blotch in the past, apple orchards in Shimla district of Himachal Pradesh were again threatened by a hitherto unknown problem of brown spotting on the leaves. This disease was characterized by the appearance of 1-3 mm diameter reddish brown necrotic spots on young apple leaves, later coalescing to form bigger blotches and giving blighted appearance to the foliage followed by defoliation. Dark brown to black elongated lesions also appeared on leaf petioles and mid ribs in early part of the season causing severe yellowing and dropping of the infected leaves. Apart from Kotkhai, this problem has also been encountered in some of the orchards in Jubbal and Baghi areas and caused up to 40 per cent defoliation in some of the orchards even by the month of August. All the commercially grown Delicious cultivars viz., Red Delicious, Royal Delicious and Red Chief were found highly susceptible; while pollinizers Golden Delicious, Red Gold and Tydeman's Early Worcester showed tolerance to this disease in the field. Microscopic studies showed the consistent association of *Alternaria alternata* with the infected leaves. The fungus was isolated in pure culture and its pathogenicity was proved in controlled conditions. The identity of this fungus was also confirmed by Global Plant Clinic Diagnostic and Advisory Service, UK (No. W7919). Protective sprays of dodine, mancozeb, zineb, iprodione and mancozeb + hexaconazole exhibited effective control in the field.

INTRODUCTION

With the intensification of apple cultivation, there has always been an increased threat of diseases affecting this fruit crop. Its monoculture has resulted in various disease problems in the past and the outbreak of apple scab and premature leaf fall due to *Marssonina coronaria* are well known (Gupta, 1983; Sharma, 1999). It seems that the spray schedules being adopted by the apple growers against apple diseases have contributed for amelioration of the latter and have fire-backed by giving way to yet another disease problem. The apple plantations in some well sprayed orchards in Kotkhai area have been known to suffer from yet another problem of leaf spotting on the terminal expanding leaves of apple during May- June months for the last three years. This phenomenon was earlier ignored and thought to be a physiological disturbance due to tank mixing of various pesticides. However, the constant vigil in such orchards by the senior author in the last few years and diagnostic field trials led its correct diagnosis as Alternaria leaf spot due to *Alternaria alternata* (Fr.) Keissler, hitherto thought to be a weak pathogen. Earlier, Gupta and Agarwala (1968) also reported similar disease in Kotgarh apple belt of the state. In the present paper, results on its prevalence, diagnosis and management are described.

MATERIALS AND METHODS

Disease prevalence and severity

After the report of this disease in the previous years, surveys were conducted in 2003 season to record its prevalence and severity in Solan, Sirmour and Shimla districts of Himachal Pradesh including Kotkhai by using modified Horsfall - Barratt 0-5 scale as follows :

0 No symptoms
1 1-3 per cent of leaf area covered with lesions
2 4-6 per cent of leaf area covered with lesions
3 7-12 per cent of leaf area covered with lesions
4 13-25 per cent of leaf area covered with lesions
5 26-50 per cent of leaf area covered with lesions

Per cent disease index and severity was calculated according to McKinney (1923).

Disease diagnosis, fungal culture and pathogenicity

Infected leaf samples were collected periodically from the affected orchards and the disease was diagnosed on the basis of foliage symptoms. Repeated isolations were carried out and pure culture was obtained by following the standard procedure in the laboratory. The pathogenic fungus *Alternaria alternata* was identified by studying its morphology and compared with the standard descriptions. The fungal culture was sent to the Global Plant Clinic Diagnostic and Advisory Service CABI Bioscience, UK Centre for the correct identification of the culture at the species level. The disease specimen and culture were also deposited at IARI, New Delhi. The pathogenic nature of the fungus was confirmed by inoculating apple plants under semi-controlled conditions. Royal Delicious apple plants were inoculated by the standardized conidial suspension (50,000 conidia/ml) by a hand held atomiser. Inoculated plants were individually bagged to produce high humidity (> 90%) and were kept at ambient temperature (20-25°C) for 24 hours and were further incubated in ambient conditions after removal of the bags. The inoculated plants were regularly inspected for disease symptoms and severity. Suitable control was maintained side by side where only sterile water was sprayed.

Disease management

To assess the efficacy of different fungicides viz., mancozeb (0.3% Indofil M-45 75 WP), zineb (0.3 % Indofil Z-78 75WP), dodine (0.075% Noor 65WP), iprodione + carbendazim (0.15% Quintal 50WP), carbendazim (0.05% Carbestin 50WP), carbendazim + mancozeb (0.25% Companion 75WP) and hexaconazole + mancozeb (0.35% Sitara MZ 34.25SC) were sprayed on three apple trees each (replications) in the field at Kotkhai at an elevation of 1950 m amsl, where the disease had been found consistently prevalent in the last 3-4 years. Six applications of fungicides were made as high volume spray from pink bud stage of apple till fruit harvest at 21 day intervals. Suitable control plants were maintained without the fungicide application. Observations were recorded on per cent disease severity according to modified Horsfall and Barratt's 0-5 scale as described earlier. Per cent disease control was also calculated.

RESULTS AND DISCUSSION

Disease prevalence and severity

Periodic disease surveys conducted in Solan, Sirmour and Shimla districts of Himachal Pradesh revealed that the Alternaria leaf spot was present in almost every apple nursery/orchard at Nauni in 2003 with 4-5 per cent disease severity and the orchards in Nohradhar area of Sirmour district were almost free from the disease. However, this disease was more severe in Chaithla village in Kotkhai fruit belt of the state, where the disease was encountered in all the orchards inspected in 2003 with 22 per cent overall disease severity. In one of the worse affected orchards, as high as 40 per cent defoliation was recorded in the month of August. At least one apple orchard each in Mohli and Ratnari areas was found severely infected by Alternaria leaf spot disease. Earlier, Alternaria blight of apple was reported to be caused by *Alternaria mali* from the orchards of Kotgarh area of the state by Gupta and Agarwala (1968). Elsewhere, unknown disease characterized by the severe defoliation of the spotted leaves in mid-summer was detected in Japan in 1956 and was supposed to be caused by *Altertnaria mali* and was named Alternaria blotch and distinguished from Alternaria leaf spot of apple in

United States which is apparently caused by a biotype of *Alternaria alternata* (Filajdic and Sutton, 1991; Sawamura *et al.,* 1993).

Disease diagnosis and symptomatology

A leaf spot causing up to 30-40 per cent defoliation of the apple (*Malus* x *domestica* Borkh.) occurred during the months of June to August, 2002 and 2003 in the apple growing areas of Shimla district of Himachal Pradesh. Initial symptoms appeared on the leaves in the mid of the May as small circular reddish brown spots, gradually enlarging up to 2-5 mm in diameter and turning to blackish brown spots bordered by dark brown to purple margins. With advancement of age the lesions turned into dark reddish brown spots. Initially, the lesions were regular with distinct margins, later they exhibited secondary expansion, becoming irregular and darker in colour. Petioles showed elongated purple to blackish brown lesions. Lesions with infected petioles often turned yellow and abscised early in the season. Typical leaf blight symptoms were usually seen in the rainy season i.e. July and August, where large coalescing brown to dark brown spots were visible on the leaves particularly on the leaf margins. The infected leaves withered and pronounced defoliation resulted. The disease specimen was deposited in IARI, New Delhi (HCIO 45556). These disease symptoms were different from Marssonina blotch *(Marssonina coronaria)* in which brown leaf spots appear on the mature whorl leaves accompanied by yellowing and excessive premature defoliation. *Alternaria mali* has also been reported to cause apple blight in Kotgarh apple belt of Himachal Pradesh (Gupta and Agarwala, 1968).

Fungal culture and pathogenicity

The pathogenic fungus was isolated on potato dextrose agar medium and was identified as *Alternaria* sp. Two different isolates have been found which showed characteristic differences from one another in their morphology. Isolate-1 showed sparse aerial mycelium and produced abundant olivaceous conidia on a dark mycelial mat on the surface of the medium. Isolate-2 had abundant grey mycelium and produced fewer number of conidia. The conidial size varied from 7-39 x 5-12 μm in both the isolates. The fungus was finally identified as *Alternaria alternata* by the mycologists of Diagnostic and Advisory Service, CABI Bioscience, UK Centre. *Alternaria* species had also been reported from several countries attacking apple leaves and fruit. Roberts (1924) identified the fungus as *Alternaria mali,* while Tailor (1996) considered it to be *Alternaria tenuis. Alternaria mali* was regarded as a weak parasite on apple leaves and did not originate the leaf spot on apple leaves but got entry by areas killed by *Sphaeropsis malorum* (Roberts, 1924). Although *Alternaria mali* and *Alternaria tenuis* have been reported to be associated with the leaf spot, blight and defoliation symptoms on apple, recent investigation indicated that all *Alternaria* pathogens known to produce host specific toxin should be classified as *Alternaria alternata* because of the morphological similarity of their asexual state. Thus *Alternaria mali* has been referred as apple pathotype of *Alternaria alternata* (Sawamura, 1990).

The pathogenicity of *Alternaria alternata* (Table 1) was proved under semi-controlled conditions. Typical brown spot symptoms were observed after 5-7 days of inoculation on all the inoculated apple plants. The disease increased with time and maximum disease severity (73 %) was achieved on young expanding leaves of Royal Delicious after 21 days. The infected leaves withered and fell down from the inoculated plants thereafter. The corresponding Golden Spur plants which were also inoculated side by side showed high degree of tolerance with 9 per cent disease severity and the lesions were limited in size and devoid of fungal spores. No disease symptoms appeared on the control plants maintained with sterile distilled water.

Table 1. Pathogenicity test of *Alternaria alternata* on apple leaves

Cultivar	Disease incidence (%)					Disease severity (%)				
	11/8	18/8	25/8	1/9	Mean	11/8	18/8	25/8	1/9	Mean
Royal Delicious	20.0	35.0	60.0	100.0	53.7	30.0	44.0	54.0	73.0	50.7
Golden Spur	0.0	3.0	5.0	12.0	5.0	0.0	3.0	7.0	9.0	4.5

Field reaction of cultivars

Different apple cultivars grown in the farmers' fields were evaluated for their reaction to brown spot disease due to *Alternaria alternata* at Kotkhai under natural epiphytotic conditions.

Table 2. Field reaction of different cultivars to brown spot disease due to *Alternaria alternata* in the field

Cultivar	Disease severity (%)	Reaction class
Royal Delicious	55.0	HS
Red Delicious	43.9	HS
Vance Delicious	52.5	HS
Red Chief	30.3	S
Red Gold	8.5	MR
Golden Spur	5.0	MR
Golden Delicious	3.5	MR
Malus floribunda	0.0	I

HS = Highly susceptible; MR = Moderately resistant; S = Susceptible; I = Immune

Data given in Table 2 indicated that the Delicious cultivars were highly susceptible showing up to 55.0 per cent disease severity in Royal Delicious followed by Vance Delicious (52.5%), Red Delicious (43.9%) and Red Chief (30.3%). Whereas the pollinizing cultivars viz., Red Gold, Golden Spur and Golden Delicious were found moderately resistant showing 3.5 to 8.5 per cent disease severity. Crab apple (*Malus floribunda*) remained free from disease and was rated as immune to Alternaria leaf spot in the field. Starking Delicious and Indo have earlier been reported to be highly susceptible to Alternaria blotch pathogen *A. mali (Alternaria alternata)* in Japan

(Sawamura and Yanase, 1963). Sawamura (1990) reported Indo and Delicious highly susceptible; Ralls Janet as moderately susceptible; and Golden Delicious, Yellow Newton, McIntosh and Black Ben Davis as relatively resistant cultivars to Alternaria blotch.

Disease management

Seven different fungicides viz., mancozeb, zineb, dodine, iprodione + carbendazim, carbendazim, carbendazim + mancozeb and hexaconazole + mancozeb were tried in the field to assess their efficacy against Altrenaria leaf spot in the field. Protective sprays of dodine (0.075%) controlled the disease completely and was followed by iprodione + carbendazim (0.15%), hexaconazole + mancozeb (0.35%), zineb (0.3%) and mancozeb (0.3%) which provided 94.8 to 98.0 per cent disease control, all of which were statistically at par in this respect.

Table 3. Bio-efficacy of different fungicides in controlling Alternaria leaf spot of apple in the field

Fungicide	Dose (%)	Disease severity (%)	Disease control (%)
Mancozeb	0.3	4.0	95.8
Zineb	0.3	3.5	96.4
Dodine	0.075	0.0	100.0
Iprodione + carbendazim	0.15	1.0	98.0
Carbendazim	0.05	94.8	5.0
Carbendazim + mancozeb	0.75	5.0	94.8
Hexaconazole + mancozeb	0.35	2.0	97.9
Control	—	96.6	—
$CD_{0.05}$	—	4.9	—

Six different spray schedules were also adopted by alternating the commonly available fungicides during the season and were evaluated against Alternaria leaf spot in the field. The observations recorded on per cent disease severity in each treatment revealed that in spray schedule involving mancozeb + carbendazim (0.25%), propineb (0.3%), hexaconazole + mancozeb (0.35%), mancozeb flowable (0.35%) and ziram (0.3%) given at pink bud, fruit set and successive fruit development and pre harvest stages at 21 day intervals was the most effective and provided up to 98.3 per cent disease control and was followed by spray schedule wherein above fungicides were rotated at different stages. Earlier, Sawamura *et al.* (1993) reported Polyoxin AC (an antibiotic) and iprodione to be superior to copper and quinolinol, captan and dithiocarbamates in controlling Alternaria blotch of apple in Japan. Iprodione provided 69.1-79.8 and 71.6-75.0 per cent disease reduction in Alternaria blotch for higher (0.3%) and lower application rate (0.15%), respectively under natural infection conditions on different stages (Filajdic and Sutton, 1992). In India also, Sud and Agarwala (1971) found ziram (0.3%) the most effective against *Alternaria mali*, causing Alternaria blight of apple which was closely followed by ferbam, zineb and captan (0.15%) in controlling the disease.

REFERENCES

Filajdic, N. and Sutton, T.B. (1991). Identification and distribution of *Alternaria mali* on apple in North Carolina and susceptibility of different varieties of apple to Alternaria blotch. *Plant Dis.* 75: 1045-1048.

Filajdic, N. and Sutton, T.B. (1992). Chemical control of Alternaria blotch of apples caused by *Alternaria mali. Plant Dis.* 76: 126-130.

Gupta, G.K. (1983). Behaviour of fungicides and different spray schedules in the control of apple scab (*Venturia inaequalis*). *Int. J. Trop. Plant Dis.* 1: 181-185.

Gupta, G.K. and Agarwala, R.K. (1968). Alternaria blight of apple. *FAO Plant Prot. Bull.* 16: 32-33.

McKinney, H.H. (1923). Influence of soil temperature and moisture on infection of wheat seedlings by *Helminthosporium sativum. J. Agric. Res.* 26: 195-207.

Roberts, J.W. (1924). Morphological characteristics of *Alternaria mali* Roberts. *J. Agric. Res.* 27: 699-708.

Sawamura, K. (1990). Alternaria blotch . In : *Compendium of Apple and Pear Diseases* (Eds. Jones, A.L. and Aldwinckle, H.S.). APS Press, St Paul USA, pp. 24-25.

Sawamura, K., Harada, Y. and Kikuchi, T. (1993). The apple industry in Japan : A historic sketch and diseases specific to the region. *Plant Dis.* 77: 546-552.

Sawamura, K. and Yanase, H. (1963). Studies on spotted diseases of apples. 2. On *Alternaria* sp., causal organism of apple blotch and Japanese name of the disease. *Bull. Hortic. Res. Stn. Japan Ser. C* (Morioka) 1: 77-94.

Sharma, J.N. (1999). Marssonina blotch- A new disease causing premature leaf fall in apple. *Indian Phytopathol.* 52: 101-102.

Sud, V.K. and Agarwala, R.K. (1971). Laboratory and field evaluation of fungicides for the control of Alternaria blight of apple. *Indian Phytopathol.* 24: 201-204.

Tailor, J. (1996). Ghost spot of apple leaves by *Alternaria tenuis. Phytopathology* 56: 553-555.

Integrated Plant Disease Management (2005)
Eds. Sharma, R.C. and Sharma, J.N.
Scientific Publishers (India), Jodhpur, pp. 33-38.

5

A rapid method for evaluation of apple germplasm for resistance to *Dematophora necatrix* causing white root rot

SATISH K. SHARMA, Y.P. SHARMA[1], D.K. KISHORE and K.K. PRAMANICK

Indian Agricultural Research Institute, Regional Station, Amartara Cottage, Shimla - 171 004 (H.P.)
[1]ICAR Research Complex for NEH Region, Barapani, Shillong -793 103 (Meghalaya)

ABSTRACT

Dormant excised twigs, root segments and detached leaves of apple cv. Royal Delicious were inoculated with *Dematophora necatrix* to standardize a rapid technique for evaluation of apple germplasm. These plant parts were put in specially prepared culture jars and incubated at 22 ± 1°C for two weeks. The lesion length and extent of mycelial growth on leaves, shoot and root segments varied among different genotypes. Of the fifteen rootstocks tested, *Malus baccata* Simcoe, *M. baccata* Khrot and *M. orientale* exhibited very restricted mycelial growth and were least susceptible. The response of all these rootstocks was almost similar with respect to leaf, root and shoot colonization. The *in vitro* reaction matched with the already known field reaction of rootstocks. Hence, the technique can be used for quick evaluation of apple germplasm against *D. necatrix*.

INTRODUCTION

White root rot (*Dematophora necatrix* Hartig) is an important soil borne disease of apple nurseries and grown up trees. The fungus is widely distributed in temperate and sub-temperate regions of the world attacking more than 160 plant species belonging to 45 families and serious on apricot, pear, strawberry, plum, peach and grapes (Behdad, 1975; Ito and Nakamura, 1984). Almost all the rootstocks are susceptible to the pathogen; however, the degree of susceptibility varies (Gupta and Verma, 1978). Most of the efforts made by earlier workers have resulted in the development of screening techniques to identify sources of resistance against the pathogen under field conditions, which are difficult and require lot of time and energy to come out with the results. Moreover, sometimes the plant material is very less to be spared for screening.

To date no quick method is available in the literature for rapid evaluation of apple germplasm for resistance against the pathogen. Therefore, leaves, shoot and root segments of apple cv. Royal Delicious were used to standardize the technique and fifteen *Malus* spp./rootstocks were selected to record their response against *D. necatrix* for validation of the technique.

MATERIALS AND METHODS

During investigations of root rot infected *Pyrus calleryana*, it was noticed that fallen buried leaves in the vicinity of infected roots and crown showed fungus colonization. These leaves were surface washed and incubated in the moist chamber at 25 ± 1°C for 48 h. Microscopic observations revealed the colonization by root rot fungus *Dematophora necatrix*. It gave an indication that fungus could also infect leaves. Therefore, detached apple leaves, dormant shoot twigs and root segments were used to see the growth of the pathogen on these plant parts to standardize the screening method. The leaves were washed with sterilized distilled water, dried under shade and inoculated with 7 mm fresh culture bits of *D. necatrix* in pre-sterilized Petri plates lined with double layer of blotting paper and incubated at 22±1°C in moist chamber for 7 days. The mycelial growth of the pathogen on upper (Fig. 1a) and lower leaf surface (Fig. 1b) of various genotypes was measured as per cent leaf area covered.

The dormant excised shoot and root segments (10 cm long) were washed with distilled water; air dried and surface disinfected with mercuric chloride (0.1% for 1 min) and subjected to vacuum in order to remove trapped air between the bud scales. The 5 mm portion of these segments was further excised from both the ends to eliminate any residual effect of the disinfectant. Surface sterilized pieces were again washed three times with sterile water before putting in specially prepared culture jars.

The glass jars (12 cm long and 5 cm dia) were half filled with rough sand, supplemented with 50 ml potato dextrose broth and autoclaved at 15 lbs for 30 min. Each jar was inoculated with 20 spawn grains of *D. necatrix* (Sharma, 1993). These jars were closed and incubated at 22 ± 1°C for 7 days. Control jars were kept without culture. Subsequently, four pieces of shoot/root segments were placed vertically in each jar and incubated at 22±1°C for 2 weeks. Each treatment was replicated three times. The segments were taken out and the observations on the extent of mycelial growth (Fig. 1 c,d) and necrotic lesions induced by infection on each segment were measured. Subsequently, leaves, scion wood thick shoot twigs and root segments of fifteen apple rootstocks available namely, *Malus baccata* Shillong, *M. baccata.* Khrot, *M. baccata* Giabung, *M. baccata* Srinagar, *M. baccata* Dhack, *M. baccata* USA, *M. baccata* Rohru, *M. baccata* Mandschurica, *M. sargentii, M. simcoe, M. orientale,* M.E.M Wilsom, MM106, MM26 and MM27 were evaluated with this method for their response to *D. necatrix* in jars and Petri plates to validate the technique. The observations are presented in Fig.1 and Fig. 2 for comparative analysis of the results.

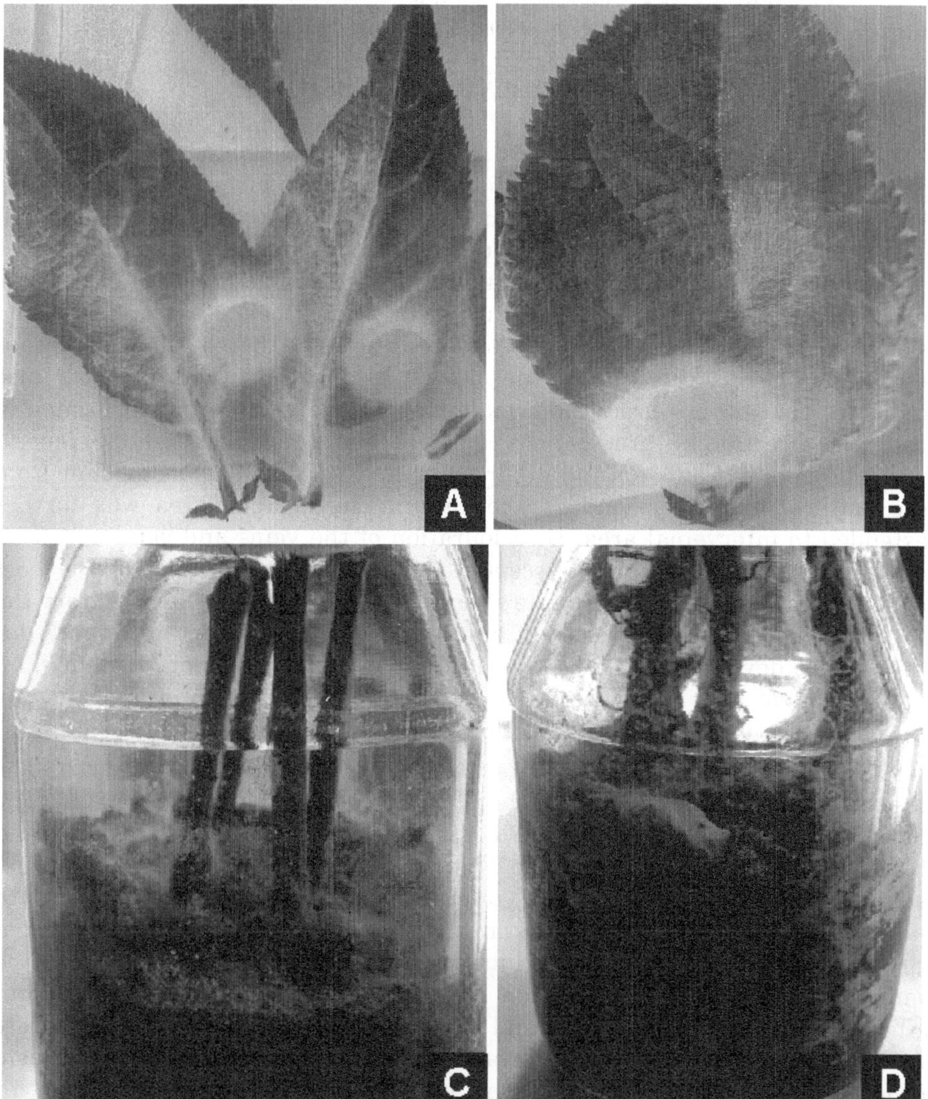

Fig. 1. Growth of *D. necatrix* on lower (a) and upper leaf surface (b) Growth of *D. necatrix* on shoots (c) and roots in culture jars (d)

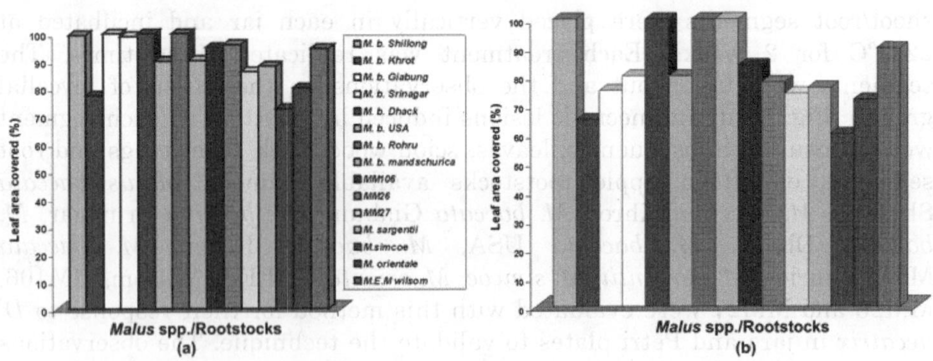

Fig. 2. Response of different *Malus* spp. / rootstocks leaf inoculations to *D. necarix* on lower (a) and upper (b) surface.

RESULTS AND DISCUSSION

The apple leaves were easily infected by the fungus and mycelium growth from culture bit slowly radiated on the leaf surface. Corresponding leaf surface showed light brown necrosis. With the increase of mycelial growth, necrosis was observed on both the surfaces and the colour changed to dark brown to black. Rate of mycelial growth and spread of necrotic area was more on the lower leaf surface resulting into quick colonization. The lesion size was generally larger than the mycelial growth, thereby indicating that the fungus first kills the tissue before establishing. Observations also indicated that fungus produced some metabolites which were translocated in basipetal direction towards petiole point and again redistributed in acropetal direction through mid rib and other veins and veinlets as was evident from the pattern of necrosis and discolouration. The necrotic area was further extended to interveinal area. Discolouration of the veins and veinlets varied with the species. In highly susceptible varieties, rapid killing of the host tissues leads to blackish discolouration of the leaves. It is obvious from the Fig. 2 (a & b) that leaf surface of different rootstocks reacted differently for the disease development and lower and upper leaf surfaces of *M. simcoe*, *M. baccata* Khrot and *M. orientale* allowed the least disease progress, thereby indicating some resistance to the colonization of the fungus.

The shoot and root segments also showed the mycelial growth. The lesion size was generally larger than the mycelial growth as in leaves. The bark on twigs and roots turned dark brown followed by drying and blackening. It is evident from the Fig. 3 (a & b) that shoot segments of *M. simcoe*, *M. baccata* Khrot and *M. orientale* have minimum mycelial growth and necrosis whereas growth was maximum on M27 followed by *M. baccata* Shillong. Root segments also exhibited almost similar reaction where the minimum mycelial growth was on *M. simcoe*, *M. orientale* and *M. baccata* Khrot. Root segments of M27 and *M. baccata* Shillong and MM106 were highly susceptible. Ram (1982) evaluated the temperate fruits germplasm collected from Himachal Pradesh, Kashmir and north-east India, for resistance to *Dematophora necatrix* and reported that *Cydonia oblongata*, *M. baccata* Japan, *M. baccata* Khrot, *M. floribunda*, *Prunus persica*, *Prunus cornuta*, *Prunus cerasoides* and *Pyrus pashia* var. Kumabni survived in an infected field for 6 years. The cultural filtrate of *D. necatrix* are toxic to apple seedlings and *in vitro* testing of rootstocks for resistance against cultural filtrate revealed that M9, MM101 and 102 were less susceptible whereas Rus Pippin was resistant (Gupta and Singh, 1985).

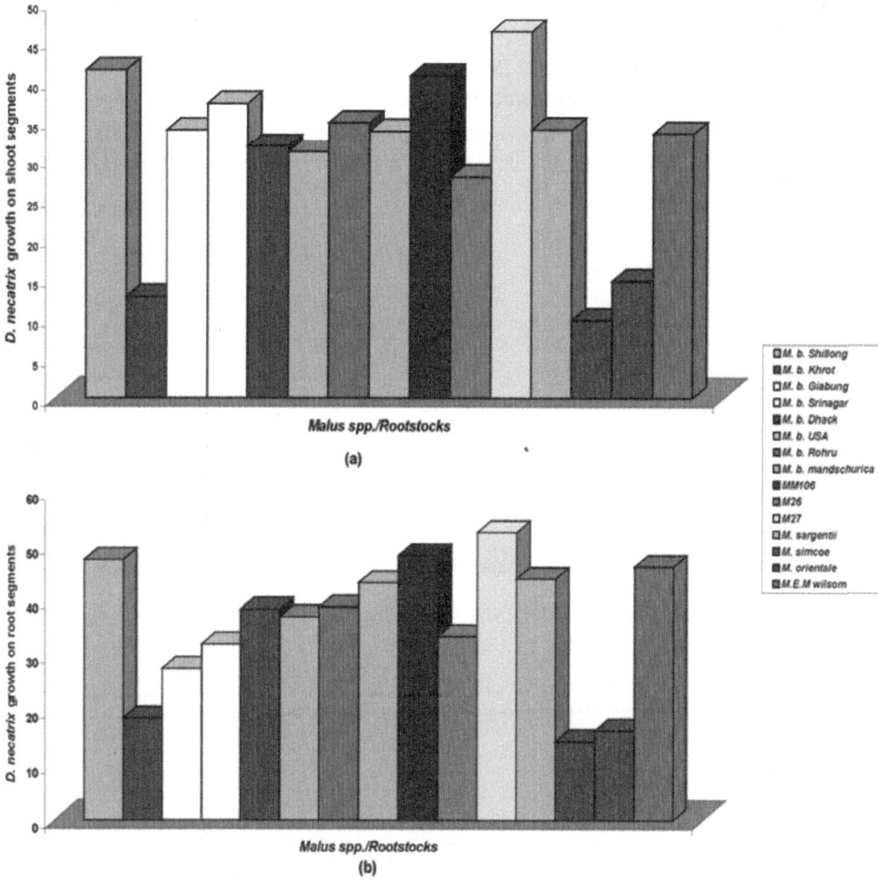

Fig. 3. Response of different *Malus* spp./ rootstocks shoots (a) and root (b) inoculations to *D. necatrix*

The leaves, shoot and roots of *M. simcoe, M. baccata* Khrot, and *M. orientale* exhibited almost similar response to *D. necatrix.* This indicated that these segments could be used for screening the genotypes for resistance to *D. necatrix* and the results of the *in vitro* tests matched with the known field reaction of these rootstocks. Hence, the technique will be very useful for screening large number of genotypes and particularly in those cases where the availability of plant material is very limited.

REFERENCES

Behdad, E. (1975). Morphology, distribution, importance and hosts of *Rosellinia necatrix* (Hart.) Berl. the cause of white root rot in Iran. *Iran. J. Plant Pathol.* 11(1/2): 13-17, 30-71.

Gupta, V.K. and Singh, N. (1985). Studies on toxic compound produced in cultural filtrate of *Dematophora necatrix*. *J. Tree Sci.* 40: 39-43.

Gupta, V.K. and Verma, K.D. (1978). Comparative susceptibility of apple rootstocks to *Dematophora necatrix*. *Indian Phytopathol.* 31: 377-378.

Ito, S.I. and Nakamura, N. (1984). An outbreak of white root rot and its environmental conditions in the experimental arboretum. *J. Jpn. For. Soc.* 66: 262-267.

Ram, R.D. (1982). Evaluation of wild germplasm of temperate fruits against root rot of apple caused by *Dematophora necatrix* Hartig. *Prog. Hortic.* 14: 4, 249-251.

Sharma, S.K. (1993). *Studies on Management of White Root Rot of Apple.* Ph.D. Thesis, Dr. Y.S. Parmar Univ. of Hort. and Forestry, Nauni, Solan, HP. 146 p.

Integrated Plant Disease Management (2005)
Eds. Sharma, R.C. and Sharma, J.N.
Scientific Publishers (India), Jodhpur, pp. 39-46.

6

Role of bacterial antagonists in the management of white root rot of apple

S.K. SHARMA, MAMTA SHARMA and MEENU GUPTA

Department of Mycology and Plant Pathology, Dr.Y.S. Parmar University of Horticulture and Forestry, Nauni, Solan - 173 230 (H.P.)

ABSTRACT

White root rot (*Dematophora necatrix* Hartig) is one of the most important soil borne diseases of apple. In the present study, three strains of *Pseudomonas fluorescens* i.e. IISR-8, $T_1R_2K_4$ and 81 and *Bacillus subtilis* were evaluated against *Dematophora necatrix* under *in vitro* as well as pot culture conditions. *P. fluorescens* IISR-8, $T_1R_2K_4$ and *Bacillus subtilis* were found more inhibitory giving 61.29, 59.62 and 56.11 per cent inhibition, respectively, under *in vitro* conditions. In pot culture experiments, maximum disease reduction with all the antagonists was achieved when bacterial antagonists were applied 6 and 8 days prior to pathogen inoculation in comparison to 2 and 4 days and improved the health of apple seedlings significantly.

INTRODUCTION

White root rot of apple (*Dematophora necatrix* Hartig) causes severe losses both in nursery as well as in grown up orchards and is a limiting factor in raising healthy nursery of apple. Seedlings are killed within a short period depending upon the age of the host and environmental factors. The control of soil borne pathogens is difficult as these pathogens are deep in the soil and the fungicides when applied as a soil drench fail to reach the target pathogen due to adsorption on soil particles. The interest in alternatives that reduce our dependence on agrochemicals has increased during the recent years. The present study was, therefore, aimed to manage the disease using bacterial antagonists and to see their effect on the health of apple seedlings.

MATERIALS AND METHODS

In vitro evaluation of antagonists

Three strains of *Pseudomonas fluorescens* i.e. IISR-8, $T_1R_2K_4$ and 81 procured from IISR, Calicut and *Bacillus subtilis* were screened for their inhibitory effect against *Dematophora necatrix* under *in vitro* conditions. A loopful of 48 h old culture of the antagonists was streaked in the centre of the Petri plates containing potato dextrose agar medium (PDA) while mycelial discs of test fungus (7 mm diameter), cut with the help of cork borer from the seven days old culture, were placed simultaneously on either side of the Petri plates. Medium inoculated only with test pathogen served as control. Each treatment was replicated four times. The inoculated plates were kept in BOD incubators at 25±1°C. The observations on radial growth were recorded after 14 days and per cent inhibition was calculated (Vincent, 1947).

$$\text{Growth inhibition (\%)} = \frac{C-T}{C} \times 100$$

Where,

C = Linear growth of fungus in control

T = Linear growth of fungus in treatment

Mass culture of antagonist

The mass culture of bacterial antagonists was prepared in nutrient broth medium in which a loopful of bacterial culture was transferred aseptically into test tubes containing 5 ml of sterilized nutrient broth at 27°C for 48 h. Subsequently, the actively growing culture was transferred to 100 ml nutrient broth at 27°C for 48 h.

Preparation of inoculum of the test fungus

The mass culture of *Dematophora necatrix* was prepared on wheat grains. Wheat grains were soaked in water for 12 h and thereafter boiled for 30 minutes. Boiled wheat grains were supplemented with sucrose (50 g) and sand (200 g) per 2 kg of grains after excess water had been drained off and mixed thoroughly. The mixture was sterilized in autoclave at 1.5 kg cm^{-2} pressure for 30 minutes in heat resistant polypropylene bags plugged with non-absorbent cotton. The medium was inoculated with 3 to 4 fungal bits of two weeks old culture of test fungus under aseptic conditions and incubated at 25±1°C.

In vivo evaluation of antagonists

Bacillus subtilis and three strains of *P. fluorescens* (IISR-8, $T_1R_2K_4$ and 81) which showed antagonistic activity against *D. necatrix* under *in vitro* conditions were further evaluated under *in vivo* conditions. Three one-year-old apple seedlings were planted in each pot (15 cm diameter) in February containing sterilized sandy loam soil. The soil was amended with 100 ml of bacterial suspension of each bacterial culture separately having a population density of $10^6 - 10^8$ c.f.u. ml^{-1}. Wheat grain inoculum of *D. necatrix* was added in each pot after 0, 2, 4, 6 and 8 days of bacterial inoculation. Pots inoculated only with test fungus served as control. The experiment was laid out in completely randomized block design with five replications. Plants were observed regularly for symptom appearance. Data on disease incidence were recorded after 15, 30, 45 and 60 days after pathogen inoculation. Plant health parameters shoot and root length as well as fresh and dry weight of the roots of seedlings were recorded after sixty days of pathogen inoculation and per cent increase over control was calculated.

RESULTS AND DISCUSSION

In vitro evaluation of antagonists

All the bacterial antagonists namely, *Pseudomonas fluorescens* IISR-8, *P. fluorescens* $T_1R_2K_4$, *P. fluorescens* 81 and *Bacillus subtilis* were antagonistic to *Dematophora necatrix*, however, *P. fluorescens* IISR-8 followed by *P. fluorescens* $T_1R_2K_4$ and *Bacillus subtilis* were more inhibitory giving 61.29, 59.62 and 56.11 per cent inhibition, respectively (Table 1). *P. fluorescens* IISR-8 and *P. fluorescens* $T_1R_2K_4$ produced maximum inhibition zone of 11.25 and 11.20 mm, respectively. Earlier several workers have also reported inhibitory effect of fluorescent pseudomonads and *Bacillus subtilis* against *Rosellinia necatrix* (Utkhede and Li, 1989; Gupta and Jindal, 1990; Cattelan, 1994; Lopez *et al.*, 1998).

Table 1. *In vitro* evaluation of bacterial antagonists against *Dematophora necatrix*

Antagonist	Growth inhibition (%)	Inhibition zone (mm)
Bacillus subtilis	56.11 (48.52)	7.00
P. fluorescens IISR-8	61.29 (51.53)	11.25
P. fluorescens $T_1R_2K_4$	59.62 (50.55)	11.20
P. fluorescens 81	55.37 (48.09)	6.85
$CD_{0.05}$	(3.43)	

Figures in parentheses are arcsine transformed values

In vivo evaluation of antagonists

Biological potential of *P. fluorescens* IISR-8, *P. fluorescens* $T_1R_2K_4$, *P. fluorescens* 81 and *Bacillus subtilis* which showed inhibitory effect against *D. necatrix in vitro* were further tested under pot culture conditions.

Effect of *P. fluorescens* IISR-8

It is evident from the data (Table 2) that disease did not appear up to sixty days when pathogen was applied 4, 6 and 8 days after the application of *P. fluorescens* IISR-8. Incidence of root rot was maximum (51.39%) when pathogen was applied simultaneously with the bacterial inoculation followed by the treatment when pathogen was added 2 days after bacterial application (40.28%). There was no disease development in any of the treatments up to 15 days after pathogen inoculation except control where 69.45 per cent disease incidence was recorded. Symptom appearance was delayed by 8-13 days in treatments when pathogen was inoculated simultaneously or 2 days after bacterial application in comparison to control where first symptom appeared 13 days after inoculation.

Table 2. Effect of *Pseudomonas fluorescens* IISR-8 on the incidence of white root rot of apple

Pathogen inoculation (after days)	Symptom appear-ance (after days)	Per cent disease incidence (after days)				Overall mean	Per cent disease control
		15	30	45	60		
0	21	0.00	50.00	72.22	83.33	51.39	44.36
		(0.00)	(45.00)	(63.25)	(75.00)	(45.81)	
2	26	0.00	27.78	55.56	77.28	40.28	56.39
		(0.00)	(26.75)	(48.25)	(66.49)	(35.37)	
4	—	0.00	0.00	0.00	0.00	0.00	100.00
		(0.00)	(0.00)	(0.00)	(0.00)	(0.00)	
6	—	0.00	0.00	0.00	0.00	0.00	100.00
		(0.00)	(0.00)	(0.00)	(0.00)	(0.00)	
8	—	0.00	0.00	0.00	0.00	0.00	100.00
		(0.00)	(0.00)	(0.00)	(0.00)	(0.00)	
Control	13	69.45	100.00	100.00	100.00	100.00	—
		(56.49)	(90.00)	(90.00)	(90.00)	(90.00)	
Overall mean		11.56	29.63	37.96	43.44	—	—
		(9.42)	(26.96)	(33.58)	(38.58)		

Figures in parentheses are arcsine transformed values

$CD_{0.05}$

Treatment = (8.12)

Interval = (6.63)

Treatment x Interval = (16.23)

Effect of *P. fluorescens* $T_1R_2K_4$

Data on per cent disease incidence (Table 3) as influenced by *P. fluorescens* $T_1R_2K_4$ show that disease incidence was minimum (20.83%) when pathogen was added 4 days after the application of bacterial antagonist, giving 79.17 per cent disease control followed by 8 and 6 days which gave

73.61 and 70.63 per cent disease control and were statistically at par with each other. Overall progress of the disease was minimum (16.67%) after 15 days of inoculation which increased to 66.67 per cent after 60 days. Symptoms were delayed by about 12–16 days in all the treatments as compared to control where first symptom appeared after 12 days.

Table 3. Effect of *Pseudomonas fluorescens* $T_1R_2K_4$ on the incidence of white root rot of apple

Pathogen inoculation (after days)	Symptom appear-ance (after days)	Per cent disease incidence (after days)				Overall mean	Per cent disease control
		15	30	45	60		
0	25	0.00	16.67	55.55	66.67	34.72	65.28
		(0.00)	(15.00)	(53.51)	(60.00)	(32.13)	
2	26	0.00	33.33	33.33	66.67	33.33	66.67
		(0.00)	(35.26)	(35.26)	(60.00)	(32.63)	
4	28	0.00	16.67	33.33	33.33	20.83	79.17
		(0.00)	(15.00)	(35.26)	(35.26)	(21.38)	
6	27	0.00	16.67	33.33	66.67	29.17	70.83
		(0.00)	(15.00)	(35.26)	(60.00)	(27.57)	
8	28	0.00	11.11	27.78	66.67	26.39	73.61
		(0.00)	(11.75)	(26.75)	(60.00)	(24.63)	
Control	12	100.00	100.00	100.00	100.00	100.00	—
		(90.00)	(90.00)	(90.00)	(90.00)	(90.00)	
Overall mean		16.67	32.41	47.22	66.67		—
		(15.00)	(30.34)	(46.01)	(60.88)		

Figures in parentheses are arcsine transformed values

$CD_{0.05}$

Treatment = (9.72)

Interval = (7.80)

Treatment x Interval = (20.50)

Effect of *P. fluorescens* 81

Data presented in Table 4 indicated that disease control varied between 80.00 to 82.85 per cent when pathogen was added 2, 4, 6 and 8 days after antagonist application and was minimum (68.57%) when antagonist and pathogen were added simultaneously. Interaction studies showed that all the treatments gave complete control of root rot up to 15 days of pathogen inoculation as compared to control where 88.89 per cent disease incidence was recorded. With the increase in number of days the disease incidence also increased in all the treatments being maximum (61.11%) in the simultaneous inoculation treatment followed by 2 days (50.00%), 4 days (38.89%), 8 days (38.89%) and was minimum in 6 days treatment (33.33%) after 60 days of pathogen inoculation.

Table 4. Effect of *Pseudomonas fluorescens* 81 on the incidence of white root rot of apple

Pathogen inoculation (after days)	Symptom appearance (after days)	Per cent disease incidence (after days)				Overall mean	Per cent disease control
		15	30	45	60		
0	22	0.00	16.67	44.44	61.11	30.56	68.57
		(0.00)	(15.00)	(41.56)	(51.49)	(27.06)	
2	32	0.00	0.00	27.78	50.00	19.44	80.00
		(0.00)	(0.00)	(26.75)	(45.00)	(17.94)	
4	33	0.00	0.00	27.78	38.89	16.67	82.85
		(0.00)	(0.00)	(26.75)	(33.25)	(15.00)	
6	28	0.00	16.67	16.67	33.33	16.67	
		(0.00)	(15.00)	(15.00)	(30.00)	(15.00)	82.85
8	29	0.00	16.67	16.67	38.89	18.06	81.42
		(0.00)	(15.00)	(15.00)	(33.25)	(15.81)	
Control	12	88.89	100.00	100.00	100.00	97.22	
		(73.94)	(90.00)	(90.00)	(90.00)	(85.99)	
Overall mean		14.82	25.00	38.89	53.70		
		(12.32)	(22.50)	(35.88)	(47.17)		

Figures in parentheses are arcsine transformed values

$CD_{0.05}$

Treatment = (15.22)

Interval = (11.51)

Treatment x Interval = (30.45)

Symptom appearance was delayed by 20 to 21 days when pathogen was added after 2 and 4 days of antagonist application. With the increase in number of days after pathogen inoculation, the symptom appearance was delayed by about 4 to 5 days; however, earliest symptoms were observed when pathogen and antagonist were applied simultaneously.

Effect of *Bacillus subtilis*

Symptoms did not appear up to 60 days when pathogen was applied 6 and 8 days after the application of antagonist (Table 5). Disease incidence was minimum (36.94%) when pathogen was applied 4 days after the bacterial application followed by 2 days (70.83%) and it was maximum (90.67%) when pathogen was applied simultaneously with the application of *Bacillus*. Interaction between treatment and interval show that in any of these treatments there was no disease after 15 days of inoculation as compared to 66.67 per cent disease incidence resulting from simultaneous application of antagonist. Disease incidence increased to 100 per cent after 30 and 45 days when pathogen was applied simultaneously and after 2 days of bacterial application. Symptom appearance was delayed by about 7 and 14 days when pathogen was applied 2 and 4 days after bacterial treatment, respectively, as compared to control where first symptom appeared after 12 days of inoculation.

Table 5. Effect of *Bacillus subtilis* on the incidence of white root rot of apple

Pathogen inoculation (after days)	Symptom appearance (after days)	Per cent disease incidence (after days)				Overall mean	Per cent disease control
		15	30	45	60		
0	13	66.67	100.00	100.00	100.00	91.67	8.33
		(54.74)	(90.00)	(90.00)	(90.00)	(81.19)	
2	19	0.00	83.33	100.00	100.00	70.83	29.17
		(0.00)	(65.90)	(90.00)	(90.00)	(61.48)	
4	26	0.00	33.33	50.00	64.45	36.94	63.06
		(0.00)	(35.26)	(45.00)	(53.42)	(33.42)	
6	—	0.00	0.00	0.00	0.00	0.00	100.00
		(0.00)	(0.00)	(0.00)	(0.00)	(0.00)	
8	—	0.00	0.00	0.00	0.00	0.00	100.00
		(0.00)	(0.00)	(0.00)	(0.00)	(0.00)	
Control	12	100.00	100.00	100.00	100.00	100.00	
		(90.00)	(90.00)	(90.00)	(90.00)	(90.00)	
Overall mean		23.81	45.24	50.00	52.05		
		(20.68)	(40.17)	(45.00)	(46.20)		

Figures in parentheses are arcsine transformed values

$CD_{0.05}$

Treatment = (1.55)

Interval = (1.17)

Treatment x Interval = (3.09)

Gupta and Jindal (1990) recorded reduced incidence of white root rot of apple in pots when *Bacillus subtilis* and *Enterobacter aerogenes* were applied as soil drench prior to the pathogen inoculation. Sharma (1993) have also reported that pre-inoculation of *E. aerogenes* gave good control of white root rot both in pot culture and field conditions.

Effect on plant health

The data presented in Table 6 show that all the bacterial antagonists significantly increased the growth of apple seedlings in comparison to control. Significant increase in plant growth parameters was registered in *P. fluorescens* IISR-8 giving 33.11, 214.76, 50.36 and 142.27 per cent increase in shoot length, root length, fresh root weight and dry root weight, respectively. This was followed by *P. fluorescens* $T_1R_2K_4$, *P. fluorescens* 81 and *Bacillus subtilis*. The strains of *Pseudomonas fluorescens* and *Bacillus subtilis* have also been reported to increase the plant biomass of apple significantly by various workers (Caesar and Burr, 1987; Utkhede and Li, 1989; Utkhede, 1999).

Table 6. Effect of bacterial antagonists on the growth of apple seedlings

Bacterial antagonist	Shoot length (cm)	Root length (cm)	Fresh root weight (g)	Dry root weight (g)
Bacillus subtilis	57.50 (52.32)	36.50(167.99)	7.02 (27.64)	4.12 (87.27)
P. fluorescens IISR-8	50.25 (33.11)	42.87 (214.76)	8.27 (50.36)	5.33 (142.27)
P. fluorescens $T_1R_2K_4$	49.83 (32.00)	41.25 (202.86)	8.00 (45.45)	4.26 (93.64)
P. fluorescens 81	44.75 (18.54)	36.00 (164.32)	7.12 (30.91)	4.13 (87.73)
Control	37.75	13.62	5.50	2.20
$CD_{0.05}$	11.23	13.57	4.61	2.83

Figures in parentheses are values of per cent increase over control

REFERENCES

Caesar, A.J. and Burr, T.J. (1987). Growth promotion of apple seedlings and rootstocks by specific strains of bacteria. *Phytopahtology* 77: 1583-1588.

Cattelan, A.J. (1994). Antagonism of fluorescent pseudomonads to phytopathogenic fungi from soil and soybean seeds. *Rev. Brasi. Ciencia do Solo* 18: 37-42.

Gupta, V.K. and Jindal, K.K. (1990). Management of white root rot of apple (*Rosellinia necatrix*) by biological and cultural methods. *Proc. 23rd Int. Hortic. Congr.,* Italy, August 2-11, 1990, Vol. 2. 3229 (Abstr.).

Lopez, J.C., Ropas, E.B., Rivas, P.G.G., Rivillas, O.C.A. and Perez, L.C.M. (1998). Isolation of fluorescent *Pseudomonas* potential antagonist of *Rosellinia necatrix* in solarized soils. *Eur. J. Plant Pathol.* 105: 571-576.

Sharma, S.K. (1993). *Studies on Management of White Root Rot of Apple.* Ph.D. Thesis, UHF, Nauni, Solan, 146 p.

Utkhede, R.S. (1999). Biological treatments to increase apple tree growth in replant problem soil. *Allelopathy J.* 6: 63-68.

Utkhede, R.S. and Li, T.S.C. (1989). Evaluation of *Bacillus subtilis* for potential control of apple replant disease. *J. Phytopathol.* 126: 305-312.

Vincent, J.M. (1947). Distortion of fungal hyphae in the presence of certain inhibitors. *Nature* 150: 850.

Integrated Plant Disease Management (2005)
Eds. Sharma, R.C. and Sharma, J.N.
Scientific Publishers (India), Jodhpur, pp. 47-52.

7

Effect of decontamination processes on residues of carbendazim in apple fruits following dip treatment

I.D. SHARMA and AMIT NATH

Department of Entomology and Apiculture, Dr. Y.S. Parmar University of Horticulture and Forestry, Nauni, Solan – 173 230 (H.P.)

ABSTRACT

Fruits of apple are generally given dip treatment of carbendazim (0.1%) before these are kept in storage or transported to far of distances to control rotting. Since an apple fruit is consumed as raw, the present studies were undertaken to determine the effect of decontamination processes like washing followed by peeling on the level of carbendazim residues following dip treatment. Uniform good quality apple fruits (cv. Golden Delicious) were given 4 and 8 minutes dip treatment in 500 and 1000 ppm carbendazim (Bavistin 50 WP) suspension separately. From each treatment, half of the fruits were placed at ambient temperature (8-24°C) and the other half in cold storage (0-2°C). The peeling of fruits was found effective in removing the carbendazim residues to more than 95 per cent up to third day and afterwards the residue level increased with time i.e. 0.35 – 0.91 ppm at ambient temperature (up to 30 days) and 0.48 – 1.70 ppm in cold storage (up to 120 days). Washing following by peeling removed the residues by more than 80 per cent even up to seventh day and even afterwards the residue value obtained were lower than the acceptable limits (5 ppm) by World Health Organization at different time intervals irrespective of dipping time, dosage and storage condition.

INTRODUCTION

Apples are generally given dip treatment of carbendazim (0.1%) before these are kept in storage or transported to far of distances to control their rotting. The dipping of apple fruits in carbendazim prevents the storage rot caused by *Penicillium expansum*, *Aspergillus niger*, *Haplospora mali* and *Lasidiplodia theobromae* (Sharma and Kaul, 1999). However, the data available on the effect of decontamination process on the residual level of fungicide that persists in the fruits at the time of consumption is scanty. Therefore, the present studies were undertaken to determine the effect of decontamination processes like washing followed by peeling on the level of carbendazim residues at ambient temperature and cold storage following dip treatment.

MATERIALS AND METHODS

High quality medium sized apple (cv. Golden Delicious) fruits (100 kg) were used to determine the effect of simple decontamination processes on residues of carbendazim. The major portion of fruits was divided in two equal lots. Both lots of fruits were separately plunged into their respective baths each containing 500 and 1000 ppm carbendazim. From each bath half of the apples were removed immediately after 4 minutes and the rest were left for 8 minutes. The treated fruit lots were allowed to dry at room temperature (23°C) and from each lot half of the air dried fruits were placed in refrigerated storage (2-4°C) with a mean relative humidity of 90-95 per cent for 120 days. Similarly, the other half of each lot was kept at ambient temperature (8-24°C) with a relative humidity 60$\pm$2 per cent for 30 days. For control, the apples were given water treatment and for each treatment separate controls were maintained. Each treatment was replicated thrice and from each replication six apples were drawn randomly for residue analysis. The apples were chopped, mixed and 100g representative sample was extracted directly with ethyl acetate and concentrated ammonia by blending in a blender. The clean-up and residue estimation was done by the spectrophotometric method described by Nath *et al.* (1993). The residues in whole and peeled fruits were estimated simultaneously at different time intervals over a period beginning from 0 day after dip to 30 and 120 days at ambient temperature and in cold storage, respectively. Pulp samples were obtained by removing the peel (not more than 1 mm thick). During peeling, precautions were taken to ensure that pulp remained unaffected from contamination.

RESULTS AND DISCUSSION

The average recoveries of carbendazim in apple fruits fortified at two levels (0.5 and 1 ppm) varied from 82-85 per cent. Following dipping for 4 minutes in 500 and 1000 ppm carbendazim suspension, the apple contained 4.87 and 6.58 ppm initial deposits while for 8 minutes dipping, the deposits were 6.24 and 7.68 ppm at the corresponding concentrations (Fig. 1, 2). This indicated that the amount of deposits recorded on the treated fruits depend on the contact time. Persistence of carbendazim was observed more in cold storage than at ambient temperature. At ambient temperature, 50 per cent of the residues lost with in 10 days and on 30th day residues were found in traces (about 1 ppm) and similar results have been reported earlier by Jain and Agnihotri (1987). In cold storage, half of the residues dissipated with in 60 days and these gradually lower down to about 1 ppm within 120 days after treatment. These studies are in agreement with the findings of Monico and Xiram (1987). The rate of dissipation in the beginning was dependent upon the initial deposits but in the later stages, it was found almost same irrespective of treatments and time.

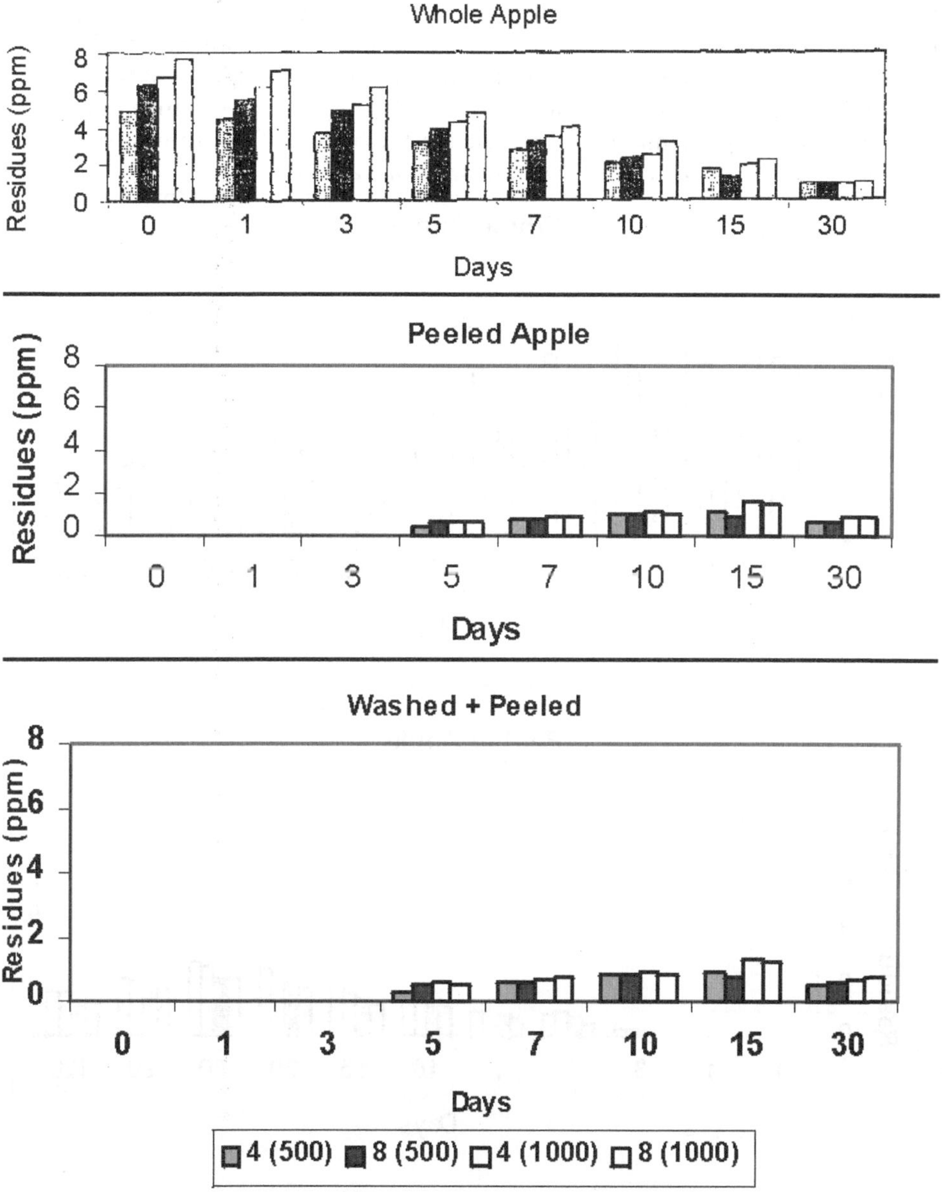

Dip time 4 and 8 min.; Doses 500 and 1000 ppm

Fig. 1. Persistence of carbendazim residue in/on apple at ambient temperature

The simple decontamination process like peeling alone was effective in removing the carbendazim residues to a great extent, providing a relief of 84-88 per cent at ambient temperature and 83-88 per cent in cold storage after 5 days of treatment (Table 1). Washing followed by peeling was comparatively more effective as it was found to remove the residues by more than 90 per cent on the same day of treatment. The effect of this simple decontamination process was found to be in decreasing trend with the passage of time, as on 30th day at ambient temperature the relief provided by peeling and washing plus peeling reduced to 11-33 and 69-47 per cent, respectively. In cold storage, residues dissipated slowly and after 120 days of treatment, relief noticed was about 8-13 per cent. The decrease in efficacy may be attributed to the fact that traces of carbendazim might have translocated to pulp with the passage of time. In earlier study, the relief provided by peeling alone and washing + peeling for mancozeb residues after 5 days of treatment was 53-58 and 90-91 per cent, respectively (Sharma *et al.*, 2001).

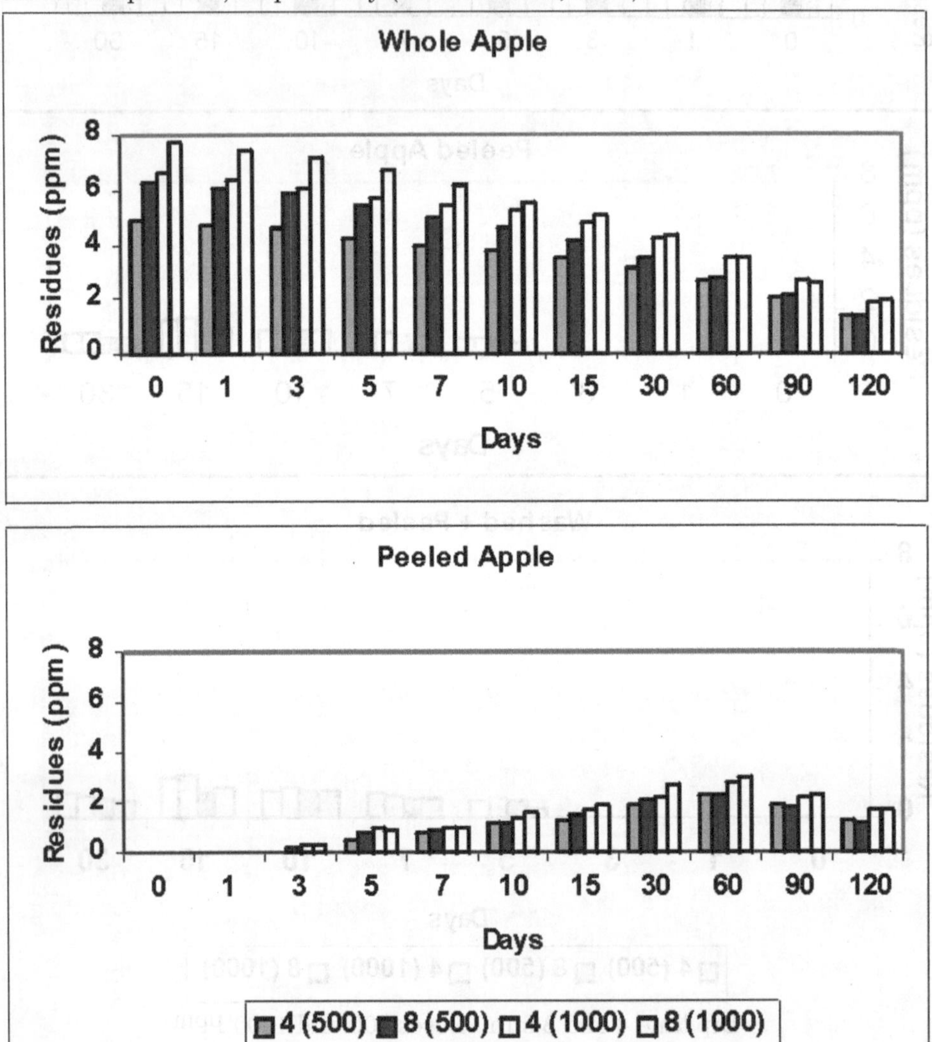

Fig. 2. Persistence of cabendazim residues on apple in cold storage

From this study, it can be concluded that more than 80 per cent residues remained on the peel for 4-5 days and afterwards these penetrated into the pulp. The residues detected in the peeled fruits were almost similar to that of whole fruits during the last days of storage which indicated that skin/peel of fruits has become free of carbendazim and what so ever traces of pesticide are detected these are present in the pulp of fruits. Even at recommended dose (500 ppm), the residues obtained following 4 min dipping were lower than those considered acceptable (5 ppm) by the World Health Organization (1974). From the consumption of fresh fruits and processing point of view, the simple decontamination process like peeling and washing assume considerable significance.

Table 1. Per cent relief from carbendazim residues in processed apple fruits

Treatment	Dose (ppm)	Dip time (min)	Per cent relief at days after treatment									
			1	3	5	7	10	15	30	60	90	120
Ambient temperature	500	4			88.81 (91.05)	73.31 (78.57)	52.47 (61.88)	30.86 (44.44)	33.71 (47.19)			
		8			84.90 (88.02)	76.89 (81.65)	53.81 (63.23)	33.85 (42.30)	22.35 (37.65)			
	1000	4			84.24 (87.29)	75.57 (80.46)	52.48 (61.98)	19.39 (35.71)	13.68 (30.53)			
		8			86.05 (89.47)	77.81 (81.79)	68.12 (74.61)	29.30 (43.26)	11.65 (29.13)			
Cold storage	500	4			88.54	80.71	68.36	61.27	39.10	15.67	9.27	8.03
		8			84.20	82.16	72.87	64.56	38.40	18.70	15.71	11.76
	1000	4			83.01	81.30	72.33	64.58	47.64	19.77	16.60	11.23
		8			86.21	83.22	70.91	63.33	34.49	15.25	12.31	13.71

Figures in parentheses are per cent relief after washing followed by peeling

REFERENCES

Jain, H.K. and Agnihotri, N.P. (1987). Presence of thiophanate methyl and MBC residues in apple following post harvest dip. *Pesticides* 21: 31-32.

Monico, P. and Xiram, V. (1987). Monitoring residues of carbendazim (applied as benomyl) and thiabendazole in Wellspur apples. *J. Assoc. Official Anal. Chem.* 70: 596-598.

Nath, A., Patyal, S.K. and Dubey, J.K. (1993). A new protocol for rapid sample preparation for spectrophotometric estimation of carbendazim residues in apple, tomato and mushroom. *J. Food Sci. Technol.* 30: 239-240.

Sharma, I.D., Patyal, S.K., Nath, A. and Joshi, V.K. (2001). Effect of processing on mancozeb residues in apple. *J. Food Sci. Technol.* 38: 23-25.

Sharma, R.C. and Kaul, J.L. (1999). Postharvest diseases of temperate fruits and their management. In: *Diseases of Horticultural Crops: Fruits* (Eds. Verma, L.R. and Sharma, R.C.), Indus Publishing Company, New Delhi, pp. 583-623.

World Health Organization (1974). *Technical Report Series No. 545,* WHO, Geneva, pp. 1-48.

Integrated Plant Disease Management (2005)
Eds. Sharma, R.C. and Sharma, J.N.
Scientific Publishers (India), Jodhpur, pp. 53-58.

8

Susceptibility of pear cultivars to stem end rot in relation to fruit qualitative characters

R.L. SHARMA, DEEPIKA SOOD and MANJU CHOUDHARY

Department of Mycology and Plant Pathology, Dr.Y.S. Parmar University of Horticulture and Forestry, Nauni, Solan –173 230 (H.P.)

ABSTRACT

Fruits of 23 pear cultivars were found susceptible to stem end rot pathogen. Cultivars Luhimitsa, Napoleon, Autumn of Yakolove and Smart pear were highly susceptible, while Bargument, Beurre Hardy, Max Red Bartlett, Kashmir pear and Beurre Baltira Giffard showed some tolerance. There existed a significant negative simple correlation between the stem end rot pathogen and fruit firmness, total and OD phenols whereas for the remaining factors (total soluble solids, total, reducing and non-reducing sugars, pH and titratable acidity) the correlation was positive. The positive correlation was significant only with pH and titratable acidity. The regression coefficient of variables on the prevalence of stem end rot indicated that fruit firmness and pH were the main variables affecting disease development significantly.

INTRODUCTION

Stem end rot caused by *Phacidiopycnis piri* (Fuckel) Weindlmayr is responsible for considerable post-harvest losses of pears in Himachal Pradesh (Sharma, 1991). Information, however, is not available on the effect of various qualitative characters of fruit in relation to post-harvest rotting due to *P. piri*. The present investigations were therefore, initiated to know the comparative behaviour of fruits of various pear cultivars towards *P. piri* and influence of qualitative characters individually and in combination on the development of stem end rot.

MATERIALS AND METHODS

The pathogen *P. piri* was isolated from the naturally infected pear fruit and maintained on potato dextrose agar medium. Fresh and apparently healthy pear fruits of 23 cultivars with uniform ripeness and size were inoculated (Grainger and Horne, 1924) with 24 h active culture of *P. piri*. Each cultivar was replicated thrice having 20 fruits in each replication. Inoculated and uninoculated fruits were kept at 25±0.5°C and percentage rot was calculated after five days (Srivastava and Tandon, 1968).

The qualitative characters in terms of total soluble solids (TSS), acidity, flesh firmness, pH, reducing sugars, non reducing sugars, total phenols, orthodihydric (OD) phenols and ascorbic acid were assayed from healthy fruits of each cultivar. The healthy fruit tissue of each cultivar was extracted with 80 per cent ethanol in hot water bath and total phenols, orthodihydric (OD) phenols, reducing sugars, non reducing sugars and ascorbic acid were estimated (Mahadevan and Sridhar, 1982). Fresh fruit flesh was grated for each treatment and juice extracted. The pH of juice was determined using "Elico" digital pH meter. The total soluble solids content was determined with "Erma" hand refractometer by putting a drop of juice and taking the reading. Firmness of flesh was measured by Effigi F-310 penetrometer with 9/16-inch plunger by removing a thin slice of the skin and expressed as firmness in pounds. Titratable acidity was determined in terms of malic acid by titrating pulp extract against 0.1 N NaOH using phenolphthalein as indicator.

Statistical analysis

In order to assess critically the contribution of variables individually towards rot causing pathogen, data were analyzed statistically and simple correlations worked out (Golden, 1952). The combined effect of all the six qualitative characters on rot development was estimated using multiple regression technique. The least square method of estimation of parameters of specified relationship between independent and dependent variables assumes the following set up:

$$Y= b_0 + b_1x_1 + b_2x_2 + b_3x_3 + b_4 x_4+ b_5x_5 + b_6x_6 + b_7x_7 + b_8x_8 + b_9x_9 + b_{10}x_{10} + e$$

Where, dependent variable Y being disease and independent variables x_1, x_2, x_3, x_4, x_5, x_6, x_7, x_8, x_9 and x_{10} being flesh firmness, total soluble solids (TSS), pH, acidity, total sugars, reducing sugars, non reducing sugars, ascorbic acid, total phenols and orthodihydric (OD) phenols, respectively; b_1, b_2, b_3, b_4, b_5, b_6, b_7, b_8, b_9 and b_{10} represent their partial regression coefficients and b_0 the intercept with Y-axis and e is error term. Significance of simple correlation coefficient and regression coefficient between the variables was tested by t-test.

RESULTS AND DISCUSSION

Susceptibility of pear fruit cultivars to *Phacidiopycnis piri*

The behaviour of 23 pear fruit cultivars revealed (Table 1) that none was resistant to *P. piri*. Cultivar Luhimitsa Kalapa was most susceptible as it accounted for more than 87 per cent rot after five days of inoculation, followed by Napoleon, which revealed insignificant differences with Napoleon, Autumn of Yakolove and Smart pear varieties. This might be due to suitable range of pH (3.72-3.80) of the fruit juice resulting in fast growth and development of the pathogen (Sood, 2001). Beurre Capiaumont, Starking Delicious, Beurre Bosc and Russian EC were more susceptible as compared to Ishiiwase, Jargonelle, Laxton's Superb and William Bone Chretien. Varieties, Manning Elizabeth and Sentya Braskaya showed higher susceptible reaction than Baggugosha and Flemish Beauty. Seckel and Lisnova Karasvitsa were less susceptible. Bargument, Beurre Hardy, Max Red Bartlett, Kashmir pear and Beurre Baltira Giffard showed less rot and could be rated as somewhat tolerant varieties.

Table 1. Susceptibility of pear cultivars to stem end rot (*P. piri*) in relation to their fruits qualitative characters

Pear cultivar	Mean per cent rot*	Fruit qualitative characters									
		Fruit firmness (kvs.sq inch)	TSS Brix	pH	Titrat-able acidity (%)	Total sugar (mg/g)	Reducing sugars (mg/g)	Non-reducing sugars (mg/g)	Ascorbic acid (mg/g)	Total phenol (mg/g)	OD phenols (mg/g)
		X_1	X_2	X_3	X_4	X_5	X_6	X_7	X_8	X_9	X_{10}
Autumn of Yakolove	75.93 (60.69)[bc]	19.50	14.20	3.72	0.54	80.00	68.80	10.60	2.25	1.77	0.78
Baggugosha (Nihari)	33.58 (35.24)[h]	25.00	12.40	3.55	0.35	81.20	67.10	13.40	3.00	1.92	1.08
Bargument	24.94 (29.95)[hi]	26.00	13.40	3.48	0.48	75.20	63.00	11.60	3.00	2.80	1.15
Beurre Baltira Giffard	20.86 (27.12)[i]	26.50	10.20	3.43	0.40	80.20	59.80	19.40	2.40	2.35	1.16
Beurre Bosc	67.31 (55.33)[cd]	21.50	13.00	3.71	0.58	80.90	68.60	11.70	2.25	1.82	0.98
Beurre Capiaumont	72.38 (58.32)[c]	19.50	11.40	3.74	0.48	76.50	63.80	12.10	2.40	1.83	0.84
Beurre Hardy	23.79 (28.99)[hi]	25.00	13.25	3.41	0.60	66.20	47.30	18.00	2.40	2.90	1.70
Flemish Beauty	33.57 (35.38)[h]	23.50	12.80	3.47	0.33	85.00	71.20	13.10	2.70	1.90	0.75
Ishiiwase	57.11 (49.13)[de]	22.00	14.90	3.60	0.49	78.90	66.30	12.00	2.85	1.85	1.20
Jargonelle	56.18 (48.61)[def]	22.00	14.60	3.56	0.59	77.80	62.50	14.50	2.10	1.88	0.78

Continue.....

Pear cultivar	Mean per cent rot*	Fruit qualitative characters									
		Fruit firmness (kvs.sq inch)	TSS Brix	pH	Titratable acidity (%)	Total sugar (mg/g)	Reducing sugars (mg/g)	Non-reducing sugars (mg/g)	Ascorbic acid (mg/g)	Total phenol (mg/g)	OD phenols (mg/g)
		X_1	X_2	X_3	X_4	X_5	X_6	X_7	X_8	X_9	X_{10}
Kashmir Pear	21.04 (27.22)[I]	26.50	12.80	3.46	0.44	80.60	63.20	16.50	2.25	2.70	1.39
Laxton's Superb	53.62 (47.09)[ef]	22.50	14.10	3.52	0.43	78.60	64.30	13.60	2.55	2.15	1.14
Lisnova Karasvitsa	26.55 (30.92)[hi]	25.00	13.00	3.48	0.43	73.50	56.80	15.90	2.40	1.92	0.88
Luhimitsa Kalapa	87.92 (69.76)[a]	16.50	12.60	3.80	0.60	77.80	58.60	18.20	2.15	1.70	0.88
Manning Elizabeth	45.22 (42.24)[fg]	24.00	14.40	3.51	0.46	79.20	66.60	12.00	3.00	2.30	1.05
Napoleon	84.63 (66.92)[ab]	19.00	12.20	3.80	0.61	79.40	56.70	21.60	2.10	2.70	1.39
Russian EC	64.44 (53.42)[c]	22.00	11.20	3.68	0.60	76.80	54.80	20.50	3.00	1.80	0.95
Seckel	29.55 (32.83)[hi]	25.00	13.20	3.43	0.33	71.00	56.20	14.10	2.70	2.32	1.22
Sentya Braskaya	38.81 (38.53)[g]	24.00	14.80	3.50	0.56	74.30	60.00	13.80	2.45	1.90	0.88
Smart Pear	74.19 (59.63)[bc]	20.00	11.80	3.75	0.47	74.40	60.60	13.10	2.85	1.90	0.95
Starking Delicious	69.90 (56.75)[c]	21.00	15.40	3.76	0.52	78.80	60.30	17.60	2.25	1.75	0.90
William Bon Chretien	53.47 (47.00)[ef]	23.00	14.20	3.55	0.33	80.90	68.60	11.70	3.00	1.88	0.85
$CD_{0.05}$	(6.82)										

Figures in parentheses are arcsine transformed values

*Figures represented by different alphabets differ significantly

Correlation between fruit quality characters and stem end rot

The simple correlation of stem end rot with fruit firmness, total phenols and OD phenols, as one of the factors was significantly negative (Table 2). Correlation with ascorbic acid was non significantly negative. Non-significant but positive correlations of stem end rot with other factors namely TSS, total, reducing and non-reducing sugars was observed. Significantly positive correlation of fruit rot severity was noticed with pH and titratable acidity only.

Partial correlation coefficients between per cent rot and TSS, pH, titratable acidity, total sugars, ascorbic acid and OD phenols were positive and non-significant except pH which was significantly positive whereas reducing and non reducing sugars and total phenols indicated negative and insignificant correlation. The correlation obtained with fruit firmness was significantly negative.

Table 2. Correlation coefficients (simple, partial and multiple) between per cent rot and fruit firmness (X_1), TSS (X_2), pH (X_3), acidity (X_4), total sugars (X_5), reducing (X_6) and non-reducing sugars (X_7), ascorbic acid (X_8), total phenols (X_9) and OD phenols (X_{10})

Simple correlation

	X_1	X_2	X_3	X_4	X_5	X_6	X_7	X_8	X_9	X_{10}
Per cent rot(Y)	-0.9657*	0.0584	0.9590*	0.5611*	0.2011	0.1041	0.0583	-0.2496	-0.74578*	-0.6075*

Partial correlation

	X_1	X_2	X_3	X_4	X_5	X_6	X_7	X_8	X_9	X_{10}
Per cent rot (Y)	-0.7854*	0.3340	0.7144*	0.2492	0.1514	-0.1410	-0.1412	0.2649	-0.0870	0.1271

Regression coefficient and coefficient of multiple determination (R^2)

	Constant (a)	X_1	X_2	X_3	X_4	X_5	X_6	X_7	X_8	X_9	X_{10}	R^2(%)
Coefficient	-172.64	-4.30*	1.16	72.23*	14.35	6.34	-5.94	-6.33	3.99	-1.86	3.40	97.75
S.E.	96.50	0.98	0.94	20.42	16.10	11.94	12.04	12.81	4.19	6.14	7.67	

*Significant at 1% level of significance

The regression coefficients of variables TSS (x_2), pH (x_3), titratable acidity (x_4), total sugars (x_5), ascorbic acid (x_8) and OD phenols (x_{10}) were positive indicating their greater impact on fruit rot. The regression coefficients of variables namely fruit firmness (x_1), reducing (x_6) and non-reducing sugars (x_7) and total phenols (x_9) were negative indicating their negative impact on fruit rot. However, fruit firmness and pH were the only statistically significant variables. It shows that fruit firmness and pH were the main factors which increased the rot significantly, whereas other factors also increased the disease but to an insignificant level.

Coefficient of multiple determination (R^2) revealed that 97.75 per cent of the total variation in stem end rot was explained by the included variables. The analysis indicate that simple negative correlation existed between different variables namely, fruit firmness, total and OD phenols and ascorbic acid versus *P. piri*, while with the remaining factors the correlation was positive. The multiple linear regression equations on the prevalence of stem end rot indicate that higher levels of pH and titratable acidity increased the susceptibility whereas firmer fruit having higher concentration of phenols increased the tolerance to disease. Increase or decrease in other independent variables had non-significant effect. Kaul (1984) has also observed a negative correlation between acidity and five fungal pathogens of apple.

REFERENCES

Golden, C.H. (1952). *Methods of Statistical Analysis*. Asia Publishing House, New Delhi, 467 p.

Grainger, K. and Horne, A.S. (1924). A method of inoculating the apple. *Ann. Bot.* 38: 213-215.

Kaul, J.L. (1984). Fruit susceptibility in relation to nutritional status of different apple cultivars of fungal rots. *Indian Phytopathol.* 37: 449-452.

Mahadevan, A. and Sridhar, R. (1982). *Methods in Physiological Plant Pathology*. Shivakami Publications, Madras, India, 199 p.

Sharma, R.L. (1991). Prevalence of storage rots of China pear in Himachal Pradesh. *Plant Dis. Res.* 6: 86-88.

Sood, Deepika. (2001). *Studies on Stem End Rot of Pears Caused by Phacidiopycnis piri (Fuckel) Weindlmayr*. Ph.D. Thesis. Dr. Y.S. Parmar University of Horticulture and Forestry, Solan (H.P.) 108 p.

Srivastava, M.P. and Tandon, R.N. (1968). Influence of temperature on Botryodiplodia rot of citrus and sapodilla. *Indian Phytopathol.* 21: 195-197.

Integrated Plant Disease Management (2005)
Eds. Sharma, R.C. and Sharma, J.N.
Scientific Publishers (India), Jodhpur, pp. 59-63.

Novel approach for screening different antagonists against *Agrobacterium tumefaciens* causing crown gall in stone fruits

A. K. GUPTA, BRIJESH KAMAL, SANDEEP KUMAR and R.C. SHARMA

Department of Mycology and Plant Pathology, Dr. Y.S. Parmar University of Horticulture and Forestry Nauni, Solan – 173 230 (H.P.)

ABSTRACT

Screening of different fungal and bacterial isolates from soil against isolates of *Agrobacterium tumefaciens* causing crown gall in stone fruit plants under dual culture could not be achieved properly. As *A. tumefaciens* is a slow growing pathogen and grows only on the specific media (Kado and Heskett and Schroth *et al* medium), which do not favour the growth of other microorganisms. Therefore, in the present investigation screening of known antagonists viz., *Trichoderma viride*, *Bacillus* sp., *Pseudomonas flourescens* and a soil isolate of *A. tumefaciens* were done by prick inoculation of the stem of 25 days old tomato plants. Subsequently pathogenic isolate of *A. tumefaciens* was applied on the same wounds after 24 h of inoculation of different antagonists. Development of galls on tomato stem was recorded as negative whereas no development of galls or retarded growth of galls was considered as positive sign towards the antagonistic activity of tested microorganisms.

INTRODUCTION

Crown gall caused by *Agrobacterium tumefaciens* (Smith and Townsend) Conn. is a serious bacterial disease of stone fruits in nurseries (Kaul and Sharma, 1988). The present status of biological control of crown gall has been recently reviewed by Sharma *et al.* (2001). Although nonpathogenic strains of *A. tumefaciens* = (*A. radiobacter*) have been reported highly effective against pathogenic *A. tumefaciens* yet the non-availability of these in India and their ineffectiveness against agrocin producing strains of *A. tumefaciens* warrants the dependence on other native antagonists which can take care of crown gall disease under field conditions. However, the screening of different known native antagonists under dual culture has been proved a failure, as *A. tumefaciens* is a slow growing and the media, which support the growth of *A. tumefaciens* did not support the growth of other microorganisms. Therefore, in the present investigations, native fungal and bacterial isolates were screened on tomato plants for their antagonistic activity against *A. tumefaciens* by cross protection method.

MATERIALS AND METHODS

Isolation of antagonists

Different fungi and bacteria were isolated from the rhizosphere soil of healthy and infected peach plants for their antagonistic activity. Samples were collected from March to August, 2001 from stone fruit nurseries of districts Solan and Sirmour of Himachal Pradesh that had recurrent high incidence of crown gall for the past several years on fruit especially on peach. The peach seedlings uprooted with the help of a crow bar with rhizosphere soil intact were brought to the laboratory. Rhizosphere soil was collected on a sterilized butter paper and shade dried for 24 h. The soil was diluted to 10^{-1}, 10^{-2} and 10^{-3} with a sterile distilled water (serial dilution method) and 0.1 ml of each dilution was spread over the solidified surface of 25 ml nutrient agar (NA) medium contained in each Petri plate with the help of sterilized L-shaped glass rod. The Petri plates were incubated at 25± 1°C and observed for growth at 24 h interval. After 72 h of growth, different fungal and bacterial colonies were picked with the help of sterilized inoculating needle and were sub-cultured/ sub-streaked to obtain pure progeny of fungi and bacteria. The cultures so obtained were maintained in slants of PDA and NA medium, respectively for fungi and bacteria and were kept in refrigerator. Periodic sub- culturing of these isolates was done after every two months to maintain different isolates.

Method of wounding and inoculation

The method used by New and Kerr (1972) was employed for wound inoculation on tomato seedlings under pot culture conditions. Stem of 4 weeks old tomato plants were wounded at crown, stem and tip portions to a depth of 3 mm with the help of a sterilized blunt cylindrical sterilized steel rod of 2 mm (dia) and 0.004 ml suspension of pathogen containing 10^8 c.f.u./ml was then deposited in each wound by an autoclaved micropipette. A control was kept where only 0.004 ml of sterilized distilled water was deposited in the wounds at crown, stem and tip portion of tomato plants. The wounds were wrapped with sterilized cotton to avoid secondary invasion by saprophytes. The pots were kept at room temperature for development of galls at the wounded and inoculated portions. The observations for any gall development were made everyday after third day of inoculation.

Evaluation of antagonistic activity

Evaluation of antagonistic activity of different isolated fungi and bacteria was made on four weeks old tomato seedlings under pot culture conditions by cross inoculation method by causing 3 mm deep injuries with a sterilized blunt cylindrical steel rod. With the help of sterilized micropipette, 0.004 ml of rhizobacterial suspension (having 10^8 c.f.u./ml) was deposited in each wound. After 24 h, 0.004 ml pathogen suspension (having 10^8 c.f.u./ml) was applied to the same wound. Observations on gall development over inoculated wounds on tomato seedlings were made after third day and on each successive day.

RESULTS AND DISCUSSION

Three bacterial and one fungal isolate viz., *Pseudomonas flourescens*, *Bacillus* sp., a soil isolate of *A. tumefaciens* and *Trichoderma viride* were selected for evaluation against *Agrobacterium tumefaciens.*

The *in vitro* evaluation of rhizosphere microflora could not be achieved by dual culture technique as neither test antagonist grew well on *Agrobacterium* specific media nor did *A. tumefaciens* on nutrient agar and vice-versa. However, upon their cross inoculation on tomato, a significant effect was observed on gall development. *Bacillus* sp. *P. fluorescens* and *T. viride* inhibited gall development significantly. While evaluating these microbes against *A. tumefaciens,* it was observed that out of nine wounds inoculated with the suspension of these bacteria and fungi 24 h before inoculation of wounds with pathogenic isolates of *A. tumefaciens,* minimum number of wounds (3) showed gall development in *T. viride* followed by *Bacillus* sp. (3.25 wounds out of 9) as compared to control where all nine inoculated wounds developed galls (Table 1). The number of wounds showing galls was five in *A. tumefaciens* (soil isolate), as compared to control (9/9). However, the numbers of galls per wound were minimum (one) in *P. fluorescens, A. tumefaciens* (soil isolate), *T. viride* and 1.5 galls/wound in *Bacillus* sp. as compared to control (4.25 galls/wound). The other test antagonists showed results statistically at par with the control. The gall size was small in *Bacillus* sp., *P. fluorescens* and *T. viride* inoculated wounds but tiny in *A. tumefaciens* (soil isolate) as compared to medium large size in control. Other test antagonists too had medium sized galls. On seeing the overall effect, what so ever, exhibited by these test antagonists on gall development, a minimum value of 4.33 was obtained in *T. viride* followed by 4.58 in *Bacillus* sp. as compared to control (7.42).

Cross inoculation of soil isolate of *A. tumefaciens* resulted in tiny galls and hence was pathogenic but hypovirulent. Kerr and Panagopoulos, (1977) and Garrett (1979) also observed similar inhibition in gall development by different isolates of *A. tumefaciens* and soil microbes.

Table 1. Evaluation of test antagonists against *Agrobacterium tumefaciens* on tomato plants

Antagonist	No. of wounds inoculated	No. of wounds showing galls	No. of galls/ wound	Gall size	Mean
Bacillus sp.	9.00 (3.00)	3.25 (1.79)	1.50 (1.20)	Small	4.58 (1.99)[de]
Pseudomonas fluorescens	9.00 (3.00)	6.00 (2.45)	1.00 (1.00)	Small	5.33 (2.15)[c]
Agrobacterium tumefaciens (soil isolate)	9.00 (3.00)	5.00 (2.23)	1.00 (1.00)	Tiny	5.00 (2.08)[cd]
Trichoderma viride	9.00 (3.00)	3.00 (1.72)	1.00 (1.00)	Small	4.33 (1.91)[e]
Control	9.00 (3.00)	9.00 (3.00)	4.25 (2.05)	Medium large	7.42 (2.68)[a]
Mean	9.00 (3.00)[a]	6.16 (2.43)[b]	2.62 (1.53)[c]		

Figures in parentheses are square root transformed values

Figures represented by same letter do not differ significantly

	$CD_{0.05}$
Antagonist (A)	= (0.12)
Interval (I)	= (0.07)
A x I	= (0.22)

Gall size Tiny = 1 mm

Small = 1 to 2 cm

Medium = 2 to 5 cm

Large = 5 cm

The success in controlling crown gall disease world over by *A. radiobacter,* a non-pathogenic strain of *A. tumefaciens* (Kerr, 1972; Kerr and Htay, 1974; Grimm *et al.,* 1982) provided a great impetus for search of other native antagonists. In our quest for suitable antagonist, we also succeeded in isolating *Bacillus* sp. *P. fluorescens* and *T. viride* from rhizosphere soil of cherry plants which resulted in production of less number of galls on tomato plants as compared to untreated. Unfortunately, *A. tumefaciens* (soil isolate) isolated from soil was pathogenic nevertheless hypo-virulent. *Bacillus* sp., *P. fluorescens* and *T. viride* were highly effective, and were selected for their evaluation in field conditions against crown gall. Utkhede and Smith (1993) also observed *Bacillus subtilis* strain ABB root dip as highly effective against crown gall disease on apple under field conditions. Zhang *et al.* (1991) achieved 95 per cent control of crown gall disease on peach with *P. fluorescens* strain. Deep and Young (1965) observed increased incidence of crown gall in fungicide treated soils. From these findings they concluded that non-pathogenic soil microflora is a potent source of competition for pathogens. They also suggested the presence of natural fungal competition and obtained 25 per cent reduction in gall development by using suspension of unidentified fungi as pre-plant treatment.

REFERENCES

Deep, I.W. and Young, R.A. (1965). The role of pre planting treatments with chemicals in increasing the incidence of crown gall. *Phytopathology* 55: 212-216.

Garrett, C.M.E. (1979). Biological control of crown gall in cherry rootstock propagation. *Ann. Appl. Biol.* 91: 221-226.

Grimm, R., Sule, S. and Lozano, J.C. (1982). Control of crown gall (*A. tumefaciens* Smith and Townsend) Conn., in nurseries. *Plant Pathogenic Bacteria. Swiss Federal Res.* St. Proc. Vth Int. Conf. pp. 531-537.

Kaul, J.L. and Sharma, R.C. (1988). Stone fruit diseases and their management. In: *Tree Protection* (Eds. Gupta, V.K. and Sharma, N.K.). ISTS, Solan, pp. 26-41.

Kerr, A. (1972). Biological control of crown gall. Seed inoculation. *J. Appl. Bacteriol.* 35: 493-497.

Kerr, A. and Htay, A. (1974). Biological control of crown gall through bacteriocin production. *Physiol. Plant Pathol.* 4: 37-44.

Kerr, A. and Panagopoulos, C.G. (1977). Biotypes of *Agrobacterium radiobacter* var. *tumefaciens* and their biological control. *Phytopathol. Z.* 90: 172-179.

New, P.B. and Kerr, A. (1972). Biological control of crown gall: Field measurements and glasshouse experiments. *J. Appl. Bacteriol.* 35: 279-287.

Panagopoulos, C.G., Psallidas, P.G. and Alivizatos, A.S. (1978). Studies on biotype 3 of *Agrobacterium radiobacter* var. *tumefaciens. Proc. 4th Int. Conf. Plant Path. Bact. Angers.*1: 221-228.

Sharma, R.C., Sharma, S. and Gupta, A.K. (2001). Present status of biological control of crown gall. In: *Plant Pathology* (Ed. Trivedi, P.C.) Pointer Publication, Jaipur, pp. 19-33.

Utkhede, R.S. and Smith, E.M. (1993). Evaluation of biological control and chemical treatments for control of crown gall on young apple trees in the Kopotnay Valley of British Columbia. *J. Phytopathol.* 137: 265-271.

Zhang, C.M., Lu, S.H., Wei, C.M. and Zhao, J.H. (1991). Bacterium strain 1-1-4 *Pseudomonas fluorescens* for biocontrol of peach crown gall disease. *Acta Agric. Shanghai* 7: 65-69.

Integrated Plant Disease Management (2005)
Eds. Sharma, R.C. and Sharma, J.N.
Scientific Publishers (India), Jodhpur, pp. 65-69

10

Widespread occurrence of biovar - 1 of *Agrobacterium tumefaciens* affecting stone fruits in Himachal Pradesh

SANDEEP KUMAR, BRIJESH KAMAL and A.K. GUPTA

Department of Mycology and Plant Pathology, Dr. Y. S. Parmar University of Horticulture and Forestry Nauni, Solan – 173 230 (H.P.)

ABSTRACT

Crown gall disease caused by *Agrobacterium tumefaciens* (Smith and Townsend) Conn. is one of the most destructive diseases in stone fruit nurseries in Himachal Pradesh. Extensive surveys were carried out in the four stone fruit growing districts of the state and isolations of the causal organism was carried out to ascertain the prevalence of different biovars. Different isolates were subjected to biochemical tests viz., 3-ketoglycoside test, acid production from erythritol, growth on Schroth *et al* medium, growth at 35^{0}C and reaction to litmus milk. All the isolates tested positive to biovar – 1 tests indicating that biovar–1 of *Agrobacterium tumefaciens* is widely prevalent in Himachal Pradesh infecting stone fruit plants in nurseries.

INTRODUCTION

Crown gall caused by *Agrobacterium tumefaciens* in stone fruits results nearly 40 per cent rejection of nursery plants. Sometimes it cause the hundred per cent rejection of the nursery plants. The disease has been reported on cherry, peach, plum and almond from India (Durgapal, 1974; Jindal and Sharma, 1988a, 1988b). Biovar-I and II of the pathogen have been reported responsible for infecting stone fruit plants. However, the occurrence of biovar-I of *A. tumefaciens* on almond in India has been first reported by Jindal and Sharma (1988a). The present studies were carried out to know the occurrence of different biovars of *A. tumefaciens* infecting stone fruits in different regions of Himachal Pradesh.

MATERIALS AND METHODS

Isolation and identification

The pathogen *Agrobacterium tumefaciens* was isolated on nutrient agar (NA), soil extract agar, yeast extract mannitol agar (YEMA) supplemented with Congo Red (500 ppm), Schroth *et al.* and Kado and Heskett (K&H) medium from crown gall infected stone fruit plants- peach and cherry rootstock Colt and rhizosphere of diseased plants from different stone fruit nurseries located at Nauni, Sanahari and Rajgarh in Himachal Pradesh. The Petri plates were incubated at 25 ± 1^0C and examined after every 24 h for development of colonies. A single colony was further streaked to obtain pure progeny of single cell colony.

The growth obtained on the selective medium was critically studied for its cultural and morphological characteristics. Among the colony characteristics, shape (round, circular, oblong, ovate etc.), colour (white, transparent, opaque, or coloured), size (cm) and margins (smooth, rough, zigzag etc.) were studied. Cell shape (rod, coccus, vibrio, helical etc.), size (under PC based fluorescent microscope: x 1000) and colour (hyaline or coloured) were also studied

Gram staining was employed for establishing the identity of the bacterium under compound microscope (x 1000). Also the pure cultures of the causal agent were subjected to different biochemical tests namely, 3-ketoglycoside test, acid production from erythritol, growth on Schroth *et al* medium, growth at 35^0C and litmus milk test to ascertain the identity of the test bacterium as descried by Panagopoulos *et al.* (1978).

Pathogenicity

The method used by New and Kerr (1972) was employed for wound inoculation on tomato seedlings under pot culture conditions to prove the pathogenicity of all the isolates. Stem of 4 weeks old tomato plants were wounded at crown, stem and tip portions to a depth of 3 mm with the help of a sterilized blunt cylindrical sterilized steel rod of 2 mm (diameter) and 0.004 ml suspension of pathogen containing 10^8 c.f.u./ml was then deposited in each wound by an autoclaved micropipette. A control was kept where only 0.004 ml of sterilized distilled water was deposited in the wounds at crown, stem and tip portion of tomato plants. The wounds were wrapped with sterilized cotton to avoid secondary invasion by weak parasites. The pots were kept at room temperature for development of galls at the wounded and inoculated portions. The observations for any gall development were made everyday after third day of inoculation.

RESULTS AND DISCUSSION

All the isolates viz., I_1 (Rajgarh), I_2 (Sanahari), both from peach, I_3 (Nauni) from cherry and a soil isolate of *Agrobacterium* (I_4) when subjected to Gram staining retained the pink-red colour of safranine when examined under compound microscope (X 1000). Colony shape in I_1, I_3 and I_4 was circular to ovate with the exception of I_2 where colonies were observed slightly elongated and hence designated as circular to oblong. The colony colour of all the isolates was off-white. Colony diameter in I_3 on YEMA ranged from 0.9 to 1.3 cm after 5 days of incubation. In other isolates, colony diameter ranged from 0.8 to 1.4 cm on YEMA and 1.2 to 2.2 cm on K&H medium. However, as compared to YEMA, the size of the bacterial colonies was larger in K&H medium after 5 days of incubation being maximum in I_3 (2.3 cm). Interestingly, no variations were observed in colony margins and all isolates produced colonies with smooth margins. Similarly, in other colony characteristics, transparent to translucent colonies were observed in all the isolates and these colonies were protruding towards outside and were squishy-squashy (Table 1).

Table 1. Cultural and morphological characteristics of *Agrobacterium tumefaciens* isolated from different stone fruit plants and soil

Cultural character	Isolates from different locations			
	I_1 (Rajgarh Peach)	I_2 Nauni (Cherry)	I_3 Sanahari (Peach)	I_4 Soil isolate
Gram reaction	Negative, cells stain pink red	Negative, cells stain pink red	Negative, cells stain pink red	Negative, cells stain pink red
Colony shape	Circular to ovate	Circular to ovate	Circular to ovate	Circular to ovate
Colony colour	Off white	Off white	Off white	Off white
Colony size (cm) on YEMA	(After 5 days) 0.8-1.2	(After 5 days) 0.7-1.4	(After 5 days) 0.9-1.3	(After 5 days) 0.8-1.3
Kado & Heskett	1.2 to 2.2	1.3 to 2.2	1.2 to 2.3	1.1 to 2.2
Colony margins	Smooth	Smooth	Smooth	Smooth
Other characters	Colonies translucent, convex and squishy-squashy	Colonies transparent to translucent, protruding towards outside, and squishy-squashy	Colonies translucent, convex and squishy-squashy	Colonies translucent, protruding towards outside, convex squishy-squashy
Morphological characters				
Cell shape	Single rods sometimes in chains of 2-5	Single rods sometimes in chains of 2-4	Single rods sometimes in chains of 3-5	Single rods sometimes in chains of 2-3
Cell size (µm)	1-4 x 0.2 to 0.4	1-4 x 0.2 to 0.35	1-4 x 0.2 to 0.4	1-3 x 0.3 to 0.45
Cell colour	Hyaline	Hyaline	Hyaline	Hyaline

All the isolates upon microscopic examination appeared typical rod shaped bacterial cells occurring generally singly but occasionally in chains of 2-5 cells. The cell size ranged from 1-4 x 0.2-0.4 μm. Hyaline cells were observed in all the four isolates. Jindal *et al.* (1990) too observed small, circular and smooth, glistening, opaque to translucent colonies of *A. tumefaciens* on nutrient agar. Morphological studies of the causal agent revealed hyaline rod shaped cells measuring 1-1 x 0.2- 0.45 μm when viewed under compound microscope. These findings are in conformity with those of Romanenko and Perepnikhatka (1984) who reported similar morphological characters of *A. tumefaciens.*

The reactions of different isolates of *A. tumefaciens* to different biochemical tests have been mentioned in Table 2. It is evident that all the isolates showed positive reaction to 3-ketoglycoside test, growth on Schroth *et al.* medium, growth at 35°C and alkaline reaction to litmus milk test. No isolate was observed producing acid from erythritol. On the basis of these reactions, all the isolates were identified as *Agrobacterium tumefaciens* belonging to biovar –I group.

Table 2. Characteristics of standard biovars of *A. tumefaciens* and isolates infecting stone fruits collected from Himachal Pradesh

Diagnostic test	Characteristics of biovars of *Agrobacterium tumefaciens* (Panagopoulos *et al.*, 1978)			Reaction of different isolates to biovar tests			
	Biovar -I	Biovar -II	Biovar -III	I_1	I_2	I_3	I_4
3-Ketoglycoside test	+	-	-	+	+	+	+
Growth on Schroth *et al.*	+	-	-	+	+	+	+
Growth at 35°C	+	V⁻	+	+	+	+	+
Acid from erythritol	-	+	-	-	-	-	-
Litmus milk test	alk.	acid.	alk.	alk.	alk.	alk.	alk.

V⁻= Variable; + = Positive reaction; - = Negative reaction; I_1 = Rajgarh (Peach); I_2 = Nauni (Peach); I_3 = Nauni (Cherry); I_4 = Isolate from soil

Earlier occurrence of biovar-I of *A. tumefaciens* on pecan and all the commercial cultivars of almond in Himachal Pradesh has been reported (Jindal and Sharma, 1988a; Jindal *et al.*, 1990). However, Al-Karablieh and Khlaif (2002) observed the occurrence of biovar-I (60.30%), biovar-II (23.30%) and biovar-III (1%) in addition to an intermediate biovar (15%) which clearly showed the dominance of biovar-I populations of *A. tumefaciens* on all temperate fruits in Jordan. All isolates upon artificial inoculation produced small creamy galls on tomato plants, which grew in size as earlier observed by Kerr (1969) and New and Kerr (1972). However, *A. tumefaciens* soil isolate as compared to other isolates induced tiny galls on tomato. It is presumed that the soil isolate was no doubt pathogenic but hypovirulent as compared to rest three.

REFERENCES

Al- Karablieh, N. and Khalaif, H. (2002). Occurrence and distribution of crown gall disease in Jordan. *Phytopathol. Mediterr.* 41: 226-234.

Durgapal, J.C. (1974). A preliminary note on bacterial diseases of temperate plants in India. II. Bacterial diseases of stone fruits. *Indian Phytopathol.* 24: 379-382.

Jindal, K.K. and Sharma, R.C. (1988a). Occurrence of biovar-I of *Agrobacterium tumefaciens* on almond in India. *Indian Phytopathol.* 41: 614-615.

Jindal, K.K. and Sharma, R.C. (1988b). Crown gall and hairy root in fruit plant nurseries in Himachal Pradesh. *Proc. 5th Int. Conf. Plant Pathol.*, Kyoto, Japan (Abstr.) pp. 98.

Jindal, K.K., Ved Ram and Gupta, G.K. (1990). Occurrence of crown gall on pecan nut plants. *Plant Dis. Res.* 5: 216-217.

Kerr, A. (1969). Crown gall on stone fruits. I. Isolation of *Agrobacterium tumefaciens* and related species. *Aust. J.Biol. Sci.* 22: 111.

New, P.B. and Kerr, A. (1972). Biological control of crown gall; Field measurements and glasshouse experiments. *J. Appl. Bacteriol.* 35: 279-287.

Panagopoulos, C.G., Psallidas, P.G. and Alivizatos, A.S. (1978). Studies on biotype 3 of *Agrobacterium radiobacter* var. *tumefaciens. Proc. 4th Int. Conf. Plant Pathogenic Bacteria.- Angers.*1: 221-228.

Romanenko, V.M. and Perepnikhatka, V.I. (1984). Flagellae of *Agrobacterium tumefaciens* cells. *Mikrobiol. Zh.* 46: 66-68.

Integrated Plant Disease Management (2005)
Eds. Sharma, R.C. and Sharma, J.N.
Scientific Publishers (India), Jodhpur, pp. 71-74.

11

Association of sooty mould development with citrus whitefly, *Dialeurodes citri* Asmead, infestation on different citrus rootstocks

RAVINDER KUMAR, R.C. BATRA and D.R. SHARMA

Department of Horticulture, Punjab Agricultural University,
Ludhiana -141 004 (Punjab)

ABSTRACT

The present studies were carried out in the Department of Horticulture, Punjab Agricultural University, Ludhiana and Regional Fruit Research Station, Abohar during 2000 and 2001. It was observed that there was direct and positive relation between sooty mould development and varying levels of *Dialeurodes citri* nymphal infestation. Among different citrus rootstocks, sweet orange (*Citrus sinensis* Osbeck.) and chakotra (*Citrus grandis* Macf.) were found to be highly susceptible to sooty mould and required less *D. citri* nymph population (9.87 ± 0.37 and 5.07 ± 0.39) as compared to other cultivars like rough lemon (*Citrus jambhiri* L.), Cleopatra (*Citrus reticulata* Blanco.) and Carrizo (*P. trifoliata* × *C. sinensis*), the later requiring comparatively more population for the development of same level of sooty mould incidence (14.03 ± 0.47, 14.16 ± 0.40 and 14.25 ± 0.40), respectively.

INTRODUCTION

In India, as many as 12 species of whiteflies are reported to be of economic importance and are widely distributed throughout the country (Konar, 1997; Batra *et al.*, 1990). The nymphs and adults of whiteflies suck sap and excrete copious amounts of honeydew, especially the nymphs. This honeydew is a suitable medium for colonisation by the sooty mould complex including *Capnodium citri, Alternaria* sp., *Botryodiplodia* sp., *Cladosporium* sp. and *Colletotrichum* sp., etc. (Srivastava and Thakre, 2000). Infection of sooty moulds and excessive desapping reduces the vigour and subsequently fruit bearing capacity of the plants (Kapoor *et al.*, 1992). Screening of wide ranging germplasm of citrus and its association with sooty moulds offer a good scope for identifying strains resistant to this pest.

MATERIAL AND METHODS

Investigations were carried out during peak activity period of the pest (April-May and September-October) during 2000-01 in the orchards of the Department of Horticulture, Punjab Agricultural University, Ludhiana. In all, 42 scion and rootstock collections from California and Florida including group of sweet orange (*Citrus sinensis* Osbeck.), mandarin (*Citrus recticulata* Blanco.), grapefruit and pummelo (*C. paradisii* Macf., *C. grandis* Osbeck.), hybrids and citranges planted in old and new orchards were observed at weekly intervals for whitefly infestation during periods of peak infestation. Hundred leaves were collected from each group of citrus genotypes and observed for population of *D. citri* nymphs and sooty mould development. The leaf area covered by sooty mould was measured with the help of a graph paper and per cent intensity of sooty mould was worked out using the following relationship:

$$\text{Per cent sooty mould intensity} = \frac{\text{Area covered by sooty mould}}{\text{Total leaf area}} \times 100$$

The data on sooty mould incidence were correlated with the nymph population of *D. citri* on infested leaves.

RESULTS AND DISCUSSION

The data on population of whitefly nymphs revealed that none of the citrus species/cultivars was free from whitefly infestation (Table 1). Although citrus groups registered varying degree of infestation yet there was a direct and positive correlation between nymph population and sooty mould development. Varying levels of sooty mould colonization were also observed on different cultivars. However, sooty mould growth did not show any relationship with adult population of the pest.

The incidence of sooty mould with regard to different citrus groups revealed that nymph population of *D. citri* had a direct and positive correlation ($r = 0.67$) with sooty mould development. Whereas the adult population played a negligible role ($r = 0.10$) in determining sooty mould incidence. Further, it was observed that 1-20 per cent colonies of sooty mould on chakotra and sour orange were associated with low nymph population i.e. 2.08 and 1.95 per 25 leaves, respectively, compared with mould growth on Cleopatra and Carrizo which had significantly higher nymph population (8.04 and 7.55 per 25 leaves) while kinnow, trifoliate orange and sweet orange had moderate nymph population of 2.73, 3.17 and 3.76 per 25 leaves, respectively. The results showed that for the same level of sooty mould colonies hosts being more susceptible to *D. citri* has similar degree of tolerance to sooty mould incidence and vice versa. A similar trend was observed with Cleopatra and sour orange for sooty mould incidence range of 20.1-40 and 80.1-100 per cent. The results indicated that chakotra and sour orange with low nymph population provided some degree of resistance to sooty mould on them. Conversely, Carrizo, Cleopatra and rough lemon being more susceptible to whitefly, provided suitable condition for the growth of sooty moulds. This study also pointed out that nymph population of *D. citri* was a major contributing factor for sooty mould development on citrus plants which is in consonance of the earlier report by Dhiman *et al.* (2000).

Table 1. Association of whitefly *Dialeurodes citri* population with sooty mould incidence on citrus germplasm during 2000-2001

Whitefly population per 25 leaves	Sooty mould incidence (%)				
	1-20	20.1-40	40.1-60	60.1-80	80.1-100
Trifoliate orange					
Nymph	3.17±0.25	6.33±0.20	9.84±0.34	-	-
Adult	0.84±0.08	2.04±0.10	3.30±0.09	-	-
Rough lemon					
Nymph	4.72±0.19	6.62±0.22	13.2±0.18	13.73±0.65	14.03±0.47
Adult	1.35±0.12	2.93±0.09	2.88±0.10	2.84±0.10	3.51±0.17
Kinnow					
Nymph	2.73±0.17	7.11±0.23	9.71±0.30	13.18±0.55	13.60±0.55
Adult	1.97±0.09	2.81±0.12	3.33±0.14	3.93±0.18	4.22±0.09
Sweet orange					
Nymph	3.76±0.13	6.81±0.15	7.97±0.29	8.19±0.16	9.87±0.34
Adult	1.84±0.07	2.55±0.18	2.95±0.66	3.19±0.03	4.0±0.13
Chakotra					
Nymph	2.08±0.06	3.35±0.09	3.82±0.11	4.71±0.42	5.07±0.39
Adult	0.91±0.04	1.44±0.09	1.83±0.07	2.0±0.09	2.01±0.08
Sour orange					
Nymph	1.95±0.44	3.37±0.06	4.84±0.02	-	-
Adult	0.66±0.03	0.94±0.06	1.26±0.05	-	-
Kagzi lime					
Nymph	4.22±0.10	5.44±0.29	6.75±0.18	7.70±0.90	-
Adult	1.77±0.15	2.17±0.44	2.53±0.15	3.25±0.12	-
Cleopatra					
Nymph	8.04±0.42	10.15±0.30	12.50±0.13	13.15±0.37	14.16±0.40
Adult	2.59±0.06	2.48±0.12	2.57±0.18	3.25±0.15	3.75±0.67
Carrizo					
Nymph	7.55±0.29	10.66±0.56	12.75±0.13	13.15±0.75	14.25±0.40
Adult	2.17±0.12	2.15±0.19	2.50±0.12	2.62±0.15	2.98±0.23

REFERENCES

Batra, R.C., Sharma, D.R. and Singh, S.N. (1990). Status of citrus insect pests and their natural enemies in different agro-climatic zones of Punjab. *Indian J. Hortic.* 47: 33-36.

Dhiman, J.S., Batra, R.C. and Sharma, D.R. (2000). Synchronous attack of insect pests and diseases in kinnow: hitech citrus management. *Proceedings of the International Symposium on Citriculture,* Nagpur, 23-27 November 2000, pp. 810-817.

Kapoor, S., Cheema, S.S., Kapoor, S.P. and Kapoor, S. (1992). Some of the biological changes in citrus leaves infected with sooty mould (*Capnodium citri*). *J. Res. Punjab Agric. Univ.* 29: 354-356.

Konar, A. (1997). Incidence pattern of whiteflies on orange *Cirus reticulata* Blanco. in different orchards of Darjeeling district of West Bengal. *Environ. Ecol.* 15: 797-799.

Srivastava, V.K. and Thakre, R.P. (2000). Synergism and antagonism among the fungi grown in honeydew secretion on leaf surface of Nagpur Mandarin. *Indian Phytopathol.* 53: 193-196.

Integrated Plant Disease Management (2005)
Eds. Sharma, R.C. and Sharma, J.N.
Scientific Publishers (India), Jodhpur, pp. 75-81.

12

Studies on *Phytophthora* spp. attacking strawberry in Himachal Pradesh

MEENU GUPTA and L.N. BHARDWAJ

Department of Mycology and Plant Pathology, Dr Y.S. Parmar University of Horticulture and Forestry, Nauni, Solan 173 230 (H.P.)

ABSTRACT

In Himachal Pradesh, strawberry is grown in valley areas of Solan, Sirmour, Kangra, Mandi and Kullu districts. Amongst various diseases, the fungal diseases are the most troublesome and reduce the yield and quality of fruits considerably. Among major fungal pathogens attacking strawberry, *Phytophthora fragariae*, *P. cactorum* and *P. nicotianae* causing red stele, leather rot and fruit rot of strawberry, respectively, are the most damaging ones. Of these *Phytophthora* spp., *P. fragariae* is the most devastating followed by *P. cactorum* and *P. nicotianae*. These associated *Phytophthora* spp. were isolated and their morphological characters were studied. In pathogenicity tests conducted, *P. fragariae* infected only roots, while *P. cactorum* and *P. nicotianae* were pathogenic to fruits only.

INTRODUCTION

Strawberry is one of the most important soft fruit crops grown in varied agro-climatic conditions of the world. In Himachal Pradesh, it is cultivated in valley areas of Solan, Sirmour, Kangra, Mandi and Kullu districts. Since the crop is highly perishable and water loving, so it is highly prone to *Phytophthora* diseases. The present investigations were, therefore, undertaken to record the prevalence and incidence of *Phytophthora* spp. attacking strawberry. The study revealed the association of three *Phytophthora* species namely, *P. fragariae*, *P. cactorum* and *P. nicotianae* causing red stele, leather rot and fruit rot, respectively.

MATERIALS AND METHODS

Periodic surveys were conducted in Solan, Sirmour, Kullu and Kangra districts of Himachal Pradesh for recording the incidence of red stele, leather rot and fruit rot of strawberry during 1999 through 2000. *Phytophthora* affected plant parts i.e. fruits, leaves, stems and runners were brought to the laboratory and the associated fungi were isolated on potato dextrose agar medium (PDA) and bean agar medium. The different fungal species responsible for causing red stele, leather rot and fruit rot of strawberry were identified on the basis of morphological characters documented in standard authentic descriptions and taxonomic keys (Waterhouse *et al.*, 1983; Stamps *et al.*, 1990). Pathogenicity tests were conducted on healthy fruits, leaves, stems and runners of strawberry. The fruits of different stages i.e. immature (green stage), mature (pale red coloured) and ripened (dark red coloured) of cv. Chandler were inoculated with seven days old culture of test fungi in two sets i.e. injured and uninjured and kept on moist chamber at 25±1°C for symptoms to appear. Similarly, healthy strawberry leaves and stems (5 cm in length) of cv. Chandler were also inoculated with 4 mm culture discs of test fungus after washing them with 70 per cent ethanol. The leaves and stems after inoculation were kept on 2 per cent agar medium in Petri plates at 25±1°C. However, the strawberry runners were inoculated with zoospore suspension and mycelial slurry as inoculum. The zoospores of *P. fragariae* are not formed on any of cultural media, however, they are formed by flooding the 2-3 days old culture of pathogen grown on bean agar in non-sterile pond water and incubating at 15±1°C for 24-48 hours. In contrast, *P. cactorum* and *P. nicotianae* formed a lot of sporangia in culture. In such cases, the aerial hyphae from 5-10 days old cultures were scrapped with the help of sterilized razor into a Petri plate containing 0.1 M KNO_3 to keep them moist but not submerged and incubated at 25±1°C for 72 h. The spornagia thus formed were harvested by flooding the culture plate with sterile distilled water. The zoospore suspension was added to the strawberry runners raised in pots. Observations on incubation period and symptom appearance were recorded in each case.

RESULTS AND DISCUSSION

Incidence

During surveys, occurrence of three diseases namely, red stele (*P. fragariae*), leather rot (*P. cactorum*) and fruit rot (*P. nicotianae*) were encountered in nature (Table 1). The average incidence of these diseases ranged between 1.26 to 19.03 per cent. The average incidence of red stele, in general, was found highest (19.03%) followed by leather rot (12.80%), while the least incidence was recorded in case of fruit rot (1.26 %). Of the various localities surveyed, maximum incidence was observed in Dhaulakuan (19.33%) followed by Paonta Sahib (18.50%) and Majra (17.66%).

Table 1. Incidence of *Phytophthora* diseases of strawberry in Himachal Pradesh during 1999-2000

District/ Location	Incidence (%)			Overall mean
	Red stele (*P. fragariae*)	Leather rot (*P. cactorum*)	Fruit rot (*P. nicotianae*)	
Solan				
Nauni	30.50	10.00	4.50	15.00
Oachghat	17.50	8.00	2.50	9.33
Kandaghat	29.50	10.00	3.00	14.17
Dedgharat	27.00	12.00	1.50	13.50
Deothi	20.50	4.50	0.00	8.33
Nalagarh	19.03	5.50	0.00	5.17
Mean	22.50	8.33	1.91	10.91
Sirmour				
Shotti	25.00	5.00	0.00	10.00
Dhaulakuan	32.50	25.50	0.00	19.33
Majra	22.50	30.50	0.00	17.66
Shambhuwala	11.00	22.50	0.00	11.17
Paonta Sahib	29.50	26.00	0.00	18.50
Mean	24.10	21.90	0.00	15.33
Kullu				
Bajoura	15.00	8.00	3.50	8.83
Jhiri	12.00	10.00	4.00	8.66
Mean	13.50	9.00	3.75	8.75
Kangra				
Jwalamukhi	1.50	6.00	0.00	2.50
Jachh	1.50	8.50	0.00	3.33
Mean	1.50	7.25	0.00	2.91
Overall mean	19.03	12.80	1.26	—

The data further envisaged that out of the three *Phytophthora* diseases, incidence of red stele was highest at Dhaulakuan (32.50%) that of leather rot at Majra (30.50%) and fruit rot at Nauni (4.5%). Amongst various districts surveyed, Sirmour had the highest average disease incidence (15.33%) followed by Solan (10.91%) and Kullu (8.75%), while the least incidence was observed in Kangra district (2.91%).

Symptomatology

Red stele (*Phytophthora fragariae*)

The symptoms of red stele appear in late spring or early summer. Affected plants showed severe dwarfing and death of plants in poorly drained fields, whereas in well-drained soils, the affected plants were stunted and produced few or no runners. The infected plants were stunted and the older leaves wilted, dried up and replaced by smaller new leaves borne on shortened petioles. Young leaves were blue-green in colour and the older ones were red, yellow or orange in colour. Severely infected plants bore little or no fruits and often wilted and died during the summer months. When young plants were attacked by the fungus, the roots started decaying from tips leading to the destruction of fibrous roots causing the primary roots to have a rat-tail appearance which is considered to be the most conspicuous symptom of the disease. On scrapping the central cortex and the exposing the stele of the root, one can observe reddish discolouration of the stele. Under favourable weather conditions also, no mycelial growth of the fungus was seen on infected roots of the plants as was observed in case of *Phytophthora cactorum* and *P. nicotianae.* Several workers have also observed similar observations while working on *P. fragariae* causing red stele of strawberry (Hickman, 1940; Lauber and Buhlmann, 1984; Bosshard and Schwindt, 1997).

Leather rot (*Phytophthora cactorum*)

Occurrence of leather rot was observed on green as well as ripe fruits between March to May. On green fruits, brown to dark brown areas with brownish margins were seen. As the disease progressed, the entire fruit became rough and leathery. On ripe fruits also, brown to purplish area developed that became tough and leathery and with the advancement of the infection, berries got mummified. Under high humid conditions, white mycelial growth of the fungus was seen covering the entire fruit. Similar kind of symptoms were also observed earlier by Ellis and Grove (1983), Madden *et al.* (1991) and Sharma (1998).

Fruit rot (*Phytophthora nicotianae*)

The affected fruits were soft, light pink and became brown to dark brown with the progress of the disease. Under warm and humid weather, fruits were seen covered with white mycelial growth of the fungus, while under dry conditions, the white mycelial growth disappeared and affected fruits showed complete rotting. Sharma (1998) also recorded similar types of symptoms on strawberry upon infection by *P. nicotianae.*

Isolation and identification

Isolation

Phytophthora spp. namely, *P. fragariae, P. cactorum* and *P. nicotianae* attacking strawberry were isolated from affected plant parts into pure culture. Pure culture of *P. fragariae* was maintained on bean agar medium whereas that of *P. cactorum* and *P. nicotianae* was maintained on oat meal medium. To maintain the virulence of the pathogenic fungi, repeated sub-culturing was done after every forty five to fifty days for *P. fragariae* and after thirty to thirty five days for *P. cactorum* and *P. nicotianae.*

Identification

The morphological and cultural characters of *Phytophthora* species attacking strawberry were recorded in each case and based on these characters, the species were identified as under:

***Phytophthora fragariae* Hickman:** The rate of growth of *P. fragariae* was much slower than that of other closely related species. The mycelium was white, non-septate but septate in older parts. Sporangia were usually inversely pyriform, nonpapillate with bluntly rounded apex, and measured 32-90 x 32-52 μm. The oospores were spherical, thick walled and measured 22-44 μm. Sporangia were not produced in culture but abundant sporangia developed when cultures were immersed in non-sterile pond water. These findings are in accordance with Hickman (1940) and Morita (1975). The fungus was identified as *Phytophthora fragariae* Hickman and the identity was further confirmed from Indian Institute of Spices Research (ICAR), Calicut, Kerala.

***Phytophthora cactorum* (Leb. & Cohn) Schroet:** The fungus grew slowly on the potato dextrose agar medium. The mycelium was white and colony was fluffy and slightly radiating. Sporangia were lemon shaped and papillate and measured 28.9 – 39.8 x 18.0 – 35 μm, while papilla was 3.63- 5 μm. The oospores were thick walled and measured 22-24 μm in size. On the basis of presence of conspicuously papillate sporangia and the production of oospores, pathogenic fungus was identified as *Phytophthora cactorum* (Leb. & Cohn) Schroet. It was distinguished from the other species by caducous sporangia with short pedicels by paragynous antheridia and plerotic oospores. The banding pattern of *P. cactorum* when compared with banding patterns of other *Phytophthora* isolates indicated that this isolate matched with *Phytophthora cactorum*. Although this pattern was identical to those of the very closely related species, *P. pseudosugae* and *P. idaei*, it was unlikely that either would be found on strawberry.

The identity of *Phytophthora cactorum* was confirmed by Dr. J.C. David, International Mycological Institute Identification Service, UK Centre (Egham), Surrey, England under IMI No. 378617.

***Phytophthora nicotianae* Breda de Haan:** The fungus grew fast on potato dextrose agar medium, with white mycelium and uniformly fluffy colony with no striations and zones. Sporangia were papillate and measured 32-44.3 x 26-34.0 μm and oospores measured 21-26 μm in size. The presence of papillated sporangia and amphigynous antheridia placed this isolate in *Phytophthora nicotianae* Breda de Haan. It was distinguished from many others by non-caducous sporangia and the arachnoid growth on PDA separates it from *P. citrophthora*, its heterothallic nature and smooth walled oogonia from *P. kartsurae* and the size of the oospores and growth at 35°C from *P. mexicana*. The banding pattern derived from digestion matched with *Phytophthora nicotianae* (Suzui *et al.*, 1980).

The identity of *Phytophthora nicotianae* was confirmed by Dr. J.C. David, International Mycological Institute Identification Service, UK Centre (Egham), Surrey, England under IMI No. 378618.

Pathogenicity

Pathogenicity of *Phytophthora fragariae*

Zoospores as well as mycelial slurry of *P. fragariae* were capable of infecting the strawberry roots; however, inoculation by zoospore suspension took 14 days for the production of symptoms of the disease whereas mycelial slurry took 25 days. However, *P. fragariae* failed to infect fruits, leaves and stems of strawberry. The early infection with zoospore suspension may be due to the fact that zoospores being the primary propagules of the pathogen came in contact with host early and caused infection.

Pathogenicity of *Phytophthora cactorum*

It is evident from the Table 2 that fruits of all the developmental stages i.e. immature (green coloured), mature (pale red coloured) and ripened (dark red coloured) were infected by the pathogen, however, immature berries were more prone to infection than mature and ripened fruits. Injury to berries resulted in rapid establishment of infection irrespective of stage of berries which reduced the incubation period to 34 hours in comparison to uninjured fruits. However, the pathogen failed to infect the leaves, stems and runners of strawberry upon inoculation.

Table 2. Pathogenicity of *Phytophthora cactorum* and *Phytophthora nicotianae* infecting strawberry

Fruit development stage	Incubation period (h)			
	Phytophthora cactorum		*Phytophthora nicotianae*	
	Injured	Uninjured	Injured	Uninjured
Immature (green)	24	32	48	60
Mature (pale red)	30	42	36	48
Ripened (dark red)	48	65	24	36
Mean	34	46.3	36	48

The results of these findings are in line with Ellis and Grove (1983) and Maltoni *et al.* (1995).

Pathogenicity of *Phytophthora nicotianae*

It is evident from data (Table 2) that immature, mature and ripened fruits of strawberry were prone to infection by *P. nicotianae*. The symptoms on injured ripened berries appeared after 24 h while on uninjured berries these appeared after 36 h. Ripened berries, in general, were more prone to infection than mature and immature berries. When leaves, stems and runners of strawberry were inoculated, *P. nicotianae* failed to infect them.

Pathogenicity tests of *Phytophthora nicotianae* causing fruit rot, conducted on immature, mature and ripened berries revealed that ripened berries were more prone to infection than mature and immature ones. Injury to fruits hastened the infection and reduced the incubation period.

REFERENCES

Bosshard, E. and Schwindt, M. (1997). Reporting bacterial and fungal causal agents of diseases in strawberries. *Schweiz. Z. Obst Weinbau* 133: 60-61.

Ellis, M.A. and Grove, G.G. (1983). Leather rot in Ohio strawberries. *Plant Dis.* 67: 549.

Hickman, C.J. (1940). The red core root disease of the strawberry caused by *Phytophthora fragariae* n. sp. *J. Pomol. Hortic. Sci.* 18: 89-118.

Lauber, H.P. and Buhlmann, G. (1984). Red core (*Phytophthora fragariae*) in our strawberry crops. *Schweiz. Z. Obst Weinbau* 120: 183-190.

Madden, L.V., Ellis, M.A., Grove, G.G., Reynolds, K.M. and Wilson, L.L. (1991). Epidemiology and control of leather rot of strawberries. *Plant Dis.* 75: 439-446.

Maltoni, M.L., Mazzara, M. and Mezzetti, B. (1995). Pathological, physiological and genetic aspects of resistance to *Colletotrichum acutatum* and *Phytophthora cactorum* in strawberry. *Riv. Frutticolturae di Ortofloricoltura* 57: 59-66.

Morita, H. (1975). Studies on red stele root disease of strawberry. *Shizuoka Agric. Exp. Stn. Bull.* No. 10. 63 p.

Sharma, A. (1998). *Studies on Phytophthora Rot of Strawberry.* M.Sc. Thesis. Dr. Y.S. Parmar University of Horticulture and Forestry, Nauni, Solan, H.P. 53 p.

Stamps, D.J., Waterhouse, G.M., Newhook, F.J. and Hall, G.S. (1990). *Revised Tabular Key to the Species of Phytophthora* (2nd Ed.). Mycol. Paper 162: 28 p.

Suzui, T., Markino, T. and Ogoshi, A. (1980). *Phytophthora* rot of strawberry caused by *Phytophthora nicotianae* var. *parasitica* in Shizuoka. *Ann. Phytopath. Soc. Jpn* 46: 169-178.

Waterhouse, G.M., Newhook, P.J. and Stamps, D.J. (1983). Present criteria for classification of *Phytophthora*. In: *Phytophthora: Its Biology, Taxonomy, Ecology and Pathology* (Eds. Erwin, D.C., Bartnicki-Garcia, S. and Tsao, P.H.). American Phytopathological Soc. St. Paul, MN. pp. 135-147.

VEGETABLE CROPS

Integrated Plant Disease Management (2005)
Eds. Sharma, R.C. and Sharma, J.N.
Scientific Publishers (India), Jodhpur, pp. 85-92.

13

Population dynamics of VAM fungi and its effect on wilt of pea caused by *Fusarium oxysporum* f. sp. *pisi*

SHALINI VERMA and N.P. DOHROO

Department of Mycology and Plant Pathology, Dr.Y.S. Parmar University of Horticulture and Forestry, Nauni, Solan -173 230 (H.P.)

ABSTRACT

Present studies were undertaken to study the population dynamics of VAM fungi in rhizosphere soil of pea plants and its role in disease suppression. The study revealed that there was enough variation in the population dynamics of VAM fungi in healthy and wilt infested soil when tested by t-test. Different species of VAM fungi viz., *Glomus* (14), *Acaulospora* (3), *Gigaspora* (2) and *Scutellospora* (2) were identified from rhizosphere soil of pea plants. The population of *Glomus* spp. (80.7%) was, however, found more in rhizosphere soil of pea plants. The study on effect of VAM inoculation on the disease suppression revealed minimum population of the pathogen in rhizosphere soil after application of *G. margarita, G. mosseae* and *G. heterosporum* thus suppressing the disease. It was also found that *G. margarita* and *G. mosseae* increased the seed germination to 87.5 and 84.1 per cent as compared to 52.5 per cent in the control and suppressed the disease to 10.8 and 15.8 per cent, respectively.

INTRODUCTION

Pea (*Pisum sativum* L.) is one of the most important cash crops of farmers of Himachal Pradesh in autumn season. It is attacked by several diseases caused by fungi, bacteria and viruses. Among these diseases, wilt caused by *Fusarium oxysporum* Schl. ex Fr. f.sp. *pisi* (J.C. Hall) Snyd. and Hanna. is one of the most destructive diseases of the crop and occurs in epiphytotic form almost every year. Heavy losses (60%) due to the disease have been reported to occur in the crop during the years of epiphytotics (Kumar, 1993). The disease is essentially soil-borne and poses a greater problem during management by using fungicides which are uneconomical. Their frequent and indiscriminate use often leads to atmospheric pollution and development of resistance in the pathogen. Keeping this in view, the studies were carried out to know the population dynamics of vesicular-arbuscular mycorrhizal (VAM) fungi in rhizosphere soil of pea plants and its role in disease suppression. The biocontrol potential of VAM against root pathogens is amply substantiated by various workers (Jalali and Jalali, 1991).

MATERIALS AND METHODS

Soil samples from the rhizosphere of healthy and wilt affected plants were collected from pea growing areas of Solan district of Himachal Pradesh. The samples of soil were collected in polythene bags and pooled separately representing healthy and disease infested areas. These samples were stored in a refrigerator at 4°C for isolation of VAM fungal spores by wet sieving and decanting method (Gerdemann and Nicolson, 1963). Twenty five g of soil was mixed with a convenient volume of water in a large beaker (500 ml) and stirred with a glass rod to make uniform suspension. The suspension was left for five minutes. In this process, the mycorrhizal debris floated on the top. The suspension was passed through different sieves (1000, 300, 250, 125, 105 and 45 MIC mesh sizes) arranged one over the another in a descending order. The residue left in each sieve was thoroughly washed under tap water and collected in a beaker. This process was repeated 8-9 times to trap all spores of VAM fungi and the final volume was made to 50 ml. A known quantity of the above suspension was transferred to the counting plate to count the number of live and dead spores under stereoscopic binocular microscope in each soil sample representing rhizosphere soil of healthy and diseased plants. The data were statistically analysed by using t-test (Gupta and Kapoor, 1970) as below:

$$t = \frac{|\overline{X} - \overline{Y}|}{\sigma_p \sqrt{\frac{1}{n_1} + \frac{1}{n_2}}}, \quad t \sim n_1 + n_2 - 2$$

where,

$$\sigma_p^2 = \frac{1}{n_1 + n_2 - 2} \left[\sum (X - X)^2 + \sum (Y - Y)^2\right]$$

The conclusions were drawn for working out significance between two means as under:

(i) Number of VAM live and dead spores in rhizosphere soil of healthy plants

(ii) Number of VAM live and dead spores in rhizosphere soil of diseased plants

(iii) Number of VAM spores in rhizosphere soil of healthy and diseased plants

(iv) Number of VAM live and dead spores.

The VAM fungi obtained as loose cluster of chlamydospores or as sporocarps in soil were picked up with Pasteur pipette and were mounted in polyvinyl lactic acid (Omar *et al.*, 1979) and their microscopic characteristics were recorded under light microscope (675 X). The taxonomic determination was made up to species level by referring to the photographic slide collections together with the synoptic keys given by Hall and Abbott (1983) and also standard keys defined by Gerdemann and Trappe (1975), Hall (1984) and Trappe (1987).

In general, the genus and species of VAM fungi were categorized by the morphology of their resting spores produced in soil. The following diagnostic characteristics were used:

(a) Size - Spores obtained by wet sieving and decanting method were placed on a clean slide with a drop of distilled water and cover slip. The length, width and wall thickness were measured by using an ocular micrometer.

(b) Subtending hyphae - The stalk attachment of the spores, whether simple or bulbous, was recorded.

(c) Wall layers -The number, type and colour of wall layers were noticed under oil immersion.

Per cent population of VAM fungi was also worked out.

Vesicular arbuscular mycorrhizal fungi viz., *G. mosseae, G. heterosporum* and *G. margarita* were cultured on the roots of fenugreek (*Trigonella foenum-graecum*) plants grown in a soil medium under laboratory conditions. Then the VAM fungi were evaluated against *Fusarium* wilt of pea under greenhouse conditions in a completely randomized block design (CRD) with six replications using variety Arkel (2.5 kg) of sterilized soil was thoroughly mixed with 50 g inoculum of the test pathogen grown on corn-sand meal and/or 50 g culture of VAM fungi, filled in trays (30 x 18 cm) and sown with pea seeds (20 seeds per tray). The data on seed germination, pre-emergence rot and wilt were recorded.

RESULTS AND DISCUSSION

The soil samples collected from the rhizosphere of healthy and wilt affected plants revealed variation in VAM population (Table 1).

In the rhizosphere soil of healthy plants, the number of VAM live spores was more (19.5/g soil) as compared to dead spores (9.5/g soil), whereas in rhizosphere soil of wilted plants, the number of VAM live spores was less (5/g soil) than VAM dead spores (12.5/g soil). The number of VAM spores (live and dead) in rhizosphere soil of healthy plants was 14.5/g soil in comparison to 8.75/g soil in wilted plants (Table 2).

Table 1. VAM population in pea rhizosphere soil

Rhizosphere soil of healthy pea plants		Rhizosphere soil of wilted pea plants		VAM spores/g soil from rhizosphere soil of		Rhizosphere soil of healthy and wilted pea plants	
Live VAM spores/g soil	Dead VAM spores/g soil	Live VAM spores/g soil	Dead VAM spores/g soil	Healthy pea plants	Wilted pea plants	Live VAM spores/g soil	Dead VAM spores/g soil
20	11	4	10	20	4	20	11
22	10	3	11	22	3	22	10
20	9	3	10	20	3	20	9
21	12	8	14	21	8	21	12
18	7	3	12	18	3	18	7
19	11	8	12	19	8	19	11
18	9	4	13	18	4	18	9
15	7	4	15	15	4	15	7
22	8	7	13	22	7	22	8
20	11	6	15	20	6	20	11
Mean				11	10	4	10
19.5	9.5	5	12.5	10	11	3	11
				9	10	3	10
				12	14	8	14
				7	12	3	12
				11	12	8	12
				9	13	4	13
				7	15	4	15
				8	13	7	13
				11	15	6	15
			Mean	14.5	8.75	12.25	11

Table 2. VAM population in pea rhizosphere soil tested by using t-test

Rhizosphere soil of	VAM spores	Mean spore number/g soil	Calculated t value	Tabulated t value
Healthy pea plants	Live	19.5	11.69*	2.101
	Dead	9.5		
Wilted pea plants	Live	5	8.604*	2.101
	Dead	12.5		
Healthy pea plants	Live and dead	14.5	3.69**	2.02
Wilted pea plants	Live and dead	8.75		
Healthy and wilted pea plants	Live	12.25	0.69	2.02
	Dead	11		

* Significant at 5 per cent and 18 degrees of freedom

** Significant at 5 per cent and 38 degrees of freedom

These results are in accordance with the findings of Caron *et al.* (1986) and Sharma and Dohroo (1996) who reported that diseases caused by soil-borne fungi were influenced by the formation of VAM in the root system. It was inferred that calculated value of t for number of VAM live and dead spores in rhizosphere soil from healthy plants (11.69), in rhizosphere soil from wilted plants (8.604), and number of VAM spores from rhizosphere soil of healthy and wilted pea plants (3.69) was much more than the tabulated value of t at 0.05 level. Hence, the results were found significant and null hypothesis might be rejected. It is obvious that variation existed among mean number of VAM live and dead spores in the rhizosphere soil of healthy and wilted pea plants and mean number of VAM spores from rhizosphere soil of healthy and wilted pea plants. In rhizosphere soil of healthy and wilted pea plants, number of VAM live spores was more (12.25/g soil) than VAM dead spores (11/g soil). The calculated value of t for number of VAM live and dead spores (0.69) was much less than the tabulated value of t (2.02) at 5 per cent level of significance and 38 degrees of freedom. Hence, the results were found nonsignificant and null hypothesis might be accepted. It showed that the mean live and dead spore number was same for both the rhizosphere soil samples. A good deal of information has been generated in recent years on VAM to suppress soil- borne diseases, especially root rots and wilts (Schenck, 1981; Dehne, 1982; Caron, 1989).

After detailed morphotaxonomic study, photomicrographs were compared with already described VAM fungi. The VAM fungi found associated with rhizosphere of pea were *Acaulospora gerdemannii, A. longula, A. nicolsonii, Gigaspora albida, G. gigantea, Glomus albidum, G. ambisporum, G. boreale, G. botryoides, G. caledonium, G. delhiense, G. fasciculatum, G. heterosporum, G. macrocarpum, G. monosporum, G. mosseae, G. pulvinatum, G. segmentatum, G. tenerum, Scutellospora calaspora* and *S. gilmorei*. The details of morphological characters were given in Table 3.

Soil samples from pea rhizosphere had maximum population of *Glomus* spp. (80.76%) followed by *Scutellospora* spp. (10.25%) and *Acaulospora* spp. (5.12%) while minimum population of *Gigaspora* spp. (1.28%) and *Sclerocsystis* spp. (2.56%) was recorded. More population of *Glomus* spp. in the rhizosphere of the crop illustrated its universal occurrence that could largely be attributed to geographical features of crop. These findings corroborate to the work of Al-Raddad (1993). So far, there is no report on the occurrence of VAM fungi in pea soil.

All the VAM fungi significantly increased the seed germination, reduced pre-emergence rot and wilt over control (Table 4).

Maximum seed germination was recorded in *G. margarita* treatment (87.50%) followed by *G. mosseae* (84.17%) whereas minimum germination was found in *G. heterosporum* (70%) which was significantly more than control (52.50%). Similarly, minimum and maximum pre-emergence rot was recorded in *G. margarita* (12.50%), and *G. heterosporum* (30.0%) treatment, respectively. However, minimum wilt was recorded in *G. margarita* (10.83%) followed by *G. mosseae* (15.83%) and maximum wilt was observed in *G. heterosporum* (19.17%) which was significantly lower than control (45.83%). The results could be related to the findings of Sharma and Dohroo (1996) who

Table 3. Morphological characteristics of VAM fungi associated with pea rhizosphere soil

VAM fungi	Morphological characters		
	Diameter (μm)	Wall number	Wall thickness (μm)
Acaulospora gerdemannii	243.2x250	Bilayer	3
	229.4x229.4	Bilayer	3
A. longula	81.4x85.	Bilayer	2.78
A. nicolsonii	114x123.5	Bilayer	4
	144.3x159.1	Bilayer	4
	177.6x196.1	Bilayer	4
Gigaspora albida	89.4x144	Bilayer	7.4
G.gigantea	183x291	Bilayer	2.78
Glomus albidum	105.6x114	Bilayer	3.8
	96.2x96.2	Bilayer	3.7
G. ambisporum	155.4x159.1	Hexalayer	7.4
	74x77.7	Bilayer	2.78
G. boreale	111x116.55	Bilayer	8
G. botryoides	160.95x162.8	Single layer	7.4
G. caledonium	240.5x266.4	Bilayer	6
	130x133	Bilayer	6
G. delhiense	118.4x120.25	Bilayer	11.1
G. fasciculatum	86.2x92.3	Single layer	3
G. heterosporum	103.6x107.3	Bilayer	9.25
	151x206	Bilayer	3.8
	138.75x146.15	Bilayer	3.52
G. macrocarpum	115x117.5	Bilayer	7.6
G. monosporum	207.2x225.7	Bilayer	4
G. mosseae	111x111	Bilayer	2.78
	118.4x133.2	Bilayer	2
	133.2x144.3	Bilayer	2
	140.6x157.25	Bilayer	3.7
	110x114.7	Bilayer	2.78
	136.9x157.25	Bilayer	3.7
	159.1x207.2	Bilayer	3.52
	114.7x114.7	Bilayer	3.7
	196.1x207.2	Bilayer	3.7
	136.9x144.3	Bilayer	3.7
G. pulvinatum	90.65x92.5	Single layer	2
G. segmentatum	55.5x59.2	Bilayer	6.48
G. tenerum	77.7x86.95	Bilayer	1.85
Scutellospora calospora	181.3x185	Bilayer	1.85
	170.2x185	Bilayer	3.7
	185x188.7	Bilayer	3
S. gilmorei	310.8x318.2	Bilayer	3

found that the plants were benefitted by inoculation of VAM fungi in their root system. The VAM fungi were known to impart resistance against soil borne plant pathogens causing root rot and wilt diseases (Dehne, 1982). Dehne and Schonbeck (1975, 1979) reported that *G. mosseae* reduced the damage to tomato by *F.o.* f.sp. *lycopersici*. Dehne (1977) reported decrease in damage in mycorrhizal plants of cucumber infected with *F.o.* f.sp. *cucumerinum*.

Table 4. Effect of VAM fungi on seed germination, pre-emergence rot and wilt of pea

Treatment	Seed germination (%)	Pre-emergence rot (%)	Wilt (%)
Gigaspora margarita	87.50 (69.39)[a]	12.50 (20.61)[a]	10.83 (18.97)[a]
Glomus heterosporum	70.0 (56.82)[c]	30.0 (33.19)[c]	19.17 (25.88)[c]
Glomus mosseae	84.17 (66.59)[b]	15.83 (23.42)[b]	15.83 (23.32)[b]
Control	52.50 (46.44)[d]	47.50 (43.56)[d]	45.83 (42.60)[d]
$CD_{0.05}$	(1.08)	(1.08)	(1.65)

Figures in parentheses are arcsine transformed values

Figures with different alphabets differ significantly

REFERENCES

Al-Raddad, A. (1993). Distribution of different *Glomus* species in rainfed areas in Jordan Dirasat. *Pure Appl. Sci.* 20: 165-182.

Caron, M (1989). Problems in the use of endomycorrhizal fungi as biological control agents. *Phytoprotection* 70: 43-49.

Caron, M., Richard, C. and Fortin, J.A. (1986). Effect of pre-infestation of the soil by a VAM fungus, *Glomus intraradices* on *Fusarium* crown and root rot of tomatoes. *Phytoprotection* 67: 15-19.

Dehne, H.W. (1977). Untersuchungen Uber den Einfluss der Endotrophen Mycorrhiza auf die Fusarium Welke on Tomate and Gurke. Dissertation, Univ. Bonn, W. Germany, 150 p.

Dehne, H.W. (1982). Interaction between vesicular-arbuscular mycorrhizal fungi and plant pathogens. *Phytopathology* 72: 1115-1119.

Dehne, H.W. and Schonbeck, F. (1975). Investigation on the influence of endotrophic mycorrhiza on the Fusarial wilt of tomato. *Z. Pflanzenkr. Pflanzenschutz.* 82: 630-632.

Dehne, H.W. and Schonbeck, F. (1979). The influence of endotrophic mycorrhizae on plant diseases and colonization of tomato plants by *Fusarium oxysporum* f.sp. *lycopersici*. *Phytopathol. Z.* 95: 105-110.

Gerdemann, J.W. and Nicolson, T.H. (1963). Spores of mycorrhizal *Endogone* spp. extracted from soil by wet-sieving and decanting. *Trans. Br. Mycol. Soc.* 46: 235-244.

Gerdemann, J.W. and Trappe, J.M. (1975). Taxonomy of the endogonaceae. In: *Endomycorrhizas* (Eds. Sanders, F.E., Mosse, B. and Tinker, R.B.). Academic Press, London. pp. 35-51.

Gupta, S.C. and Kapoor, V.K. (1970). *Fundamentals of Mathematical Statistics.* Sultan Chand & Sons, Educational Publishers, New Delhi, 1266 p.

Hall, I.R. (1984). Taxonomy of VA mycorrhizal fungi. In: *VA Mycorrhiza* (Eds. Powel C.L. and Bagyaraj, D.J.). CRC Press, Florida. pp. 57-94.

Hall, I.R. and Abbott, L.K. (1983). Photographic slide collection illustrating features of endogonaceae. *Bull. Br. Mycol. Soc.* 17: 54.

Jalali, B. L. and Jalali, Indu. (1991). Mycorrhizae in plant disease control. In: *Handbook of Applied Mycology Vol. I. Soil and Plants* (Eds. Arora, D.K., Rai, B., Mukerji, K.G. and Knudson, G.R.). Marcel Dekker, Inc. New York, USA. pp.131-154.

Kumar, A. (1993). *Studies on Resistance to Fusarium Wilt in Some Cultivars of Peas (P. sativum* L.). M.Sc. Thesis, Dr. Y.S. Parmar Univ. of Hort. and Forestry, Solan, HP, 71 p.

Omar, M.B., Bolland, L. and Heather, W.A. (1979). A permanent mounting medium for fungi. *Bull. Br. Mycol. Soc.* 13: 31.

Schenck, N.C. (1981). Can mycorrhiae control root diseases? *Plant Dis.* 65: 230-324.

Sharma, S. and Dohroo, N.P. (1996). Vesicular-arbuscular mycorrhizae in plant health and disease management. *Int. J. Trop. Plant Dis.* 14: 147-155.

Trappe, J. (1987). Phylogenetic and ecological aspects of mycotrophy in the angiosperms from an evolutionary standpoint. In: *Ecophysiology of VA Mycorrhizal Plants.* (Ed. Safir, G.). CRC Press, Boca Raton. pp. 5-25.

Integrated Plant Disease Management (2005)
Eds. Sharma, R.C. and Sharma, J.N.
Scientific Publishers (India), Jodhpur, pp. 93-99.

14

Comparative efficacy of biocontrol agents against *Fusarium* wilt of pea

SHALINI VERMA and N.P. DOHROO

Department of Mycology and Plant Pathology, Dr.Y.S. Parmar University of Horticulture and Forestry, Nauni, Solan -173 230 (H.P.)

ABSTRACT

Fusarium wilt of pea caused by *Fusarium oxysporum* f.sp. *pisi* is one of the most destructive diseases of autumn pea in Solan, Sirmour and Shimla districts of Himachal Pradesh. The efficacy of different antagonists was ascertained against the pathogen *in vitro* and the disease under field conditions during 1998 and 1999. *Trichoderma* spp. viz., *Trichoderma harzianum* and *T. viride* showed the maximum growth inhibition of the pathogen *in vitro* and disease suppression under field conditions. *T. viride* and *T. harzianum* resulted in increased seed germination in comparison to control, decreased disease incidence to 22.8 and 25.5 per cent with AUDPC of 4.16 and 4.60 and minimum apparent infection rate (r) of 0.046 and 0.046, respectively, as compared to 0.063 in the control treatment.

INTRODUCTION

Pea (*Pisum sativum* L.) is one of the most important cash crops of farmers of Himachal Pradesh. Wilt caused by *Fusarium oxysporum* Schl. ex Fr. f.sp. *pisi* (J.C. Hall) Snyd. and Hanna is one of the most destructive diseases of the crop and occurs in an epiphytotic form almost every year. Heavy losses (60%) due to the disease have been reported to occur in the crop during the years of epiphytotics (Kumar, 1993). The disease is essentially soil-borne and poses a greater problem during management by using fungicides which are uneconomical, and their frequent and indiscriminate use often leads to atmospheric pollution and development of resistance in the pathogens. In this context, biological control is an alternative strategy of disease management which is eco-friendly. Several organisms have been successfully used as biocontrol agents, such as species of *Trichoderma* (Raguchander *et al.*, 1997), *Pseudomonas*, *Gliocladium*, *Penicillium*, (Castejon-Munoz and Oyarzun, 1995), *Pythium* (Bradshaw-Smith *et al.*, 1991) etc. Based on the review of literature, it appeared that biological seed treatment was feasible and attempts were made during the present study to select effective antagonists which could be applied to pea seeds and evaluated by laboratory, greenhouse and field tests.

MATERIALS AND METHODS

The soil samples were collected from pea growing areas of Himachal Pradesh in polythene bags. The samples were taken from the rhizosphere of healthy pea plants, pooled and then isolation of associated microbiota was made by dilution plate method (Johnson, 1957). The cultures were identified on the basis of morphological characters documented in taxonomic keys (Von Arx, 1981).

In vitro evaluation of antagonists

The efficacy of different antagonists was ascertained against *F.o.* f.sp. *pisi* by using dual culture method (Falck, 1907). The radial growth of antagonists and pathogen were observed besides recording zone of inhibition or interaction. The per cent growth inhibition of the test pathogen was calculated as per Vincent (1947).

Slurry seed treatment

Ten days old culture of biocontrol agents (BCA) growing on potato dextrose agar medium was taken for their evaluation against wilt of pea. The hyphae were scrapped with the help of sterilized scalpel, put in a beaker containing small quantity of molasses and properly blended for 5 min aseptically. Slurry treatment to 60 g (approximately 420 seeds) pea seeds of variety Arkel was applied properly with different BCAs. The treated seeds were dried on the blotting papers and transferred to paper bags for greenhouse and field evaluation.

Greenhouse evaluation

The sterilized soil was mixed thoroughly with the inoculum of test pathogen @ 50 g/2.5 kg soil by growing on corn-sand meal (Dohroo, 1988), filled in trays (30x18 cm) soaked with water and allowed for a week for the pathogen to establish. Twenty BCA treated seeds of different lots were then sown separately in different trays and watered regularly. Inoculated sterilized soil which was sown with untreated seeds, was kept as control. The data on seed germination, pre-emergence rot and wilt were recorded and per cent disease control for each treatment was also calculated.

Field evaluation

The comparative efficacy of six biocontrol agents viz., *Gliocladium virens* (4 x 10^6 spores/ml), *Laetisaria arvalis* (7 x 10^5 spores/ml), *Paecilomyces lilacinus* (1 x 10^6 spores/ml), *Pseudomonas fluorescens* (1% PF-1, Deptt. of Plant Pathology, TNAU, Coimbatore), *Trichoderma harzianum* (3 x 10^6 spores/ml) and *T. viride* (2 x 10^6 spores/ml) was tested by giving slurry treatment to pea seeds against *Fusarium* wilt under field conditions during 1998 and 1999. The seeds were treated with BCAs as per the method given above. Field experiment was laid out in randomized block design with four replications using the susceptible variety Arkel. Ninety seeds were sown per plot (1 m^2) at a spacing of 30 x 3.5 cm. The data on seed germination, pre-emergence rot, wilt and green pod yield (g/m^2) were recorded. The apparent infection rate(r) and area under disease progress curve (AUDPC) described by Van der Plank (1963) and Shanner and Finney (1977), respectively, were also calculated for each treatment.

$$r = \frac{2.303}{t_2 - t_1} \log_{10} \frac{x_2 (1 - x_1)}{x_1 (1 - x_2)}$$

Where,

r	= Apparent infection rate (per unit per day)
t_2-t_1	= Time interval between first and last observation
x_1	= Proportion of diseased plant parts at time t_1
x_2	= Proportion of diseased plant parts at time t_2
1-x_1	= Proportion of healthy plant parts at time t_1
1-x_2	= Proportion of healthy plant parts at time t_2

$$AUDPC = \sum \frac{(Y_i + Y_{i+1})}{2} x (t_{i+1} - t_i)$$

Where,

Y_i	= Disease incidence at time t_i
Y_{i+1}	= Disease incidence at time t_{i+1}
t_i	= Time when disease incidence was Y_i
t_{i+1}	= Time when disease incidence was Y_{i+1}.

RESULTS AND DISCUSSION

In vitro evaluation

The data (Table 1) indicated that all the antagonists inhibited the mycelial growth of *F.o.* f.sp. *pisi*. *T. harzianum* (51.55%), *T. viride* (51.08%) showed maximum inhibition and were statistically at par with each other. They were significantly superior to other antagonists followed by *T. hamatum* (48.64%) and *T. longibrachiatum* (46.98%). *Beauveria bassiana*

(21.69%) and *Aspergillus tamarii* strain 2 (22.55%) were, however, found least effective.

Table 1. *In vitro* evaluation of antagonists against *Fusarium oxysporum* f.sp. *pisi*

Antagonist	Mycelial growth inhibition (%)	Zone of inhibition (+)/ Interaction (-)
Aspergillus tamarii strain 1	32.28(34.59)[d]	—
A. tamarii strain 2	22.55(28.35)[h]	—
A. wentii	29.75(33.06)[ef]	—
Beauveria bassiana	21.69(27.63)[h]	—
Fusarium oxysporum strain 1	34.55(36.00)[c]	—
F. oxysporum strain 2	31.08(33.88)[de]	—
Gliocladium virens	26.67(31.09)[g]	—
Laetisaria arvalis	27.36(31.54)[g]	—
Paecilomyces lilacinus	29.25(32.72)[f]	—
Pseudomonas fluorescens	26.39(30.90)[g]	+
Trichoderma hamatum	48.64(44.22)[b]	—
T. harzianum	51.55(45.89)[a]	—
T. longibrachiatum	46.98(43.27)[b]	—
T. viride	51.08(45.62)[a]	—
$CD_{0.05}$	(1.059)	—

Figures in parentheses are arcsine transformed values
Figures with different alphabets differ significantly

The zone of inhibition was observed only in case of *Pseudomonas fluorescens*, whereas other treatments revealed zone of interaction. The inhibitory effect of *Trichoderma* spp. against *F.o.* f.sp. *pisi* was probably due to hyperparasitism/ mycoparasitism, competition for space and nutrition and antagonistic chemicals produced and released into the environment. *Trichoderma* spp. have been reported to produce antibiotic compounds (trichodermin, acetaldeyde), extracellular enzymes (chitinase, cellulase, (1,3)-glucanase), unsaturated monobasic acid (dermadine) and peptides (alamethicine, suzukacillin) that either damage plant pathogens or enhance their population in maintaining favourable balance as a portion of the biota (Mukerji and Garg, 1988). *Trichoderma* spp. have also been employed in biological control of diseases of vegetable crops caused by soil borne plant pathogens (Baker and Cook, 1974; Elad *et al.*, 1980; Chet and Baker, 1981).

Greenhouse evaluation

The data (Table 2) revealed that all the BCAs increased seed germination, reduced pre-emergence rot and wilt as compared to control.

Table 2. Effect of biocontrol agents on seed germination, pre-emergence rot and wilt of pea in greenhouse conditions

Biocontrol agent	Seed germination (%)	Pre-emergence rot (%)	Wilt (%)	Disease control (%)
Gliocladium virens	70	30	21.42	60.73
Laetisaria arvalis	90	10	10	81.66
Paecilomyces lilacinus	90	10	11.11	79.63
Pseudomonas fluorescens	85	15	11.76	78.44
Trichoderma harzianum	100	0	5	90.83
T. viride	100	0	5	90.83
Control	55	45	54.55	-

Cent per cent germination of pea was recorded in *T. harzianum* and *T. viride* treated seeds with no incidence of pre-emergence rot. However, minimum germination (70%) and maximum pre-emergence rot (30%) was observed in *Gliocladium virens* treated seeds. *T. harzianum* and *T. viride* were found to be highly effective in reducing the incidence of wilt to 5 per cent as compared to 54.55 per cent in control followed by *L.arvalis* (10%), *P. lilacinus* (11.11%), *P. fluorescens* (11.76%) and *G. virens* (21.42%). *T. harzianum* and *T. viride* provided 90.83 per cent disease control whereas minimum disease control of 60.73 per cent was recorded in *G. virens*. Various workers employed biocontrol agents for the disease control which reduced the number of target organisms; replaced pathogens in diseased debris with saprophytes through competition, antibiosis, predation or parasitism and imparted cross protection and host resistance (Chet, 1987).

Field evaluation

The pooled analysis of data (Table 3) showed that BCA seed treatments had a significant effect on germination, pre-emergence rot, wilt and yield of pea. *T. viride* treated seeds showed maximum germination (55.42%) followed by *T. harzianum* (49.44%). However, *G. virens* (35.69%) showed least germination of pea but was significantly more in comparison to control (28.06%). The pre-emergence rot was observed minimum in *T. viride* (44.58%) and maximum in *G. virens* (64.31%) which was significantly less than control (71.94%). During crop season 1998, less pre-emergence rot of 57.82 per cent was recorded as compared to 1999 (58.65%). *T. viride* was found highly effective in reducing the *Fusarium* wilt of pea to 22.83 per cent followed by *T. harzianum* (25.52%). Further, *G. virens* was found to be the least effective BCA with 37.37 per cent wilt in the present studies, which was significantly less than control (56.16%). During crop season 1998, less *Fusarium* wilt (17.05%) was recorded as compared to crop season 1999 (51.13%). Maximum pod yield of pea was obtained in *T. viride* (618.4 g/m^2) followed by *T. harzianum* (580.5 g/m^2) while minimum was obtained in control (244.2 g/m^2). It is evident from the data illustrated in Fig. 1, that minimum incidence of wilt was noticed in *T. viride* followed by *T. harzianum* whereas maximum was in *G. virens* which was significantly lower than control. The AUDPC was 4.16 in *T. viride* followed by *T. harzianum* (4.602) and maximum in *G. virens* (7.99) and *P. fluorescens* (7.64) which was significantly lesser than control (13.17%). *T. viride* recorded minimum apparent infection rate (0.046) as compared to control (0.063).

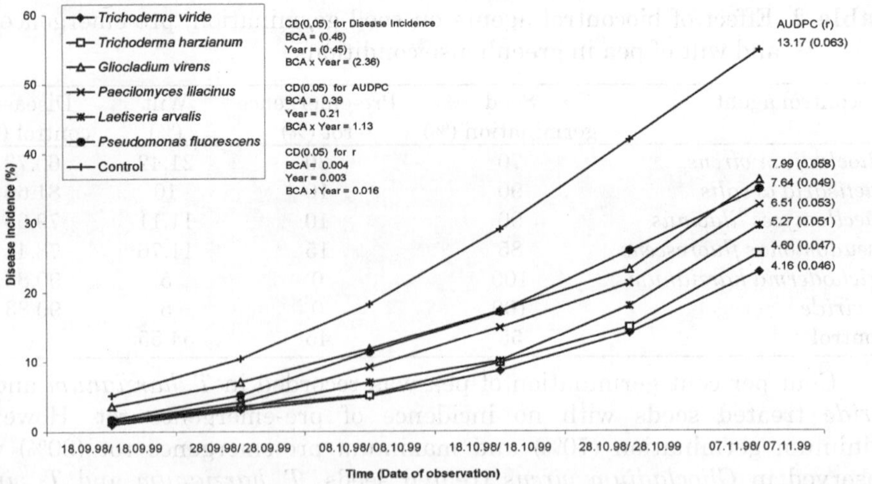

Fig. 1. Effect of biocontrol agents on disease incidence (crop season 1998 and 1999)

Table 3. Effect of biocontrol agents on seed germination, pre-emergence rot, yield and wilt of pea

Treatment	Seed germination (%)			Pre-emergence rot (%)			Wilt (%)			Yield (gm/m^2)		
	1998	1999	Pooled	1998	1999	Pooled	1998	1999	Pooled	1998	1999	Pooled
Gliocladium	31.94	39.44	35.69	68.06	60.55	64.31	18.73	56	37.37	523^{b}	462.5^{b}	492.7^{d}
virens	(34.41)^{f}	(38.91)^{d}	(36.66)^{f}	(55.59)^{f}	(51.1)^{d}	(53.34)^{f}	(25.62)^{b}	(48.45)^{d}	(37.04)			
Laetisaria	46.11	43.88	45.00	53.89	56.11	55.00	11.51	47.37	29.44	568.7^{a}	518.4^{b}	543.6^{c}
arvalis	(42.77)^{c}	(41.49)^{c}	(42.13)^{c}	(47.23)^{c}	(48.51)^{c}	(47.87)^{c}	(19.78)^{a}	(43.49)^{c}	(31.64)			
Paecilomyces	42.22	41.39	41.8	57.78	58.64	58.21	15.44	49.59	32.52	539.2^{a}	496.6^{c}	517.9^{d}
lilacinus	(40.53)^{d}	(40.04)^{d}	(40.28)^{d}	(49.48)^{d}	(49.98)^{d}	(49.73)^{d}	(23.12)^{b}	(44.77)^{d}	(33.94)			
Pseudomonas	37.5	36.39	36.95	62.50	63.61	63.05	16.23	53.48	34.85	538.5^{a}	485^{c}	511.7^{d}
fluorescens	(37.76)^{e}	(37.1)^{e}	(37.43)^{e}	(52.24)^{e}	(52.9)^{e}	(52.57)^{e}	(23.72)^{b}	(47.00)^{d}	(35.36)			
Trichoderma	52.22	46.66	49.44	47.78	53.33	50.56	10.26	40.78	25.52	574.5^{a}	586.5^{a}	580.5^{b}
harzianum	(46.27)^{b}	(43.09)^{b}	(44.68)^{b}	(43.73)^{b}	(46.91)^{b}	(45.32)^{b}	(18.67)^{a}	(39.67)^{a}	(29.17)			
T. viride	58.33	52.5	55.42	41.66	47.5	44.58	8.88	36.79	22.83	645.5^{a}	591.2^{a}	618.4^{a}
	(49.81)^{a}	(46.44)^{a}	(48.12)^{a}	(40.20)^{a}	(43.57)^{a}	(41.88)^{a}	(17.3)^{a}	(37.32)^{a}	(27.31)			
Control	26.95	29.17	28.06	73.05	70.83	71.94	38.42	73.89	56.16	263.5^{c}	225^{e}	244.2^{e}
	(31.27)^{g}	(32.68)^{f}	(31.98)^{f}	(58.74)^{g}	(57.32)^{f}	(58.03)^{g}	(38.27)^{c}	(59.65)^{e}	(48.96)			
Grand mean	42.18	41.35		57.82	58.65		17.05	51.13		521.9	480.8	
	(40.40)	(39.96)		(49.6)	(50.04)		(23.78)	(45.76)				
$CD_{0.05}$	(1.47)	(0.57)		(1.47)	(0.57)		(2.56)	(2.34)		119.82	8.36	

Figures in parentheses are arcsine transformed values

Figures with different alphabets differ significantly

Effects	$CD_{0.05}$	$CD_{0.05}$	$CD_{0.05}$	$CD_{0.05}$
BCA	(0.462)	(0.462)	(1.306)	32.39
Year	(0.247)	(0.247)	(0.698)	17.31
BCA x year	(1.323)	(1.323)	NS	NS

The seed treatment with *Trichoderma* spp. before planting in non-treated soils has been reported to apparently improve seedling vigour, increase crop yield and manage some soil borne pathogens (Mukherji and Garg, 1988). These results are in accordance with Lacicowa and Pieta (1994) who suggested that pea seeds treated with *Trichoderma* spp. gave the best disease control of *Fusarium* wilt of pea.

REFERENCES

Baker, F.K. and Cook, R.J. (1974). *Biological Control of Plant Pathogens.* W.H. Freeman and Comp., San Francisco. 433 p.

Bradshaw-Smith, R.P., Whalley, W.M. and Craig, G.D. (1991). Interactions between *Pythium oligandrum* and the fungal foot rot pathogens of peas. *Mycol. Res.* 95: 861-865.

Castejon-Munoz, M. and Oyarzun, P.J. (1995). Soil receptivity to *Fusarium solani* f.sp. *pisi* and biological control of root rot of pea. *Eur. J. Plant Pathol.* 101: 35-49.

Chet, I. (1987). *Innovative Approaches to Plant Disease Control.* John Wiley, New York, 372 p.

Chet, I. and Baker, R. (1981). Isolation and biocontrol potential of *Trichoderma hamatum* from soil naturally suppressive to *Rhizoctonia solani. Phytopathology* 71: 286-290.

Dohroo, N.P. (1988). Germplasm reaction of cauliflower to stalk rot (*Sclerotinia sclerotiorum*). *Indian J. Plant Pathol.* 6: 144.

Elad, Y., Chet, I. and Katan, J. (1980). *Trichoderma harzianum* : A biocontrol agent effective against *Sclerotium rolfsii* and *Rhizoctonia solani. Phytopathology* 70: 119-121.

Falck, R. (1907). Wachstumgesetze, Wachstumtakumtaktoren Und temperature Wertder holzersterenden. *Mycelien* 1: 143-154.

Johnson, L.F. (1957). Effect of antibiotics on the numbers of bacteria and fungi isolated from soil by dilution-plate method. *Phytopathology* 47: 630-631.

Kumar, A. (1993). *Studies on Resistance to Fusarium Wilt in Some Cultivars of Peas (P. sativum* L.). M.Sc. Thesis, Dr. Y.S. Parmar Univ. of Hort. and Forestry, Solan, HP, 71 p.

Lacicowa, B. and Pieta, D. (1994). Protective effects of microbiological dressing of pea seeds (*Pisum sativum* L.) against pathogenic fungi living in soil. *Ann. Univ. Mariae Curie-Sklodowska, Sec. EEE, Hortic.* 2: 165-171.

Mukerji, K.G. and Garg, K.L. (Eds.) (1988). *Biocontrol of Plant Diseases.* Vol. I. CRC Press Inc. Florida, 211 p.

Raguchander, T., Rajappan, K. and Samiyappan, R. (1997). Evaluating methods of application of biocontrol agent in the control of mungbean root rot. *Indian Phytopathol.* 50: 229-234.

Shanner, G. and Finney, R.E. (1977). Effect of nitrogen fertilization on the expression of slow mildewing resistance in Knox wheat. *Phytopathology* 67: 1051-1056.

Van der Plank, J.E. (1963). *Plant Diseases: Epidemics and Control.* Academic Press, New York, 349 p.

Vincent, J.M. (1947). Distortion of fungal hyphae in the presence of certain inhibitors. *Nature* 150: 850.

Von Arx, J.A. (1981). *The Genera of Fungi Sporulating in Pure Culture.* FL-9490 VADUZ, Germany, 424 p.

Integrated Plant Disease Management (2005)
Eds. Sharma, R.C. and Sharma, J.N.
Scientific Publishers (India), Jodhpur, pp. 101-105.

15

Management of pea wilt/root rot complex by integrating plant extracts and biocontrol agents

MEENA DEVI and Y.S. PAUL

Department of Plant Pathology, CSK HPKV, Palampur-176 062 (H.P.)

ABSTRACT

Fungitoxic activity of plant extracts of ten plant species was evaluated against the pea wilt/root rot complex pathogens viz. *Fusarium solani, F. oxysporum, Rhizoctonia solani* and *Sclerotinia sclerotiorum.* The plant species selected were *Melia azedarach, Eupatorium adenophorum, Eupatorium* sp., *Cannabis sativa, Ranunculus muricatus, Ocimum sanctum, Lantana camara, Vitex negundo, Camellia sinensis* and *Datura stramonium.* In addition the effects of these extracts were also studied on the four selected *Trichoderma* sp. bioagents viz., JMA-4, SMA-5, DMA-8 and JMA-11. Out of these, plant extract of *R. muricatus* completely inhibited the growth of all the pathogens as well as the bioagents. Plant extract of *C. sinensis* caused more than 70 per cent inhibition of *F. solani* and *S. sclerotiorum,* without affecting the growth of two strains only viz., JMA-4 and JMA -11. *E. adenophorum* was highly effective against *F. solani, F. oxysporum* and *R. solani* causing more than 50 per cent inhibition and did not affect the growth of any of the bioagents. *S. sclerotiorum* was inhibited to the tune of 31 per cent by *E. adenophorum* extract. Thus it can be concluded that *E. adenophorum* can safely be integrated with JMA-11 and better disease management can be achieved.

INTRODUCTION

Pea (*Pisum sativum* L.) is one of the major cash crops of Himachal Pradesh and is grown both as vegetable and pulse crop. It is threatened by a number of diseases, but in recent years root rot/wilt complex has become most destructive as not only affects the initial stand but also continues to appear throughout the cropping season. The disease is caused by a complexity of pathogens in the state. The fungi associated with the disease have been identified as *Fusarium solani* f.sp *pisi, F. oxysporum* f. sp. *pisi, Sclerotinia sclerotiorum, Rhizoctonia solani* and *Phoma medicaginis* var. *pinodella*. Two or more than two species may occur simultaneously and result in huge losses both quantitatively and qualitatively. Maheshwari *et al.* (1981) reported 13.9-93.0 per cent loss due to this disease in north India. The disease is mainly soil-borne in nature and it is difficult to think of chemical methods of disease management. Resistance is not available against the disease. In the present studies, efficacy of plant extracts of locally available species was tested against the wilt/root rot pathogens of pea. In addition, the effects of theses extracts on the locally isolated bioagents which were found effective against the pathogen complex were also evaluated with a view to develop an integrated strategy to manage the disease complex.

MATERIALS AND METHODS

Plant extracts of locally available plant species namely *Melia azedarach, Eupatorium adenophorum, Eupatorium* sp., *Cannabis sativa, Ranunculus muricatus, Ocimum sanctum, Lantana camara, Vitex negundo, Camellia sinensis* and *Datura stramonium* were prepared. Freshly collected leaves were washed with tap water. One hundred grams of leaves were mixed in 100 ml of water, macerated using pestle and mortar and filtered through muslin cloth. This was considered as the standard extract of a particular plant species. Potato dextrose agar medium was prepared in which 10 per cent extract of each of the species was incorporated. One set of the medium was autoclaved at 15 lbs per square inch for 15 min while other set was used as unsterilized. The medium was transferred in sterilized Petri plates and inoculated with the wilt/root rot pathogens (*F. solani* f. sp. *pisi, F. oxysporum* f. sp. *pisi, S. sclerotiorum, R. solani* and *Phoma medicaginis* var. *pinodella*). Also, the four selected *Trichoderma* sp. bioagents which were earlier found inhibitory to the pathogens namely JMA-4, SMA-5, DMA-8 and JMA-11 were inoculated in the plates. Three replications in each case were maintained. The plates were incubated at 22 ± 1°C. Data on mycelial growth was recorded as soon as control plate was fully covered with mycelium. Per cent inhibition was calculated as per Vincent (1947).

RESULTS AND DISCUSSION

The data on per cent inhibition of the growth of the pathogens using different plant extracts has been presented in Table 1. Under sterilized conditions, *Ranunculus muricatus* extract resulted in complete inhibition of all the pathogens tested, but under unautoclaved conditions, *F.solani* and *F.oxysporum* were inhibited only by 48.2 and 32.2 per cent, respectively. Possibly, the antifungal substance/concentration against these pathogens is released after heating only. In case of *R. solani*, maximum inhibition of 66.7 per cent under autoclaved and 52.2 per cent under unautoclaved conditions was recorded with *Eupatorium adenophorum* extract. Other extracts resulted in less than 35 per cent inhibition. More than 50 per cent inhibition of *F.*

solani was recorded in case of *Camellia sinensis* (66.7 and 81.8%), *Datura stramonium* (51.1 and 74.5 %) and *Eupatorium adenophorum* (52.2 and 72.2%) under autoclaved and unautoclaved conditions, respectively. In all these cases autoclaving reduced the fungitoxicity. *Camellia sinensis* and *Datura stramonium* gave 70.0 and 85.6 per cent inhibition of *F. oxysporum* after autoclaving; while *E. adenophorum* resulted in 72.2 per cent inhibition of this pathogen without autoclaving. In case of *S. sclerotiorum* also, *C. sinensis* gave 52.2 and 72.2 per cent inhibition with and without autoclaving, respectively; while 52.2 per cent inhibition was observed by *E. adenophorum* after autoclaving. *Datura stramonium, Cannabis sativa* and *Ocimum sanctum* were earlier also reported to inhibit various plant pathogens (Singh *et al.*, 1979; Patel *et al.*, 2000). Bansal and Gupta (2000) observed toxicity of *Ocimum bacillicum* and *Lantana camara* against *Fusarium oxysporum* and Ushamalini *et al.* (1997) had reported inhibitory effects of extracts of *Vitex negundo, Ocimum sanctum* and *Lantana indica* against *F. oxysporum* and *Macrophomina.* The fungitoxic nature of the extracts varied whether autoclaved or not and actually depended upon the plant species and the pathogen tested. The most effective extract amongst the plant species tested was found in case of *R. muricatus* followed by *E. adenophorum.* Presence of antifungal activity in extracts of members of ranunculaceae has earlier been well documented (Gilliver, 1947; Misra, 1975).

Table 1. Antifungal activity of some plant extracts against pea wilt/root rot pathogens

Plant species*	Per cent inhibition							
	R. solani		*F. solani*		*F oxysporum*		*S. sclerotiorum*	
	A**	B**	A	B	A	B	A	B
1	0.0 (0.0)	0.0 (0.0)	20.0 (26.5)	26.6 (31.1)	10.0 (18.4)	22.2 (28.1)	0.0 (0.0)	0.0 (0.0)
2	0.0 (0.0)	33.3 (35.3)	43.0 (41.0)	35.5 (36.5)	33.3 (35.3)	32.2 (34.6)	33.3 (35.3)	32.2 (34.6)
3	7.7 (15.8)	28.9 (32.5)	27.8 (31.8)	50.0 (45.0)	31.1 (33.9)	49.3 (44.6)	16.7 (23.8)	3.3 (10.4)
4	100.0 (90.0)	100.0 (90.0)	100.0 (90.0)	48.2 (43.9)	100.0 (90.0)	32.2 (34.6)	100.0 (90.0)	100.0 (90.0)
5	4.4 (12.2)	33.0 (41.0)	22.2 (28.1)	22.2 (28.1)	22.2 (28.1)	40.4 (39.4)	0.0 (0.0)	0.0 (0.0)
6	10.0 (18.4)	8.9 (17.2)	0.0 (0.0)	26.7 (31.1)	24.5 (29.6)	22.2 (28.1)	0.0 (0.0)	0.0 (0.0)
7	33.3 (35.3)	15.6 (23.2)	31.1 (33.8)	17.8 (24.9)	11.1 (19.5)	15.6 (23.0)	0.0 (0.0)	0.0 (0.0)
8	0.0 (0.0)	22.2 (28.1)	66.7 (54.7)	81.8 (64.2)	70.0 (56.8)	44.5 (41.7)	52.2 (46.3)	72.2 (68.3)
9	24.5 (29.6)	0.0 (0.0)	51.1 (45.6)	74.5 (59.7)	85.6 (67.7)	47.7 (43.7)	0.0 (0.0)	22.2 (28.1)
10	66.7 (54.7)	52.2 (46.3)	52.2 (44.3)	72.2 (58.3)	42.2 (40.4)	72.2 (58.2)	52.2 (46.2)	32.2 (34.6)
$CD_{0.05}$	3.83	2.19	2.47	6.33	6.81	7.93	4.14	3.85

**1- Melia azedarach; 2- Eupatorium sp.; 3-Cannabis sativa; 4- Ranunculus muricatus; 5- Ocimum sanctum; 6- Lantana camara; 7- Vitex negundo; 8- Camellia sinensis; 9- Datura stramonium; 10- Eupatorium adenophorum*

** A = Autoclaved; B = Non-autoclaved

The inhibitory effects of the plant extracts tested against the biocontrol agents which were locally isolated and were effective against pea wilt and root rot pathogens were also tested with a view to study their integration level. The data on per cent inhibition of the bioagents have been presented in the Table 2. A perusal of data shows that extract of *R. muricatus* completely inhibited all the bioagents under both autoclaved and unautoclaved conditions. This shows incompatibility of the botanical with the bioagents. Amongst other botanicals which were effective against the pathogens was *Eupatorium adenophorum* which did not affect three bioagents strains viz., SMA 5, JMA 11 and DMA 8, thus showing the compatibility with them. Similarly, JMA 11 and DMA 8 strains were compatible with *C. sinensis*.

Table 2. Inhibitory activity of plant extracts against different isolates of *Trichoderma* species

Plant species*	Per cent inhibition							
	JMA-4		SMA-5		JMA-11		DMA-8	
	A**	B**	A	B	A	B	A	B
1	0.0 (0.0)	0.0 (0.0)	0.0 (0.0)	0.0 (0.0)	0.0 (0.0)	0.0 (0.0)	0.0 (0.0)	13.3 (21.4)
2	0.0 (0.0)	0.0 (0.0)	0.0 (0.0)	0.0 (0.0)	0.0 (0.0)	0.0 (0.0)	0.0 (0.0)	33.3 (35.2)
3	3.3 (10.4)	41.5 (40.1)	17.8 (24.9)	36.7 (37.2)	4.4 (12.2)	41.5 (40.1)	11.1 (19.5)	53.3 (46.9)
4	100 (90.0)	100 (90.0)	100 (90.0)	100 (90.0)	100 (90.0)	100 (90.0)	100 (90.0)	100 (90.0)
5	0.0 (0.0)	0.0 (0.0)	30.0 (33.2)	23.3 (28.8)	0.0 (0.0)	20.0 (26.5)	31.1 (33.9)	26.7 (31.1)
6	43.0 (41.0)	0.0 (0.0)	20.0 (26.5)	15.6 (23.2)	0.0 (0.0)	0.0 (0.0)	22.2 (28.1)	15.7 (23.2)
7	18.9 (25.7)	4.4 (11.8)	30.0 (33.2)	28.9 (32.5)	0.0 (0.0)	0.0 (0.0)	22.2 (28.1)	11.1 (19.5)
8	11.1 (19.5)	60.0 (50.8)	0.0 (0.0)	52.2 (46.2)	0.0 (0.0)	0.0 (0.0)	0.0 (0.0)	0.0 (0.0)
9	0.0 (0.0)	74.4 (59.7)	0.0 (0.0)	22.2 (28.1)	0.0 (0.0)	24.4 (29.6)	0.0 (0.0)	24.4 (29.6)
10	0.0 (0.0)	32.2 (34.6)	0.0 (0.0)	0.0 (0.0)	0.0 (0.0)	0.0 (0.0)	0.0 (0.0)	0.0 (0.0)
$CD_{0.05}$	2.85	5.26	0.65	3.25	0.21	2.2	1.28	3.95

*1- *Melia azedarach;* 2- *Eupatorium sp.;* 3- *Cannabis sativa,* 4- *Ranunculus muricatus;* 5- *Ocimum sanctum;* 6- *Lantana camara;* 7- *Vitex negundo;* 8- *Camellia sinensis;* 9- *Datura stramonium;* 10- *Eupatorium adenophorum*

** A = Autoclaved; B = Non-autoclaved

The studies thus showed that all the botanicals may not be compatible with the biocontrol agents and proper screening must be done before integrating them. The species like *R. muricatus* though effective against the pathogens are also fungitoxic to the bioagents and such botanicals may also show allelopathic effects against the crop at higher concentrations or may affect other useful microorganisms. The present studies thus suggest the scope of integration of any of the three strains of biocontrol agents viz., SMA 5, JMA 11 and DMA 8 with application of *Eupatorium adenophorum* for the effective management of pea wilt/root rot disease.

REFERENCES

Bansal, R.K. and Gupta, R.K. (2000). Evaluation of plant extracts against *Fusarium oxysporum*, wilt pathogen of fenugreek. *Indian Phytopathol.* 53: 107-108.

Gilliver, K. (1947). The effect of plant extracts on the germination of the conidia of *Venturia inaequalis. Ann. Appl. Biol.* 34: 136-143.

Maheshwari, S.K., Gupta, J.S. and Jhooty, J.S. (1981). Effect of various cultural practices on the incidence of the wilt and root rot of pea. *Indian J. Agric. Res.* 15: 149-151.

Misra, S.B. (1975). *Antifungal Activity of Leaves of Some Higher Plants.* Ph.D. Thesis, Gorakhpur Univ. India.

Patel, K.D., Patel, P.M. and Patel, B.N. (2000). Inhibition of tobacco mosaic virus infection by some plant extracts. *J. Mycol. Plant Pathol.* 30: 229-230.

Singh, S., Goel, L.B., Sharma, S.K. and Nayar, S.K. (1979). Fungitoxicants and plant extracts in the control of hill bunt of wheat. *Indian Phytopathol.* 32: 297-299.

Ushamalini, C., Rajappan, K. and Gangadharan, K. (1997). Suppression of charcoal rot and wilt pathogens of cowpea by botanicals. *Plant Dis. Res.* 12: 113-117.

Vincent, J.M. (1947). Distortion of fungal hyphae in the presence of certain inhibitors. *Naturo* 150: 850.

Integrated Plant Disease Management (2005)
Eds. Sharma, R.C. and Sharma, J.N.
Scientific Publishers (India), Jodhpur, pp. 107-111

16

Integration of biopesticides with bioagents in the management of pea wilt/root rot complex

Y.S. PAUL, A.S. KAPOOR and MEENA DEVI

Department of Plant Pathology, CSK HPKV, Palampur-176 062 (H.P.)

ABSTRACT

Five biopesticides namely, Achook, Neemgold, Neemazal, Wanis and Spictaf were evaluated *in vitro* for their efficacy in suppressing the growth of pea wilt/ root rot pathogens viz., *Fusarium solani, F. oxysporum, Rhizoctonia solani* and *Sclerotinia sclerotiorum*. Wanis completely inhibited the growth of all the pathogens at 0.5 per cent concentration and caused more than 90 per cent inhibition in all the cases even at 0.1 per cent concentration. Other biopesticides were also found effective against these pathogens at 0.5 per cent. The effect of these biopesticides on four effective strains of biocontrol agent (*Trichoderma* sp.) was also tested with a view to integrate them in disease management. It was observed that Spictaf (0.1%) did not suppress the growth of the bioagents tested (JMA-4, JMA-11, SMA-5 and DMA-8) but inhibited the growth of *R. solani, F.solani, F. oxysporum* and *S. sclerotiorum* to the tune of 57,58,65 and 73 per cent, respectively. The mycelial growth of bioagents was moderately affected by Neemgold, Neemazal and Achook. Thus, Spictaf (0.1%) can safely be used in combination with biocontrol agents in the management of wilt/ root rot complex of pea.

INTRODUCTION

The use of natural products for the control of fungal diseases in crop plants is an interesting alternative to synthetic fungicides due to their lower negative impact on the environment. Neem is known for its pesticidal values since long and a number of neem based insecticides are in market and used commercially (Vijayalakshmi *et al.*, 1995). In the present studies, five such neem based pesticides viz., Achook, Neemgold, Neem Azal, Wanis and Spictaf were tested at two concentrations (0.1 and 0.5 %) *in vitro* against pea wilt root rot pathogens viz., *Fusarium solani, F. oxysporum, Rhizoctonia solani* and *Sclerotinia sclerotiorum.* Their effect on four bioagents which were found effective against these pathogens was also tested with a view to integrate biopesticides and the bioagents in the management of pea wilt and root rot disease, which is soil-borne in nature and attaining serious proportion in pea growing areas of Himachal Pradesh.

MATERIALS AND METHODS

Five biopesticides namely, Achook, Neemgold, Neemazal, Wanis and Spictaf at two concentrations viz., 0.1 and 0.5 per cent were tested *in vitro* against four pea wilt/root rot pathogens namely, *Fusarium solani, F. oxysporum, Rhizoctonia solani* and *Sclerotinia sclerotiorum* as well as four local isolates of *Trichoderma* viz., JMA 4, SMA 5, DMA 8 and JMA 11 which have been found effective against the pathogens. Double strength PDA was prepared and the biopesticide solutions were added as per the concentrations under sterilized conditions. The culture discs (5 mm) of the test organisms were placed in the poisoned medium and incubated at $22\pm1°C$. Control plates without the biopesticide were also maintained. The data on growth of the organisms were recorded when the control plates were full with mycelial growth and per cent growth inhibition was calculated by the formula (Vincent, 1947).

$$I = \frac{C - T}{C} \times 100$$

Where,

I = Per cent inhibition in mycelial growth

C = Growth of mycelium in control (mm)

T = Growth of mycelium in treatment (mm)

In a pot experiment, isolated biocontrol agents and biopesticides were tested against the disease under glasshouse condition. A mixture of all the four pathogens associated with wilt/root rot complex (1:1:1:1) was added to the pot mixture @ 20 g/kg soil. The biocontrol agents were applied @ 510 mg/kg soil while seed treatment with the Spictaf was done by dipping the seed in 0.1 and 0.5 per cent suspension for 15 minutes. Commercially available biofungicides viz., Ecoderma and *Pseudomonas fluorescens* were also tested by treating the seed @ 4 g/kg seed. The seed was sown and the data with reference to infected seedlings were recorded up to 45 days of sowing.

RESULTS AND DISCUSSION

The data with respect to the per cent inhibition of the pathogens as well as the bioagents have been presented in Table 1. Amongst the biopestcides tested, Wanis (0.5%) completely inhibited all the pathogens tested; whereas at 0.1 per cent the inhibition of the pathogens, *Rhizoctonia solani, F. solani, F. oxysporum* and *S. sclerotiorum* was 97.8, 97.0, 97.8 and 97.2 per cent, respectively. The other formulation Spictaf inhibited these pathogens to the tune of 97.8, 76.7, 97.8 and 82.4 per cent, respectively. Neemazal and Neemgold inhibited these pathogens to the degree of 59.3, 85.4. 83.3, 83.2 per cent and 80.7, 74.8, 52.6 and 63 per cent, respectively. Though other biopesticides were also effective against the pathogens, but their efficacy was comparatively lower. When the efficacy of the effective biopestcides was compared against the bioagents tested, Wanis inhibited more than 60 per cent all the bioagents at 0.1 per cent and more than 80 per cent at 0.5 per cent concentration. Thus the possibility of integrating this biopestcide with the bioagent was rejected. Spictaf at 0.5 per cent did not affect SMA 5 at all. Isolates, JMA 4, SMA 5 and JMA 11 were not affected at 0.1 per cent, while the isolate DMA 8 was affected only up to 5.9 per cent. Thus, Spictaf at 0.5 per cent can safely be integrated with SMA 5 in the management of wilt and root rot complex of pea and at 0.1 per cent with all the tested bioagents.

Table 1. Effect of biopesticides on pathogens and bioagents

Bio-pesticides	Conc. (%)	Per cent inhibition							
		Pathogen				Bioagent (Isolates of *Trichoderma* sp)			
		RS	FS	FO	SS	JMA-4	DMA-5	DMA-8	DMA-11
Achook	0.1	35.5	12.6	44.8	19.3	52.6	54.3	43.9	42.5
		(36.5)	(20.6)	(42.0)	(25.8)	(46.5)	(47.6)	(41.5)	(40.7)
	0.5	86.7	61.1	49.1	62.1	65.2	68.2	77.8	60.7
		(68.7)	(51.4)	(44.5)	(52.0)	(53.8)	(55.6)	(62.1)	(51.2)
Neemgold	0.1	44.4	27.1	47.7	39.6	38.5	33.3	53.3	36.1
		(41.7)	(31.3)	(43.6)	(39.0)	(38.3)	(35.3)	(47.0)	(36.9)
	0.5	80.7	74.8	52.6	63.0	48.2	43.2	59.9	65.2
		(64.0)	(60.0)	(46.5)	(52.5)	(43.9)	(41.5)	(50.8)	(53.8)
Neemazal	0.1	28.2	62.1	63.8	63.8	11.8	51.1	48.4	52.6
		(32.0)	(52.0)	(53.0)	(53.0)	(20.0)	(45.6)	(44.0)	(46.5)
	0.5	59.3	85.4	33.3	83.2	53.3	62.2	49.6	71.1
		(50.3)	(67.5)	(35.3)	(65.8)	(46.8)	(52.4)	(44.7)	(57.5)
Wanis	0.1	97.8	97.0	97.8	97.2	77.8	66.7	64.5	65.2
		(81.5)	(80.0)	(81.4)	(80.4)	(61.8)	(54.7)	(53.4)	(53.8)
	0.5	0.0	0.0	0.0	0.0	86.7	95.6	81.5	88.9
		(0.0)	(0.0)	(0.0)	(0.0)	(68.6)	(78.0)	(64.5)	(70.5)
Spictaf	0.1	57.0	58.3	71.6	66.6	0.0	0.0	5.9	0.0
		(49.0)	(49.80	(60.0)	(54.7)	(0.0)	(0.0)	(11.5)	(0.0)
	0.5	97.8	76.7	97.8	82.4	30.4	0.0	24.4	21.5
		(81.4)	(61.1)	(81.4)	(65.3)	(33.4)	(0.0)	(29.5)	(27.6)
$CD_{0.05}$	0.1	8.80	3.57	5.55	5.26	2.70	6.53	5.20	5.16
	(0.5)	(2.71)	(2.38)	(2.97)	(3.36)	(2.52)	(7.64)	(8.16)	(2.74)

Figures in parentheses are arcsine transformed values
RS = *Rhizoctonia solani;* FS = *Fusarium solani;*
FO = *Fusarium oxysporum;* SS = *Sclerotinia sclerotiorum*

The results relating to the effect of integration of biopesticides with bioagents have been presented in the Table 2. A perusal of the data suggests that maximum disease control (60.3%) was achieved when seed treatment with Spictaf (0.1%) was integrated with soil application of JMA 4 followed by integration of Spictaf (0.5%) and JMA4 without any adverse affect on seed germination. The germination per cent in case of latter was 92.1. Integration of biocontrol agents with biopesticides can be very effective in disease management when proper selections are made.

Table 2. Evaluation of biopesticides and bioagents in the management of root rot complex of pea

Treatment	Seed germination (%)	Disease incidence (%)	Disease control (%)
JMA 4 (ST)*	92.13(76.58)	43.97 (41.5)	40.3
SMA 5(ST)	91.67 (76.17)	53.6 (47.21)	27.2
DMA 8 (ST)	76.4 (65.64)	53.9 (53.9)	13.2
JMA 11 (ST)	88.4 (73.7)	44.9 (42.0)	38.9
Spictaf 0.1% (ST)	83.8 (66.7)	56.0 (48.44)	23.9
Spictaf 0.5% (ST)	87.9 (67.7)	48.1 (43.9)	34.6
Spictaf 0.1% (ST) +JMA 4 (Soil)	100 (89.9)	29.20(32.57)	60.3
Spictaf 0.5% (ST) +JMA 4 (soil)	92.1 (76.6)	39.8 (39.1)	45.9
Ecoderma (ST)	80.1 (63.7)	59.7 (50.6)	18.9
Psudomonas (ST)	91.7 (76.2)	52.8 (46.6)	28.3
Control	56.47 (48.90)	73.6 (59.7)	—
$CD_{0.05}$	18.9	13.78	—

Figures in parentheses are arcsine transformed values
*ST=Seed treatment; Soil = Soil application.

The neem based biopesticides had not been tested so far against the pea pathogens except powdery mildew of pea (Singh *et al.*, 2000), however, their fungitoxic nature has been reported earlier against other pathogens. Reh and Schlosser (1994) could control grape powdery mildew with Neemazal-S and Neemazal-F. Antifungal activity of five neem products viz., neem seed cover dust, neem oil, neem cake, neem leaf extract and Raklin against *Macrophomina phaseolina* causing charcoal rot of mungbean was reported by Ilyas *et al.* (1997).

REFERENCES

Ilyas, M.B., Iftkhar, K., Anwar, U.L. and Haq, M. (1997). Effect of different neem products on the vegetative growth and sclerotial production of *Macrophomina phaseolina*. *Pak. J. Phytopathol.* 9: 77-79.

Reh, I. and Schlosser, E. (1994). Alternative control of powdery mildew of grapevine. 46th International Symposium on Crop Protection, Gent, Belgium. 3rd May, 1994. Mededelingen faculteit and Land-bouwkundige en Toegepaste Biologische Wetenschappen Universiteit Gnt 59: 909-917.

Singh, O.P., Prithviraj, B., Singh, K.P. and Sarma, B.K. (2000). Control of powdery mildew (*Erysiphe pisi*) of pea (*Pisum sativum*) by combined applications of plant growth promoting rhizobacteria and Neem Azal TM. *Z. Pflanzenkr. Pflanzenschutz* 107: 59-66.

Vijayalakshmi, K., Radha, K.S. and Vandana, Shiva. (1995). *Neem: A User's Manual.* Centre for Indian Knowledge Systems, Chennai. 96 p.

Vincent, J.M. (1947). Distortion of fungal hyphae in the presence of certain inhibitors. *Nature* 150 : 850.

Integrated Plant Disease Management (2005)
Eds. Sharma, R.C. and Sharma, J.N.
Scientific Publishers (India), Jodhpur, pp. 113-120.

17

Management of root rot/wilt complex disease of pea

A .S. KAPOOR, Y.S. PAUL, K. KOTTAISAMY and MEENAKSHI

Department of Plant Pathology, CSK HPKV, Palampur-176 062 (H.P.)

ABSTRACT

Four bioagents have been isolated which have shown efficacy against *F. oxysporum* f. sp. *pisi* and *F.solani* f. sp *.pisi* causing root rot/wilt of pea. Among 5 bio-pesticides evaluated against *F oxysporum* f. sp. *pisi* and *F.solani* f. sp. *pisi* under *in vitro* condition, Wanis was found to be the most effective in inhibiting mycelial growth of above pathogens. Pea-kuth crop sequence in zone IV and pea-rice /maize in zone II was found effective in reducing pea root rot/ wilt complex disease. Integration of soil amendment with *Lantana camara* (10 tonne/ha) + bioagents Tricoguard (2.5 kg/75 kg FYM/ha) followed by Bavistin (0 .05%) spray at pre-flowering stage was most effective in reducing the root rot/wilt complex. Bioagent *T.viride* SMA 7 as soil application @ 2.5 kg/75 kg FYM/ha and seed treatment with Wanis (0.2%) was also found effective in managing the root rot/wilt complex.

INTRODUCTION

Pea occupies a premier place both as grain legume and vegetable in most countries in the world especially in temperate region. In Himachal Pradesh, garden pea (*Pisum sativum* L.) is cultivated over an area of 7956 ha with annual production of 8101 tonnes (Anonymous, 1998) and is the main off-season vegetable fetching premium price to the growers. However, pea diseases are the major constraint for pea productivity. Of late, pea root rot/wilt complex disease has been found to be the most serious disease especially in off-season crop in zone IV where generally crop rotation is not followed resulting in 5-40 per cent yield loss. Keeping this in view, studies were undertaken to find out the suitable management practices for root rot/wilt complex of pea.

MATERIALS AND METHODS

Evaluation of biocontrol agents

Regular surveys of pea growing areas of zone II and zone IV of Himachal Pradesh were conducted during 2001-2002, 2002-2003 and 2003-2004. The bioagents were isolated from infected plants and rhizosphere of pea crop on potato dextrose agar medium (PDA) and *Trichoderma* specific medium. After preliminary evaluations, the most effective strains of *Trichoderma* were used for further experiments. *In vitro* evaluation of some local strains of *Trichoderma* spp. was also done using dual culture technique. Seven isolates viz., SMA-7, SMA-4 and RMA-8. JMA-11, B and E of bioagent *Trichoderma viride*, one of *T. harzianum* and one commercial bioproduct Tricoguard (*T.viride*) were evaluated for their efficacy against the root rot/wilt complex @ 2.5 kg of wheat bran based formulation mixed in 75 kg FYM per ha or Tricoguard 2.5 kg/75 kg of FYM/ha under field condition.

Different formulations of bioagents based on wheat bran, charcoal and sodium alginate pellets were compared with commercial Tricoguard as soil application @ 2.5 kg /75kg of FYM/ ha along with a standard check.

Evaluation of biopesticides

Five neem based biopesticides namely, Wanis, Achook EC, Neemazal, Neemgold and Spictaf at two concentrations (0.1 and 0.5%) were evaluated against *Fusarium solani* f.sp. *pisi* and *F. oxysporum* f.sp. *pisi* by poisoned food technique at 25 ± 1°C. Each treatment was replicated thrice. The colony diameter of each treatment was recorded when the control plates were fully covered with fungal growth. Growth inhibition of the pathogens by biopesticides was calculated by using formula given by Vincent (1947).

Integrated management

Different components of integrated disease management (IDM) viz., biocontrol, biopesticide, amendment with *Lantana camara*, cultural practices i.e. crop rotation, chemical control etc. were evaluated individually and in different combinations for the management of pea root rot/wilt complex under field conditions.

RESULTS AND DISCUSSION

In vitro evaluation of bioagents

About twenty bioagents were isolated from different soil samples collected from different pea growing areas and four isolates of *Trichoderma viride* viz., SMA 7, SMA 4, RMA 8 and JMA 11 were found the most effective against *F. solani* f. sp. *pisi* and *F. oxysporum* f. sp. *pisi* (Table 1). Isolate RMA 8 and SMA 7 followed by SMA 4 gave 79.55, 79.55 and 78.86 per cent mycelial inhibition of *F. oxysporum* f. sp. *pisi*, respectively, whereas SMA 4 and JMA 11 were more effective against *F.solani* f. sp. *pisi*.

Table 1. *In vitro* evaluation of *Trichoderma viride* isolates against root rot/wilt pathogens of pea

Isolate	*F. oxysporum* f. sp. *pisi*		*F. solani* f. sp. *pisi*	
	Mycelial growth (mm)	Inhibition (%)	Mycelial growth (mm)	Inhibition (%)
SMA7	9.0	79.55 (63.08)	12.67	70.53 (57.10)
SMA4	9.3	78.86 (62.58)	7.67	82.16 (64.97)
RMA8	9.0	79.55 (63.08)	10.67	75.19 (60.07).
JMA11	12.0	72.73 (58.50)	8.34	80.60 (63.87)
Control	44	0.00 (0.08)	43.00	0.00 (0.08)
C.D $_{0.05}$		10.25		8.36

Figures in parentheses are square root transformed values

Biological control of root rot/wilt complex of pea under field conditions

All the bioagents significantly reduced the root rot/wilt complex disease under field condition however, the best control was obtained with isolate SMA 7 (67.2%) followed by SMA 4 (54.0%) (Table 2).

Table 2. Field evaluation of bioagents against root rot/wilt of pea

Treatment	Disease incidence (%)	Disease control(%)
Trichoderma viride SMA-7	29.63 (32.77)	67.2
T.viride SMA-4	41.52 (39.91)	54.0
T.viride RMA-8	53.91 (47.29)	40.2
T.viride JMA-11	58.27 (50.03)	35.4
T.viride-B	60.51 (51.09)	32.9
T.viride-E	67.92 (55.99)	24.7
T. harzianum	56.97 (49.07)	32.8
Tricoguard	57.52 (49.39)	36.2
Control	90.14 (71.69)	-
CD$_{0.05}$	14.70	

Figures in parentheses are square root transformed values

The commercial bioproduct Tricoguard was less effective as compared to local isolates. This is because of the ability of local isolates for adaptation under local conditions. *T.viride* and *T. koningii* have been reported to be potential biocontrol agents against pea root rot complex disease (Lacicowa and Pjeta, 1994). Saksirirat *et al.* (1994) reported *Gliocladium roseum* effective against pea wilt.

Evaluation of bioformulations in the management of root rot/wilt complex in pea

Soil application of wheat bran based *T.viride* SMA 4 @ 2.5kg mixed in 75 kg of FYM per ha resulted 14.32 per cent disease incidence of root rot/wilt as compared to 31.22 per cent in control followed by talcum powder based formulation of *Trichoderma harzianum* (15.50 %). Charcoal based formulation of *T. harzianum* as was least effective (29.17% disease incidence) in controlling root rot/wilt complex of pea (Table 3). Kapoor *et al.* (2002) reported the delivery of *T.harzianum* in the form of sodium alginate pellets in soil as most effective in reducing white rot of pea. Kapil (2002) found wheat bran based formulation of *Trichoderma viride* as soil application equally effective in managing white rot of pea in pot experiments. However, Umamaheshwari and Ramakrishnan (2000) reported seed treatment as a better method of delivery of *T.viride*.

Table 3. Efficacy of different bioformulations in controlling root rot/wilt of pea

Treatment	Disease incidence (%)	Green pod yield (q/ha)
Trichoderma viride SMA-7 wheat bran based	23.62 (29.04)	7.23
T.viride SMA-4 wheat bran based	14.32 (22.19)	4.90
Trichoderma harzianum talc based	15.50 (23.17)	7.27
T. harzianum charcoal based	29.17 (32.66)	6.15
T. viride talc based	23.09 (28.68)	5.86
T .viride in sodium alginate pellets	20.72 (27.04)	7.06
Bavistin (seed treatment @ 2g / kg) + 0.1% at pre-flowering stage	22.45 (28.26)	10.68
Tricoguard (soil application)	25.36 (30.23)	6.56
Control	31.22 (33.91)	5.65
$CD_{0.05}$	3.28	NS

Soil application @ 2.5kg /75kg of FYM/ha

Figures in parentheses are arcsine transformed values

Biopesticide evaluation

Wanis was found to be the most effective in inhibiting mycelial growth of *F. oxysporum* f. sp. *pisi* providing 80.29 per cent inhibition at 0.5 per cent concentration. Wanis was also the most effective against *F.solani* f sp. *pisi* whereas Neemazal was the least effective. (Table 4). Use of botanicals in the management of different soil-borne diseases has been reported. Neem derivatives have been reported to act as insecticides, fungicides, nematicides etc. (Shankar and Parmar, 1999). Sharma and Kapoor (1999) reported plant extracts of *Saussurea lappa* and *Lantana camara* most effective against *Sclerotinia sclerotiorum*. Kapil and Kapoor (2002) reported Wanis as most effective against *S. sclerotiorum*.

Table 4. *In vitro* evaluation of biopesticides against *Fusarium* wilt/root rot pathogens of pea

Treatment	*F. oxysporum* f. sp. *pisi*				*F. solani* f. sp. *pisi*			
	0.1%		0.5%		0.1%		0.5%	
	Mycelial growth (mm)	Inhibition (%)	Mycelial growth (mm)	Inhibition (%)	Mycelial growth (mm)	Inhibition (%)	Mycelial growth (mm)	Inhibition (%)
Wanis	24.08	71.29 (57.54)	16.19	80.29 (63.58)	11.50	86.28 (68.19)	7.05	91.57 (73.05)
Spictaf	35.75	46.64 (43.05)	32.41	60.55 (51.06)	41.08	50.94 (45.52)	33.44	60.02 (50.77)
Neemgold	35.17	47.33 (43.45)	30.64	62.70 (52.36)	41.58	50.35 (45.17)	32.44	61.22 (51.47)
Achook	37.67	55.09 (47.87)	23.11	71.87 (57.92)	44.42	46.95 (43.22)	44.81	46.43 (42.43)
Neemazal	60.25	28.16 (32.01)	42.07	48.79 (44.25)	60.67	27.55 (31.63)	47.79	42.87 (40.86)
Control	83.87	0 (0.08)	82.15	0 (0.08)	83.74	0 (0.08)	83.65	0 (0.08)
$C.D_{0.05}$		2.59		4.59		3.68		5.94

Figures in parentheses are arcsine transformed values

Integrated disease management in pea

Crop rotation of pea with kuth (*Saussurea lappa*) in zone IV, and rice and maize in zone II were found to reduce incidence of root rot/wilt complex (Table 5).

Table 5. Incidence of major soil-borne diseases of pea under different cropping sequences in zone- IV of Himachal Pradesh

Location	Crop sequence	Incidence (%)	
		RRWC	WR
Salpat	(i) Pea – Pea	50 - 80	5 - 15
	(ii) Pea – Pea – Pea – Pea	25 - 90	5 - 10
Madgran	Potato – Pea	10 - 50	5 - 15
Hinsa	(i) Potato – Pea	10 - 35	0 - 5
	(ii) Wheat – Pea	50 - 90	15 - 25
	(iii) Kuth – Pea – Pea	5 - 25	0-5
	(v) Barley – Pea	5 - 25	15 - 35
	(vi) Kuth – Potato – Pea	0 - 5	0-5
	(viii) Kuth – Potato – Pea – Potato – Pea	5 - 10	5-25
Chhating	(i) Potato – Pea	25 – 90	10 –15
	(ii) Pea – Potato – Pea	15 – 90	10 - 15
	(iii) Pea – Wheat – Potato – Pea	25 - 50	5 - 15
Shakoli	(i) Kuth – Pea	1 - 5	0-5
	(ii) Pea – Potato – Manu – Pea	25 - 30	5-25
	(iii) Kuth–Potato–Pea	5 - 10	10 - 15

RRWC = Root rot wilt complex, WR= White rot

It has been observed that wherever the farmers have rotated the pea with kuth, the incidence of root rot/wilt complex and white rot was invariably reduced and maximum incidence of root rot complex was 10 per cent, whereas in pea-pea rotation up to 90 per cent incidence was observed. This trend was consistent at all the locations surveyed. Barley – pea rotation was also effective in reducing the incidence of root rot/wilt complex disease. However, effectiveness of this crop sequence of pea- kuth or kuth – pea against white rot of pea was not consistent and disease ranged from 0 to 15 per cent. Yan` kov (1989) demonstrated best control of Fusarium pea wilt by adopting crop rotation with sugar beet and fodder grain at 4-6 years interval.

Rice - pea rotation, soil amendment with *Lantana camara* (10 tonnes/ha) + Tricoguard (2.5 kg/75kg FYM/ha) followed by Bavistin (0 .05%) spray (A2 B_1 F_1) at pre- flowering stage was most effective in controlling the root rot/wilt complex which resulted in minimum incidence of root rot complex (24.68%) as compared to 34.29 per cent in control (Table 6). In maize - pea rotation, A_2 B_1 F_1 resulted in minimum of 27.90 per cent incidence. Beans - pea rotation increased the disease (48.63%) as compared to crop rotation with cereals.

Table 6. Integrated management of root rot complex disease of pea in different crop sequences

Treatment	Rice-pea (% disease)	GPY pea (q/ha)	Maize-pea (% disease)	GPY Pea (q/ha)	Beans-pea (% disease)	GPY pea (q/ha)
$A_0B_0F_0$	34.29 (35.79)	22.61	34.37 (35.85)	33.02	48.63 (44.20)	24.79
$A_0B_0F_1$	47.14 (43.34)	18.98	27.3 (31.37)	39.35	30.23 (33.24)	31.58
$A_0B_1F_1$	38.60 (38.41)	17.24	30.82 (33.71)	42.29	38.44 (38.29)	31.40
$A_1B_0F_0$	28.40 (32.20)	15.07	30.91 (33.77)	29.79	47.50 (43.57)	28.64
$A_1B_0F_1$	27.92 (31.88)	28.80	21.88 (27.83)	43.15	35.80 (36.75)	31.18
$A_1B_1F_1$	25.97 (30.59)	27.00	24.31 (29.53)	42.23	42.98 (40.92)	36.06
$A_2B_0F_0$	26.92 (31.24)	23.57	28.52 (32.27)	31.74	29.95 (33.15)	25.88
$A_2B_0F_1$	26.94 (31.24)	33.41	25.77 (30.46)	42.94	45.38 (42.30)	30.13
$A_2B_1F_1$	24.68 (29.73)	30.22	27.90 (31.88)	46.32	40.30 (39.41)	45.62
CD(p=0.5)	16.13	NS	10.79	13.58	12.38	NS

Figures in parentheses are arcsine transformed values
A_0 = No FYM; B_0 = No biocontrol; F_0 = No fungicide; A_1 = FYM @ 10 tonnes/ha.; A_2 = *Lantana* @ 10 tonnes/ha; F_1 = Seed treatment with Bavistin @ 2.5 grams/ kg of the seed + spray of Bavistin @ 0.05 %; B_1 = Tricoguard (*Trichoderma viride*) 2.5 kg with 60 kg FYM /ha as soil application; GPY = Green pod yield

In another trial, integration of bioagent and biopesticide was studied. The results revealed that there was minimum disease incidence of root rot/wilt complex (14.33%) where bioagent was used as soil application @ 2.5kg/75 kg FYM/ha and seed treatment with Wanis (0.2%). Integration of bioagent, seed treatment with Wanis (0.2%) and amendment with *Lantana* @ 10 tonnes/ha and combination of bioagent soil application and amendment with *Lantana camara* were equally effective in reducing the root rot-wilt complex disease (Table 7).

Beniwal *et al.* (1994) reported the importance of integration of host resistance, cultural practices and biological control in the management of major diseases of faba bean, chickpea, lentil and pea. Integration of host resistance, crop rotation, fertilizer application, sowing time and depth for the management of root rot disease were suggested (Wang *et al.*, 1995). Tu (1987) developed an integrated management programme to control Fusarium wilt and root rot of pea. Kraft and Papavizas (1983) emphasized the integration of cultural practices and genetic resistance in the management of root rot/wilt of pea. Kapoor *et al.* (2002) reported crop rotation with non-leguminous crops such as cereals, kuth and onion.

Table 7. Integrated disease management of root rot/wilt complex

Treatment	Root rot /wilt complex Disease incidence (%)	Green pod yield (q/ha)
T_1= *Trichoderma viride* SMA-7 wheat bran based (soil application @ 2.5 kg/ 75 kg FYM/ha)	21.28 (27.45)	11.02
T_2 - Wanis (Seed treatment 0.2 %)	15.67 (23.26)	9.14
$T_3 = T_1 + T_2$	14.33 (22.21)	10.62
$T_4 = T_1$+ Spray with Wanis 0.1% at pre-flowering stage	16.74 (24.10)	12.76
T_5 =*Lantana camara* amendment @ 10 tones/ ha	17.32 (24.54)	8.70
T_6 = T1 +T2 +T5	16.31 (23.80)	9.3
T_7 = Bavistin (Seed treatment) + Spray 0.1%	17.94 (25.04)	10.91
T8 =Bioagent I + *Lantana camara*	16.80 (24.18)	12.27
T9 = Control	23.94 (29.25)	8.07
$CD_{0.05}$	2.75	NS

Figures in parentheses are arcsine transformed values

REFERENCES

Anonymous. (1998). *Area and Production of Vegetables in Himachal Pradesh.* Directorate of Agriculture (H.P.), Shimla.

Beniwal, S.P.S, Trapero, C. and Muehlbauer, F.J. (1994). Integrated control of diseases of cool season food legumes. In: *Proceedings of 2nd International Food Legume Research Conference on Pea, Lentil, Faba Bean, Chickpea and Grass Pea.* Cairo, Egypt. pp. 642-665.

Kapil, Renu. (2002). *Management of White Rot of Pea (Sclerotinia sclerotiorum) Using Trichoderma spp. and Biopesticides.* M.Sc. Thesis, CSK HPKV, Palampur.76 p.

Kapil, Renu, and Kapoor, A.S. (2002). Management of white rot of pea using bioagents and biopesticides. Paper presented in International conference on *"Challenges and Options for Sustainable Development of Himalaya- Beyond- 2002"* organized by GTZ , HPEDS and CSK HPKV Palampur held at CSK HPKV Palampur, Oct. 1-4, 2002.

Kapoor, A.S., Paul, Y.S., Kottaisamy, K. and Meenakshi (2002). Integrated disease management in pea. Paper presented in the Asian Congress of Mycology and Plant Pathology *"Plant Health for Food Security"*. Indian Society of Mycology and Plant Pathology held at University of Mysore. Oct. 1 – 4, 2002.

Kraft, G.M. and Papavizas, G.C. (1983). Use of host resistance, *Trichoderma* and fungicides to control soil borne diseases and increase seed yields of peas. *Plant Dis.* 67: 1234 – 1237.

Lacicowa, B. and Pjeta, D. (1994). Protective effect of microbiological dressing of pea seeds (*Pisum sativum* L.) against pathogenic fungi living in soil. *Annales-Universitatis-Mariaccurie-Sklodowska-Sectio-EEE-Horticultura* 2: 165–171.

Saksirirat, W, Sirimungkararat, S, Steinmetz, J and Schoenbeck, F. (1994). Applicability of *Gliocladium roseum* Bain for biological control of Fusarium wilt of pea. *Kaen Kaset Khon Kaen Agric. J.* 22: 37-42.

Shankar, J.S. and Parmar, B.S. (1999). Recent developments in botanicals and biopesticides. *Indian J. Plant Prot.* 27: 139–154.

Sharma, C.L. and Kapoor, A.S. (1999). *In vitro* evaluation of herbicides and plant extracts against *Sclerotinia sclerotiorum* causing white rot in pea. *Plant Dis. Res.* 14 : 79-81.

Tu, J.C. (1987). Integrated control of pea root rot disease complex in Ontario. *Plant Dis. Surv.* 71: 9-13.

Umamaheswari,C. and Ramakrishnan,G. (2000). Factors favouring the growth and multiplication of *Trichoderma viride.* In: *Proceedings of Indian Phytopathological Society Golden Jubilee International Conference on Integrated Plant Disease Management for Sustainable Agriculture*, Vol. 1, IARI, New Delhi, pp. 365-367.

Vincent, J.H. (1947). Distortion of fungal hyphae in the presence of certain inhibitors. *Nature* 150: 850.

Wang, K.C., Zhang, Z.S.,Chen, J.N., Fan, Z.Q. and Niu, B.S. (1995). A study on the occurrence of root rot in *Pisum sativum* and the techniques for its integrated control. *Ningxia J. Agric. For. Sci. Technol.* 3: 1-5.

Yan` kov, I.I. (1989). Fusarial wilt of peas. *Zashch. Rast.* (Mosko) No. 4 : 18.

Integrated Plant Disease Management (2005)
Eds. Sharma, R.C. and Sharma, J.N.
Scientific Publishers (India), Jodhpur, pp. 121-126.

18

Screening of pea germplasm for resistance against powdery mildew

KUMUD JARIAL and R.C. SHARMA

Department of Mycology and Plant Pathology, Dr. Y. S. Parmar University of Horticulture and Forestry, Nauni, Solan – 173 230 (H.P.)

ABSTRACT

One hundred five lines of pea germplasm were screened for resistance against powdery mildew using the criteria of disease incidence and severity, apparent infection rate and area under disease progress curve (AUDPC). Thirty eight lines remained disease free and were classified as highly resistant. Rest of the sixty seven showed varied degrees of apparent infection rates and AUDPC and classed as resistant, moderately resistant, susceptible and highly susceptible accordingly. Lines EC-269313, EC-341793 and NC-60749 had least AUDPC (1.26) and apparent infection rate (0.0611 per unit per day) while, EC-341774, PM-5 and PKT-89/159 had maximum AUDPC (14.03) and JP-747 had maximum apparent infection rate (0.2690 per unit per day).

INTRODUCTION

Powdery mildew caused by *Erysiphe pisi* Syd. is an important disease of pea causing heavy losses to the crop. Use of resistant varieties is one of the most successful, practical and economical methods of disease management. Among *Pisum* accessions, at least two sources of genetic resistance are available. A single recessive gene er_1 for resistance is available from the cultivar Strategem which was confirmed by Cousin (1965). Resistance conferred by a second recessive gene er_2 is present in Peruvian genotypes and is confined to leaves. Under Indian conditions, some work has also been done on screening for resistance (Sohi *et a*l., 1974; Sharma *et al.*, 1980; Krishna and Mishra, 1989; Thakur *et al.*, 1996). Present studies were conducted to screen different germplasm lines of pea against powdery mildew under mid hill conditions of Himachal Pradesh.

MATERIALS AND METHODS

One hundred five lines of pea germplasm were raised in the experimental farm, Department of Vegetable Crops, Dr. Y.S. Parmar University of Horticulture and Forestry, Nauni, Solan, Himachal Pradesh, in single rows of 2 m length with a planting distance of 45 x 5 cm. On all sides of this plot, standard susceptible cultivar Lincoln was sown, so as to provide enough inoculum for the disease to build up. Standard package of practices was followed for raising the crop. Disease in certain lines started appearing on 15.03.97 and since then observations were taken at weekly intervals based on the modified scale of Munjal *et al.* (1963). Observations were taken up to 18.04.97 when almost all plants turned yellow and started drying up. The apparent infection rate (r) for different genotypes was calculated according to Van der Plank (1963).

$$r = \frac{2.3}{t_2 - t_1} \log_{10} \frac{x_2}{1 - x_2} - \log_{10} \frac{x_1}{1 - x_1}$$

where,

r denotes the apparent infection rate per unit per day, $t_2 - t_1$ is the time interval between first and last observation, x_1 and x_2 represent the proportion of diseased plant parts at dates t_1 and t_2, respectively, $1-x_1$ and $1-x_2$ represent the proportion of healthy plant parts at dates t_1 and t_2, respectively.

From the severity data, the area under the disease progress curve (AUDPC) was calculated in accordance with Shaner and Finney (1977):

$$\text{AUDPC} = \sum_{i=1}^{n} (Y_{i+1} + Y_i)/2 \text{ x } (t_{i+1} - t_i)$$

where,

Y_i and Y_{i+1} are disease severity data (per unit) at times t_1 and t_{i+1}, respectively and n represents the total number of observations.

RESULTS AND DISCUSSION

Out of 105 lines, disease appeared in 67 lines while, 38 remained disease free. Among sixty seven lines, disease on forty seven lines was recorded on March 15, 1997. Minimum apparent infection rate (0.0841 per unit per day) and least AUDPC (4.06) in these lines were found in Local-64, PNL-1, G-2-A, IC-41104, IC-41311 and IC-41221. Highest AUDPC among the lines in which disease appeared on March 15, 1997 was found in lines PM-5 and PKT-89/159. In rest of the twenty lines, disease was first recorded on March 22,

1997. Out of twenty lines, minimum apparent infection rate (0.0611 per unit per day) and least AUDPC (1.26) were found in three lines namely, EC-269313, EC-341793 and NC-60749. Highest apparent infection rate (0.2690 per unit per day) was found in JP-747 line where disease first appeared on March 22, 1997 (Table 1).

Table 1. Initial time of powdery mildew appearance, AUDPC and infection rate in pea germplasm

Germplasm line	Date of disease appearance	AUDPC	Infection rate (r)
EC-269313	22.03.97	126.00	0.0611
EC-341793	22.03.97	126.00	0.0611
NC-60749	22.03.97	126.00	0.0611
Local - 64	15.03.97	406.00	0.0841
PNL-1	15.03.97	406.00	0.0841
G-2-A	15.03.97	406.00	0.0841
IC-41104	15.03.97	406.00	0.0841
IC-41311	15.03.97	406.00	0.0841
IC-41221	15.03.97	406.00	0.0841
PNL-52	15.03.97	441.00	0.1028
EC-269803	15.03.97	511.00	0.1028
EC-269643	15.03.97	511.00	0.1028
EC-173943	15.03.97	511.00	0.1028
RK-89-197	15.03.97	721.00	0.1028
PMR-2	15.03.97	721.00	0.1028
LMR-4	15.03.97	721.00	0.1028
IC-41222	15.03.97	721.00	0.1028
IC-41113	15.03.97	721.00	0.1028
IC-41216	15.03.97	721.00	0.1028
NC-64078	15.03.97	511.00	0.1028
PLP-565	15.03.97	511.00	0.1028
IC-41205	15.03.97	511.00	0.1028
EC-269295	15.03.97	651.00	0.1028
EC-269606	15.03.97	651.00	0.1028
EC-269644	15.03.97	651.00	0.1028
EC-269573	15.03.97	651.00	0.1028
PH-1	15.03.97	651.00	0.1028
IC-41105	15.03.97	651.00	0.1028
IC-41201	15.03.97	651.00	0.1028
EC-959260	15.03.97	721.00	0.1028
EC-341790	15.03.97	721.00	0.1028

Continue....

Germplasm line	Date of disease appearance	AUDPC	Infection rate (r)
EC-269319	15.03.97	721.00	0.1028
IC-41224	22.03.97	476.00	0.1295
IC-41106	22.03.97	476.00	0.1295
EC-26940	22.03.97	616.00	0.1295
EC-173940	15.03.97	808.50	0.1352
IC-41119	15.03.97	808.50	0.1352
Selection-528	15.03.97	913.50	0.1352
PLP-528	22.03.97	371.00	0.1059
PLP-5633	22.03.97	371.00	0.1059
KS-6155	22.03.97	371.00	0.1059
NDP-7	22.03.97	371.00	0.1059
NC-60743	22.03.97	371.00	0.1059
IC-41111	22.03.97	371.00	0.1059
Bridger	22.03.97	336.00	0.1295
EC-269424	22.03.97	476.00	0.1295
EC-269640	22.03.97	476.00	0.1295
IC-41204	15.03.97	1001.00	0.1675
IC-41208	15.03.97	1001.00	0.1675
EC-269384	15.03.97	1001.00	0.1675
EC-269677	15.03.87	1041.00	0.1675
IC-41307	15.03.97	1071.00	0.1675
IC-41209	15.03.97	1071.00	0.1675
IC-41212	22.03.97	738.50	0.1702
GL-195	15.03.97	913.50	0.1352
IC-41103	15.03.97	913.50	0.1352
Batari Pawan	15.03.97	1158.50	0.1352
IC-41217	15.03.97	1333.50	0.1352
EC-341774	15.03.97	1403.50	0.1352
PM-5	15.03.97	1403.5	0.1352
PKT-89/159	15.03.97	1403.5	0.1352
EC-174436	15.03.97	1001.00	0.1675
EC-269315	15.03.97	1001.00	0.1675
IC-41317	22.03.97	878.50	0.1702
JP-169	22.03.97	826.00	0.2109
LIVEM-15	22.03.97	966.00	0.2109
JP-747	22.03.97	266.00	0.2690

Different lines were categorised on the basis of disease incidence on the last date of observation. Out of 105 lines, 38 remained disease free till last date of observation and so were categorised as highly resistant. Rest 67 lines showed variable degrees of disease incidence out of which 3 lines had up to 5 per cent infection, 12 lines had up to 15 per cent infection, 30 lines had up to 25 per cent infection, 22 lines had up to 50 (12 lines) and 75 (10 lines) per cent infection and so were categorized as resistant, moderately resistant, susceptible and highly susceptible, respectively (Table 2).

Table 2. Reaction of pea germplasm to powdery mildew

Disease incidence (%)	Germplasm lines	Category
0	EC-269416, EC-101374, EC-101512, EC-269589, EC-269802, EC-269395, EC-959261, EC-333, NC-60747, NC-60756, NC-64072, NC-60759, NC-64098, MC-64081, Sugar Giant, Sugar Giant-II, VR-12, VG-213, KFER-I, DPH-96, DPH-19, DPH-13, DPH-64, P-87, P-37, HPF-9, UN-53-6W-11-HR, JP-20, JP-501, Selection-35, Selection-8, PLP-521, PLP-582, LMR-20, PMRKSM-1, 8-12-1, IC-41308, IC-41320.	Highly resistant
5	EC-269313, EC-341793, NC-60749	Resistant
15	NC-60743, NDP-7, KS-6155, Local-64, PNL-1, PLP-528, PLP-5633, G-2-A, IC-41221, IC-41111, IC-41104, IC-41311	Moderately resistant
25	EC-959260, EC-341790, EC-269295, EC-269319, EC-269606, EC-269803, EC-269643, EC 269644, EC-269573, EC-173943, EC-269424, EC-269640, EC-26940, NC-64078, RK-89-197, PLP-565, PNL-52, PH-1, PMR-2, LMR-4, Bridger, JP-747, IC-41222, IC-41205, IC-41113, IC-41216, IC-41105, IC-41201, IC-41224, IC-41106.	Susceptible
50 and > 50	EC-341774, EC-173940, PKT 89/159, PM-5, GC-195, Selection-528, Batari Pawan, IC-41217, IC-41103, IC-41119, IC-41212, IC-41317, EC-174436, EC-269677, EC-269384, EC-269315, JP-169, LIVEM-15, IC-41307, IC-41204, IC-41208, IC-41209	Highly susceptible

Screening of pea germplasm for resistance in the present studies has been done on the basis of natural field infections using a sowing time which encourages the maximum disease severity. For each germplasm line date of initial disease appearance, AUDPC and apparent infection rate were studied. Out of 105 lines, 38 were disease free and rest 67 had variable degrees of disease. Number of lines had similar apparent infection rate but revealed variable AUDPC. Various workers from time to time have screened pea germplasm to locate the sources of resistance against powdery mildew (Sohi *et al.*, 1974; Sharma *et al.*, 1980; Kalia, 1985; Mathur *et al.*, 1991; Thakur *et al.*, 1996). Certain lines reported resistant from one place have been found susceptible when screened at another location. This variability in their reaction may be due to variation in the pathogen genotype and to some extent in the environmental conditions. However, on large number of Mexican and Peruvian germplasm lines, resistance is governed by single recessive gene (Cousin, 1965).

REFERENCES

Cousin, R. (1965). Study of resistance to powdery mildew in peas. *Ann. Amelior. Plant* 15: 93-97.

Kalia, P. (1985). Pea varieties resistant to powdery mildew. *Indian J. Agric. Sci.* 55: 206-208.

Krishna, A. and Mishra, S.P. (1989). Mechanism of slow mildewing in pea. *Indian Phytopathol.* 42: 103-107.

Mathur, K., Shekhawat, K.S. and Bhatia, L.K. (1991). Performance of pea varieties against powdery mildew. *Indian Phytopathol.* 44: 137.

Munjal, R.L., Chenulu, V.V. and Hora, T.S. (1963). Assessment of losses due to powdery mildew (*Erysiphe polygoni* DC.) on pea. *Indian Phytopathol.* 16: 268-270.

Shaner, G. and Finney, R.E. (1977). Effect of nitrogen fertilization on the expression of slow mildewing resistance in Knox wheat. *Phytopathology* 67: 1051-1056.

Sharma, L.C., Mathur, A.K. and Kaur, A. (1980). Performance of pea varieties against powdery mildew in Rajasthan. *Madras Agric. J.* 67: 483.

Sohi, H.S., Sokhi, S.S. and Rawal, R.D. (1974). Varietal reaction of peas to powdery mildew (*Erysiphe polygoni*) and rust (*Uromyces fabae*). *Mysore J. Agric. Sci.* 8: 529-532.

Thakur, B.R., Kapoor, A.S. and Jamawal, R.S. (1996). Varietal resistance of pea powdery mildew in dry temperate zone of Himachal Pradesh. *Indian Phytopathol.* 49: 92-93.

Van der Plank, J.E. (1963). *Plant Diseases: Epidemics and Control.* Academic Press, London. 349 p.

Integrated Plant Disease Management (2005)
Eds. Sharma, R.C. and Sharma, J.N.
Scientific Publishers (India), Jodhpur, pp. 127-130.

19

Reaction of pea germplasm to powdery mildew and leaf miner under dry temperate agroclimate in Himachal Pradesh

C.L. BHARDWAJ, K.C. SHARMA and S.D. SHARMA
Hill Agricultural Research and Extension Centre, CSK HPKV, Bajaura-175 125 (H.P.)

ABSTRACT

Field evaluation of pea (*Pisum sativum* L.) germplasm against natural disease pressure of powdery mildew (*Erysiphe pisi* DC. ex Saint Amans) and leaf miner insect (*Cromatomyia horticola* Goureau) at CSK HPKV, Regional Research Station, Kukumseri, Lahaul and Spiti, Himachal Pradesh showed that the field pea genotypes viz., HPCC 72, HPCC 84, EC 248952, NDP 8, L 305 and L 306 showed higher level of resistance against powdery mildew compared to garden and processing / canning category of genotypes. The genotypes, viz. Alaska, DPH 71, HPCC 8, HPCC 94, KTP4, L308, L1059, Pant 8, Sel 6, Sel 143 and SP 75 hitherto reported to possess higher degree of resistance were rated moderately susceptible and susceptible. No true sources of resistance were recorded against pea leaf miner both at Kukumseri and at Bajaura mid-hill location among the tested pea germplasm. The resistant sources of pea can be used for evolving resistant cultivars before ascertaining the pathogenicity factor amongst the genotypes identified as resistant.

INTRODUCTION

Pea (*Pisum sativum* L.) is a good source of quality protein cultivated as cash crop under cold climate for marketing fresh pods or canning/processing purposes by deploying garden/canning pea varieties as well as field pea varieties for grain-legume purpose by the growers. Pea varieties suffer heavy quantitative and qualitative yield losses due to powdery mildew caused by *Erysiphe pisi* DC ex Saint Amans (Munjal *et al.*, 1963), besides common occurrence of leaf miner insect (*Cromatomyia horticola* Goureau; Diptera: Agromyzidae) under dry and less humid agro-climatic conditions. Powdery mildew on wild and cultivated legumes are very common in Himachal Pradesh and pea powdery mildew pathogen, *Erysiphe pisi* possesses a wide host range (Bhardwaj and Singh, 1984). It has also been reported to overwinter in its perfect state under dry temperate agroclimate of the state (Kapoor and Chaudhary, 1995) with the probability of spread of the pathogen to tropical and subtropical regions during pea cultivation periods. Since a small error in observations for assessment of differences in pest severity/ incidence (Banyal and Tyagi, 2000) can attribute to a greater loss during deployment of the desirable variety, the systematic evaluation or assessment of resistant donors against all the variants of the pathogen at hot spot is of fundamental importance. The pea cultivars also respond differentially to the natural populations of pathogen at regional basis (Narsinghani and Tiwari, 1986; Mathur *et al.*, 1991; Mahmood *et al.*, 1991). Therefore, germplasm both from indigenous and exotic sources comprising garden, canning/processing and field pea genotypes were field evaluated against natural and expected race spectrum at a hot spot in Himachal Pradesh.

MATERIALS AND METHODS

A total of 145 lines of pea germplasm from diverse sources comprising garden/canning/field pea genotypes were planted at CSK HPKV, Regional Research Station, Kukumseri, Lahaul and Spiti, Himachal Pradesh during summer seasons of 2000 and 2001 under natural fertility conditions in row plots of half meter row length, keeping Lincoln and Palam Priya as checks in the beginning and after set of twenty entries. From grain filling stage onwards, the pea germplasm was assessed on the basis of visual rating scale of 0 to 9 (Munjal *et al.*, 1963; Mahmood *et al.*, 1991).

Modified key for rating pea powdery mildew

Disease severity grading parameter	Numerical 0-9 visual rating score	Reaction
No infection (score of zero) or traces of powdery mildew infection in the form of pustules or lesions covering 0 to 5% area on foliage/other plant parts with proliferation of thinner mycelial mat.	0 to 3.0	Resistant reaction assumed for recessive gene action designated as R
Powdery mildew infection in the form of pustules or lesions covering 6 to 25% area on foliage/other plant parts with proliferation of slightly to medium thick mycelial mat.	3.1 to 4.5	Moderate resistant reaction designated as MR
Powdery mildew infection in the form of pustules or lesions covering 26 to 50% area on foliage/other plant parts with proliferation of slightly to medium thick mycelial mat.	4.6 to 5.5	Moderate susceptible reaction designated as MS

Powdery mildew infection in the form of pustules or lesions covering 51 to 75% area on foliage/other plant parts with proliferation of medium or slightly higher thick mycelial mat.	5.6 to 6.5	Susceptible reaction designated as S
Powdery mildew infection in the form of pustules or lesions covering 75 to 100% area on foliage/other plant parts with proliferation of thick to highly thick mycelial mat.	6.6 up to 9.0	Highly susceptible reaction designated as HS

The showing a score of 3.0 or less were categorized as resistant, between 3.1 to 4.5 moderately resistant, between 4.6 to 5.5 moderately susceptible, between 5.6 to 6.5 susceptible and 6.6 and above up to score of 9.0 as highly susceptible. The genotypes were also assessed for susceptibility to leaf miner insect, *Cromatomyia horticola* Goureau.

RESULTS AND DISCUSSION

Out of 145 pea genotypes assessed, 4.13 per cent were rated for resistant (R) having score of 3.0 or below without absolute resistance and 6.90 per cent has a score between 3.1 and 4.5 and were categorized as moderately resistant (MR). Thus, HPCC 72, HPCC 84, EC 248952, NDP 8, L 305 and L 306 genotypes were resistant; whereas, Alaska, DPH 71, HPCC 8, HPCC 94, KTP 4, L 308, L 1059, Pant 8, Sel 6, Sel 143 and SP 75 as moderately resistant (Table 1).

Table 1. Reaction of pea germplasm to powdery mildew in dry temperate agro-climatic conditions of Kukumseri in Himachal Pradesh

Resistant (R)	Moderately resistant (MR)
HPCC 72 (3.0), HPCC 84 (3.0)[1], EC 248952 (3.0), NDP 8 (1.0), L 305 (3.0), L 306 (2.0)	Alaska (4.5), DPH 71 (4.0), HPCC 8 (4.0), HPCC 94 (4.5), KTP 4 (4.5), L 1059 (4.5), Pant 8 (4.0), Sel 6 (4.5), Sel 143 (4.5) and SP 75 (4.0)

[1]The genotype responded to hypersensitive type reaction causing severe necrosis on foliage

The remaining pea genotypes were categorized as moderately susceptible (score between 4.6 to 5.5), susceptible (score between 5.6 to 6.5) and highly susceptible (score between 6.6 up to 9.0). All the genotypes showed susceptible reaction to pea leaf miner both at Kukumseri and Bajaura locations. Among the pea genotypes rated resistant, HPCC 84 seemed to show hypersensitive reaction to the natural population of pea powdery mildew fungus. Most of the pea genotypes scored resistant or moderately resistant reaction belonged to field peas (Table 1), whereas the elite pea cultivars of garden pea viz., Arkel (6.0), Azad P1 (5.5), Bonville (5.0), Lincoln (5.0), GC 477 (5.0), VL 3 (6.0), VL (6.0) and canning/processing peas, viz., WR 1158 (7.0) and WR 1167 (6.0) were moderately susceptible, susceptible and highly susceptible. Pea cultivars Rachna (score of 5.5) and Palam Priya (score of 6.0) hitherto reported resistant were susceptible to powdery mildew fungus (Mathur *et al.*, 1991) a variability in *Erysiphe pisi*. It has been observed that resistant pea cultivars developed so far lacked desirable horticultural attributes (Kumar *et al.*, 2002). Recommended pea cultivars have also been rated highly susceptible to powdery mildew (Kumar and Singh, 1981; Sokhi *et al.*, 1979; Banyal and Tyagi, 2000). The present results are in agreement with these findings.

REFERENCES

Banyal, D.K. and Tyagi, P.D. (2000). Comparison of the Gompertz and Logistic equations to describe progress of powdery mildew of peas. *J. Mycol. Plant Pathol.* 30: 79-80.

Bhardwaj, C.L. and Singh, B.M. (1984). Host range of *Oidium* state of *Erysiphe pisi* from peas on some leguminous hosts in Kangra valley of Himachal Pradesh. *Indian Phytopathol.* 37: 732-733.

Kapoor, A.S. and Chaudhary, H.K. (1995). Mode of perpetuation of *Erysiphe pisi* in dry temperate zone of Himachal Pradesh. *Indian Phytopathol.* 48: 77-78.

Kumar, H. and Singh, R.B. (1981). Genetic analysis of adult plant resistance to powdery mildew in peas (*Pisum sativum* L.). *Euphytica* 30: 147-151.

Kumar, R., Kalloo, G. and Singh, M. (2002). Genetic analysis of resistance to powdery mildew in peas (*Pisum sativum* L.). *Indian J. Agric. Res.* 72: 182-183.

Mahmood, T., Ahmad, I., Haq, I.U. and Aslam, M. (1991). Varietal resistance of pea to powdery mildew. *Indian Phytopathol.* 44: 52-53.

Matur, P.K., Shekhawat, K.S. and Bhatia, L.K. (1991). Performance of pea varieties against powdery mildew. *Indian Phytopathol.* 44: 137.

Munjal, R.L., Chenulu, V.V. and Hora, T.S. (1963). Assessment of loss due to powdery mildew (*Erysiphe polygoni* DC) in peas. *Indian Phytopathol.* 16: 206-267.

Narsinghani, V.G. and Tiwari, A. (1986). Multi-resistant peas in 2000 A.D. *Veg. Sci.* 13: 107-112.

Sokhi, S.S. Jhooty, J.S. and Bains, S.S. (1979). Response of pea varieties against powdery mildew. *Indian Phytopathol.* 32: 571-574.

Integrated Plant Disease Management (2005)
Eds. Sharma, R.C. and Sharma, J.N.
Scientific Publishers (India), Jodhpur, pp. 131-135.

20

Role of soil hydrothermal regimes in development of bacterial wilt of tomato in Himachal Pradesh

SATISH K. SHARMA[1], RAJESH KUMAR[2] , N. P. DOHROO[2] and ANJU SHARMA[3]

[1] Indian Agricultural Research Institute, Regional Station, Amartara Cottage, Shimla - 171 004 (H. P.)
[2]Department of Mycology and Plant Pathology, [3]Distributed Information Centre, Department of Biotechnology, Dr. Y.S. Parmar University of Horticulture and Forestry, Nauni, Solan -173 230 (H.P.)

ABSTRACT

Soil hydrothermal regimes i.e. soil moisture and soil temperature showed positive correlation with bacterial wilt development in tomato. The increase in soil moisture from 8.00 to 17.53 per cent and soil temperature from 20.9 to 31.1°C resulted in maximum disease incidence (100%). Further, coefficient of multiple determination (R^2) revealed that 93.61 per cent variation in disease incidence was due to soil hydrothermal regimes. The regression equation i.e. disease incidence (Y) = $-191.48 + 7.11A_1 + 3.61A_2$ indicated that a unit change in mean soil temperature (A_1) and soil moisture (A_2) influenced the disease by 7.11 and 3.61 units, respectively.

INTRODUCTION

Bacterial wilt of tomato (*Lycopersicon esculentum* Mill.), incited by *Ralstonia solanacearum* (Smith) Yabuuchi *et al.* (Syn. *Pseudomonas solanac-*

earum E.F. Smith) is one of devastating bacterial diseases affecting vascular bundles of plants and may result in complete failure of the crop (Sood and Singh, 1993). The pathogen has a very wide host range, which include several hundred plant species representing 44 families of crop plants (Rao and Sohi, 1977; He *et al.*, 1983; Persley, 1986). In India, plant species belonging to 37 plant families are affected (Shekhawat *et al.*, 1992). In tomato, the disease is of recent origin in temperate regions of India. It has engulfed almost all the potential tomato growing localities of Himachal Pradesh (Sood and Singh, 1993; Gupta *et al.*, 1998). Therefore, present studies were conducted to study the role of hydrothermal regimes in the development of disease and subsequently framing the strategies for its management.

MATERIALS AND METHODS

The role of soil hydrothermal regimes i.e. soil moisture and temperature on the progress of the disease was studied in bacterial wilt sick soils under field conditions. The seedlings of susceptible variety Naveen were transplanted in 90 x 30 cm spacing on first March 2002. Standard package of practices were followed for raising the crop. Soil temperature was recorded at 10 cm depth with the help of L.C.D. portable digital multistem thermometer at 2.00 p.m. in the rhizosphere of tomato plant by inserting the probe of thermometer into the soil. The temperature was recorded daily and averaged for a week to work out its effect on disease development.

Soil moisture was determined on volumetric basis by collecting eight random soil samples from the field at weekly intervals, starting with the appearance of the disease. These soil samples were immediately brought to the laboratory, weighed for their moist soil weight and kept in hot air oven at 105°C for 24 h to get a constant dry weight. Per cent moisture content was worked out on dry weight basis for each sample. Subsequently, soil samples were averaged to work out the soil moisture in the field. The data on the disease incidence were also recorded at weekly intervals. Correlation and regression analysis was carried out to see the contribution of independent variables on the development and prediction of disease.

RESULTS AND DISCUSSION

The disease started appearing in the second week of March at 24.2°C mean soil temperature and 10.56 per cent mean soil moisture (Fig. 1). Later on, it progressed rapidly from 3rd week of March to 3rd week of April. Maximum disease incidence was recorded in the last week of April when almost all the plants died at 31.1°C mean soil temperature and 17.53 per cent soil moisture. Temperature is the most important factor affecting the host pathogen interaction as well as survival of the bacteria in the soil (Hayward, 1991). Earlier workers while working in tropical and subtropical regions have reported that relatively higher soil temperature and moisture are favourable for bacterial wilt development (Abdullah *et al.*, 1983; Bora *et al.*, 1996; Devi *et al.*, 1982). High temperature in the range of 30-35°C increases the disease incidence and the plants that were resistant at moderate temperature become susceptible (Krausz and Thurston, 1975). Wet and well-drained soils are conducive for survival of the pathogen whereas, it is affected adversely by soil desiccation and flooding (Buddenhagan, 1986). Soil moisture affect multiplication and survival of *R. solanacearum* irrespective of soil types as the colony counts increased at highest soil moisture (Hayward, 1991). Further, soil temperature and soil moisture have synergistic effect on disease development and high temperature or high soil moisture alone may not even induce bacterial wilt (Shekhawat *et al.*, 2000).

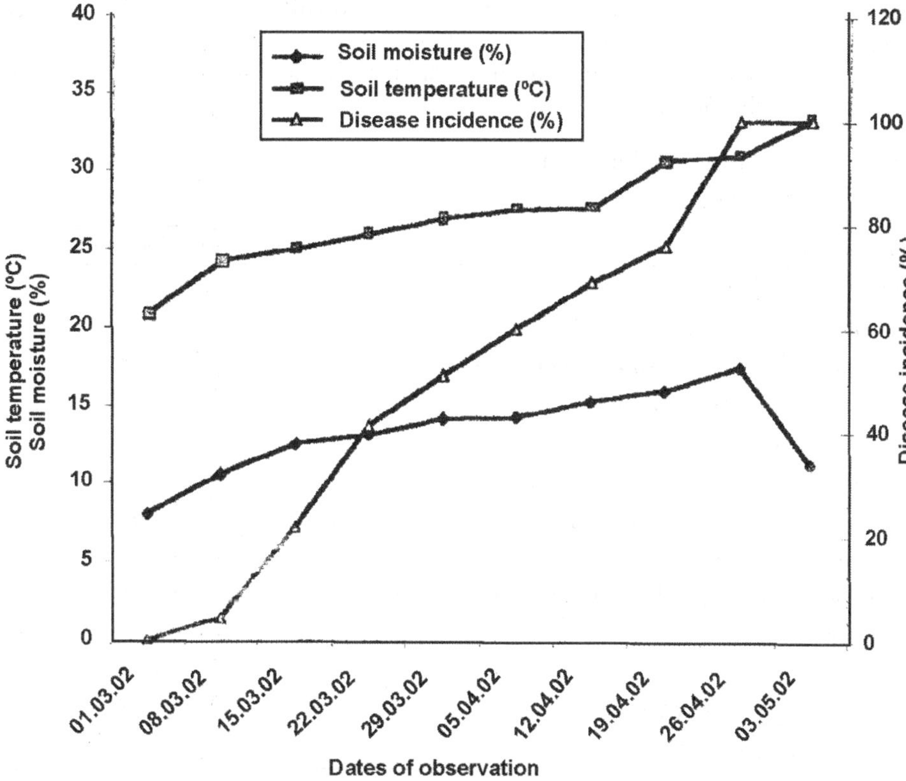

Fig. 1. Effect of hydrothermal regimes on development of bacterial wilt of tomato.

Simple correlation analysis between disease incidence and soil moisture (r = 0.8782) or soil temperature (r = 0.9553) showed correlation indicating that these independent variables influenced the disease development and with the increase in soil moisture or temperature disease incidence increased significantly (Table 1).

Soil moisture and soil temperature also showed positive correlation (r = 0.8296) with each other. When soil temperature was held constant, a non-significant positive correlation co-efficient (r = 0.5192) was recorded between disease incidence and soil temperature. Similarly, when soil moisture was kept constant disease incidence and soil temperature (r = 0.8490) were also found to be positively and significantly correlated. However, when disease incidence was kept constant the independent variables showed non-significant negative correlation (r = -0.061). Further, the coefficient of multiple determination (R^2) was also calculated to measure the contribution of linear function of independent variables, such as soil temperature (A_1) and soil moisture (A_2) on dependent variable i.e. disease incidence. The coefficient of multiple determination (R^2) between disease incidence and group of independent variables was found to be 0.9361, which indicated that 93.61 per cent change in disease incidence could be attributed to soil hydrothermal regimes included in the study collectively, whereas, rest of the variation was due to unexplained factors (error variation) and factors not included in the investigation. The computation of regression coefficients was utilized to develop regression equation ($Y = -191.48 + 7.11A_1 + 3.61A_2$), which indicated that a unit change in mean soil temperature and soil moisture influenced the disease incidence to the extent of 7.11 and 3.61 units, respectively. Hence, this equation can be utilized for prediction of disease level after its further validation in the field.

Table 1. Correlation of bacterial wilt of tomato with soil hydrothermal regimes

Correlation	Coefficient value
Simple correlation	
Disease incidence x soil moisture	0.8782*
Disease incidence x soil temperature	0.9553*
Soil moisture x soil temperature	0.8296*
Partial correlation	
Disease incidence x soil moisture	0.5192**
Disease incidence x soil temperature	0.8490*
Soil moisture x soil temperature	-0.061**
Co-efficient of multiple determination (R^2)	93.61
Regression co-efficient	
Constant	-191.48
Soil temperature (A_1)	7.11*
Soil moisture (A_2)	3.61**

* Significant at 5 per cent level of significance
** Non-significant at 5 per cent level of significance

REFERENCES

Abdullah, H., Maene, L.M.J. and Naib, H. (1983). The effect of soil type and moisture levels on bacterial wilt disease of groundnut. *Pertanitca* 6: 26-31.

Bora, L.C., Gogoi, P.K. and Das, B.C. (1996). Bacterial wilt of tomato in relation to few environmental parameters in Assam. *J. Agric. Sci. Soc. North East India* 9: 185-186.

Buddenhagen, I.W. (1986). Bacterial wilt revisited. *Proc. Int. Workshop, PCARRD, Lon Banos, Philippines 8-10 Oct. 1985, ACIAR Proc.* 13: 126-143.

Devi, L.R., Menon, M.R. and Aiyer, R.S. (1982). Population threshold of *Pseudomonas solanacearum* at the onset of first symptoms of wilt in tomato. *Indian J. Micrbiol.* 22 : 41-43.

Gupta, S.K., Dohroo, N.P. and Shyam, K.R. (1998). Occurrence of bacterial wilt of tomato in Himachal Pradesh. *Plant Dis. Res.* 13: 174.

Hayward, A.C. (1991). Biology and epidemiology of bacterial wilt caused by *Pseudomonas solanacearum*. *Annu. Rev. Phytopathol.* 29: 65-87.

He, L.Y., Sequeira, L. and Kelman, A. (1983). Characteristics of strain of *P. solanacearum* from China. *Plant Dis.* 67: 1357-1361.

Krausz, J.P. and Thurston, H.D. (1975). Breakdown of resistance to *Pseudomonas solanacearum* in tomato. *Phytopathology* 65: 1272-1274.

Persley, G.J. (1986). Bacterial wilt disease in Asia and the South Pacific. In: *Proc. Int. Workshop, PCARRD*, Lon Banos, Philippines 8-10 Oct. 1985, *ACIAR Proc.* 13: 145.

Rao, M.V.B. and Sohi, H.S. (1977). Control of bacterial wilt of brinjal. *Indian Hortic.* 22: 11-13.

Shekhawat, G.S., Gadewar, A.V. and Chakrabarti, S.K. (2000). *Potato Bacterial Wilt in India.* CPRI, Tech. Bull. No. 38: 56 p.

Shekhawat, G.S., Gadewar, A.V. and Chakrabarti, S.K. (1992). Spontaneous phenotypic reversion from afluidal to fluidal state in strains of *Pseudomonas solanacearum*. *Bacterial Wilt Newsletter ACIAR* 8 : 5-6

Sood, A.K. and Singh, B.M. (1993). Prevalence of bacterial wilt of solanaceous vegetables in mid hill sub-humid zone of Himachal Pradesh (India). *ACIAR Proc.* 45: 358-361.

Integrated Plant Disease Management (2005)
Eds. Sharma, R.C. and Sharma, J.N.
Scientific Publishers (India), Jodhpur, pp. 137-141.

21

Genetics of resistance to bacterial wilt in tomato (*Lycopersicon esculentum* Mill.)

K.C. SHARMA, J.C. GUPTA and V.K. RATHEE

Hill Agricultural Research and Extension Centre, CSK HPKV,
Bajaura - 175 125 (H.P.)

ABSTRACT

Inheritance studies on resistance to bacterial wilt (*Ralstonia solanacearum* (Smith) Yabuuchi *et al.*) in tomato (*Lycopersicon esculentum* Mill.) from six generations (parents, F_1, F_2, BC_1 and BC_2) of four crosses involving resistant and susceptible parents were conducted during summer season of 2002. The study has clearly indicated the involvement of more than one interacting genes. Further the study also indicated the role of additive and dominance components and their interactions in the expression of bacterial wilt resistance in tomato. Therefore, selection in early segregating generations would be effective in case of crosses having preponderance of additive [d] and additive x additive [i] effects while crosses having dominance effects, selection after few generations of selfing would be more fruitful for the development of bacterial wilt resistant varieties/hybrids in tomato.

INTRODUCTION

Mid hills of Himachal Pradesh have created a dent in the markets of plains by supplying fresh tomatoes from June to October thereby bringing very remunerative returns to the farmers. However, continuous cultivation of solanaceous vegetables in general and tomatoes in particular in the same fields year after year has led to the development of most serious disease bacterial wilt caused by *Ralstonia solanacearum* (Smith) Yabuuchi *et al.*, the disease causing losses up to 90 per cent to the crop during summer season (Bose and Rajan, 2000). Almost all the leading areas of tomato cultivation in Mandi, Kangra, Kullu, Hamirpur and Solan districts experiencing humid sub-temperate and sub-tropical climatic conditions are facing a serious threat from this disease. Lack of resistance in the presently grown cultivars/hybrids as well as costly chemical control (the pathogen being soil-borne) further aggravates the problem. The best possible effective, stable and economic way to overcome this menace is the incorporation of resistant genes in the hybrids/cultivars. For this, understanding of mode of inheritance of the disease resistance and identification of resistant donors is a pre-requisite. Keeping in view the above background, an investigation was undertaken to study the mode of inheritance of resistance to bacterial wilt in tomato.

MATERIALS AND METHODS

The present study was conducted on four crosses viz., Hawaii -7998 x Solan Gola, Hawaii-7998 x Roma, BT-18 x Solan Gola and TBL-4 x Solan Gola. Each cross-involved bacterial wilt resistant stable lines (Hawaii-7998, BT-18 and TBL-4) as one of the parents whereas other parent involved in the cross represented commercial but susceptible cultivars (Solan Gola, Roma). The F_1, F_2, BC_1 and BC_2 generations were developed in each cross. The seedlings of parents P_1, P_2, F_1, F_2, BC_1 and BC_2 generations were transplanted at the Research Farm of Hill Agricultural Research and Extension Centre, CSK Himachal Pradesh Krishi Vishvavidyalaya, Bajaura (representing mid hill sub humid area) during summer, 2002. The layout design followed was randomized block design with three replications maintaining a spacing of 75 x 45 cm. Each parent and F_1 had one row of 4.5 m length, BCP_1 and BCP_2 two rows and F2's at least four rows per replication. A row of susceptible check variety Solan Gola or Roma was also planted after every 12 rows to ensure the uniform inoculum throughout the experimental field.

To avoid the escape, all the plants were also inoculated artificially at the time of transplanting with the inoculum of the most prevalent boivar-3 of race-I of *Ralstonia solanacearum.* Bacterial ooze test was also carried to confirm the disease. The screening against bacterial wilt was carried out according to Mew and Ho (1976). The transformed values over replications were used for statistical analysis. The scaling test suggested by Mather and Jinks (1982) was applied for testing the adequacy of the additive dominance model. Hayman (1958) six parameter model and Jinks and Jones (1956) three parameter model were applied for the analysis depending upon the presence or absence of non-allelic interactions.

RESULTS AND DISCUSSION

All the F_1 plants of four resistant x susceptible crosses i.e. Hawaii-7998 x Solan Gola, Hawaii-7998 x Roma, BT-18 x Solan Gola and TBL-4 x Solan Gola were observed to be resistant which suggested that resistance to the disease was dominant over susceptibility. Nirmala (1987) and Sharma (1998) also observed similar observations in tomato.

Table 1. Disease reaction and mode of inheritance of bacterial wilt in tomato

Cross	F_1	F_2					BC_2				
		Observed		Expected ratio	X^2		Observed		Expected ratio		X^2
		R	S	R : S			R	S	R	S	
Hawaii-7998 x Solan Gola	Resistant	88	12	3 : 1	0. 17		41	19	1 :	1	8.06*
Hawaii-7998 x Roma	Resistant	105	15	3 : 1	10.00*		42	18	1 :	1	9.60*
BT-18 x Solan Gola	Resistant	108	12	3 : 1	14.40*		44	16	1 :	1	13.06*
TBL-4 x Solan Gola	Resistant	94	26	3 : 1	0.71		40	20	1 :	1	6.67*

* Significant at 5 per cent level

The segregation pattern in the F_2 generation (Table 1) showed a good fit of 3 resistant: 1 susceptible only in two crosses viz., Hawaii-7998 x Solan Gola and TBL-4 x Solan Gola. The other two crosses viz., Hawaii-7998 x Roma and BT-18 x Solan Gola did not follow the expected 3 resistant: 1 susceptible ratio in the F_2 generation. Further, all the four resistant x susceptible crosses also did not segregate in the expected ratio of 1 resistant : 1 susceptible in BC_1 and BC_2 generations. This clearly indicated that the disease in question is not controlled by a single gene and probably more than one gene governs this disease. Nirmala (1987) also observed quite similar results in tomato. In order to test the adequacy of additive dominance model, scaling tests A, B and C were applied. The values of A, B and C were found to be significant in all the crosses indicating the failure of additive-dominance model (Table 2).

Table 2. Scaling tests for non-allelic interactions of bacterial wilt resistance in tomato

Cross	A	B	C
Hawaii-7998 x Solan Gola	-45.40* ± 7.20	24.06* ± 6.98	-22.02* ± 13.46
Hawaii-7998 x Roma	-28.24* ± 4.97	50.72* ± 5.06	54.95* ± 7.39
BT-18 x Solan Gola	-27.50* ± 9.92	39.19* ± 9.40	56.89* ± 20.24
TBL-4 x Solan Gola	-18.43* ± 9.21	30.67* ± 9.16	29.49* ± 16.95

* Significant at 5 per cent level

The generation means were partitioned into different components like mean effect [m], additive effect [d], dominance effect [h], additive x additive [i], additive x dominance [j] and dominance x dominance [l] and presented in Table 3.

Table 3. Genetic parameters for bacterial wilt resistance in tomato

Cross	[m]	[d]	[h]	[i]	[j]	[l]
Hawaii-7998 x Solan Gola	59.49* ± 1.43	7.78* ± 2.32	37.31* ± 9.56	0.68 ± 7.38	-34.73* ± 2.81	20.67 ± 16.35
Hawaii-7998 x Roma	69.36* ± 1.26	4.62 ± 2.99	-11.41 ± 8.26	-32.47* ± 7.80	-39.48* ± 2.99	09.98 ± 14.04
BT-18 x Solan Gola	71.95* ± 2.86	3.29 ± 2.50	-11.35 ± 15.01	-45.20* ± 12.47	-33.35* ± 4.17	33.51 ± 22.57
TBL-4 x Solan Gola	62.32* ± 1.55	6.44* ± 2.99	22.32 ± 11.68	-17.25* ± 8.63	-24.55* ± 3.67	05.01 ± 20.74

* Significant at 5 per cent level

Analysis of components of means indicated that dominance effects [h] were significant and more pronounced in cross Hawaii-7998 x Solan Gola, though additive gene effects with low magnitude were also significantly important for wilt resistance. This suggested that the dominance gene effects played important role in the inheritance of resistance to wilt. The results are in line with those of Grimault *et al.* (1995). Further the relative contribution of additive and non-additive gene effects differed in crosses suggesting that the inheritance of bacterial wilt was complex. Nirmala Devi (1987) also found that the genetics of resistance to bacterial wilt is complex in nature. However additive effects were significant in cross TBL-4 x Solan Gola, thus there is good scope for improvement of this attribute by applying selection pressure for resistant plants. Among the interactions, additive x additive interaction was found significant in 3 out of the 4 crosses viz., Hawaii-7998 x Roma, BT-18 x Solan Gola and TBL-4 x Solan Gola. Additive x dominance type of interaction was observed to be significant in all the four crosses. Dominance x dominance type of interaction also played insignificant role in all the four crosses. The present study on inheritance and gene action has clearly indicated the involvement of more than one interacting genes. Further the study also indicated the role of additive and dominance components and their interactions in the expression of bacterial wilt resistance in tomato. Therefore, selection in early segregating generations would be effective in case of crosses having preponderance of additive and additive x additive effects while crosses having dominance effects, selection after few generations of selfing would be more fruitful for the development of bacterial wilt resistant varieties/hybrids in tomato.

REFERENCES

Bose, S.S.C. and Rajan, S. (2000). Peroxidase isozyme as markers for bacterial wilt resistance in tomato (*Lycopersicon esculentum* Mill.). *Veg. Sci.* 27: 136-141.

Grimault, V., Prior, P. and Anais, G. (1995). A monogenic dominant resistance of tomato to bacterial wilt in Hawaii 7996 is associated with plant colonization by *Pseudomonas solanacearum*. *J. Phytopathol.* 126: 349-352.

Hayman, B.I. (1958). The separation of epistasis from additive and dominance variation in generation means. *Heredity* 12: 371-390.

Jinks, J.L. and Jones, R.M. (1956). Estimates of the components of heterosis. *Genetics* 43: 223-224.

Mather, K. and Jinks, J.L. (1982). *Biometrical Genetics.* Ed. 3, Chapman & Hall Ltd. London.

Mew, T.H. and Ho, W.C. (1976). Varietal resistance to bacterial wilt in tomato. *Plant Dis. Rep.* 60: 264-268.

Nirmala Devi (1987). *Studies on Genetic Resistance to Bacterial Wilt* (*Pseudomonas solanacearum* E.F. Smith) *and Root Knot Nematode in Tomato* (*Lycopersicon esculentum* Mill.). Ph.D Thesis, University of Agricultural Sciences, Bangalore.

Sharma, P. (1998). *Genetic Analysis of Fruit Yield Components and Bacterial Wilt Resistance in Tomato* (*Lycopersicon esculentum* Mill.). Ph.D. Thesis, HPKV, Palampur, (H.P.).

Integrated Plant Disease Management (2005)
Eds. Sharma, R.C. and Sharma, J.N.
Scientific Publishers (India), Jodhpur, pp. 143-145.

22

Management of root-knot nematode, *Meloidogyne incognita* in tomato nursery by amendment with some organic wastes

KUM KUM WALIA

Department of Nematology, CCS Haryana Agricultural University,
Hisar - 125 004 (Haryana)

ABSTRACT

In a pot trial under greenhouse conditions, dried and powdered calendula and papaya seeds, and mango kernels were mixed in nematode infested soil @ 2 and 4 g per kg soil one week before sowing of tomato seeds (var. Sel-7, 50 seeds approx. per pot). Mango kernels @ 4 g per kg soil hampered the seed germination. Seedling height and weight were significantly improved in all amended treatments over unamended check except in mango kernel amendment where it was comparable to unamended check. Mango kernels powder was ineffective in reducing root-knot index too. Papaya seeds @ 4 g per kg soil were the most effective in improving seedling growth and reducing nematode attack with a root knot index (RKI) of 2.9 as compared to 5 in unamended check.

INTRODUCTION

The seeds of calendula (*Calendula officinalis* L.), papaya (*Carica papaya* L.) and mango kernels (*Mangifera indica* L.) have been reported to possess anthelmintic properties in literature (Dastur, 1977). The nematicidal potential of these plant materials was assessed against root-knot nematode, *M. incognita* in a pot trial.

MATERIALS AND METHODS

The seeds or kernels were air dried and powdered. The powdered material was mixed @ 2 or 4 g per kg soil filled in 15 cm diameter pots. Soil infested with nematodes was obtained from culture pots and initial population estimated. Autoclaved soil (nematode free) was used in the second set to compare the manurial effect of the amendments. Seeds of tomato cv. Sel-7 (50 per pot) were sown the next day. Each treatment was replicated four times and randomized completely on a screen-house bench. Sixty days after germination, ten seedlings were removed randomly from each pot and data recorded on the seedling height, seedling weight and root knot index on 1-5 scale (Johnson, 1985). The data were analyzed statistically.

RESULTS AND DISCUSSION

The seedling height and weight were significantly improved in calendula or papaya seed powder amended nursery soil (Tables 1, 2).

Table 1. Effect of nursery soil amendment with some organic wastes on seedling height (cm) of tomato cv. Sel-7 (average of four replications)

Treatment	With nematodes	Without nematodes	Mean
Calendula seeds @ 2g/kg soil	3.8	9.3	6.5
Calendula seeds @ 4g/kg soil	4.5	9.7	7.2
Papaya seeds @ 2g /kg soil	4.5	11.1	7.7
Papaya seeds @ 4g /kg soil	6.7	10.4	8.5
Mango kernels @ 2g /kg soil	2.7	7.4	5.0
Mango kernels @ 4g /kg soil	2.2	6.1	4.1
Unamended check	3.1	5.8	4.4
Mean	3.9	8.5	

$CD_{0.05}$

Amendment = 1.17

Nematodes = 0.62

Amendment x Nematodes = 1.66

Maximum gain in plant height was observed in higher level (4 g per kg soil) of papaya seed powder followed by its lower level and calendula powder at 4 g per kg soil. Mango kernel powder was not effective as the seedlings obtained at both the levels of the amendment had lesser height than even unamended check in the nematode infested soil. However, the differences were statistically non-significant. As far as seedling weight is concerned, maximum weight was recorded in papaya seed powder @ 4 g per kg infested soil followed by the lower level of papaya seed powder, and calendula seed powder at both the levels. Mango kernels at both the levels increased the seedling weight in autoclaved soil but not in nematode infested soil may be due to high inoculum of nematodes (10,000 juveniles per kg soil) or some detrimental organisms might have established themselves in presence of organic amendment or due to some biochemical changes in soil which need to be explored.

Table 2. Effect of nursery soil amendment with some organic wastes on seedling weight (g) of tomato cv. Sel-7 (averages of 4 replications)

Treatment	With nematodes	Without Nematodes	Mean
Calendula seeds @ 2g /kg soil	2.3	6.4	4.4
Calendula seeds @ 4g /kg soil	2.8	8.0	5.4
Papaya seeds @ 2g /kg soil	3.1	6.6	4.9
Papaya seeds @ 4g /kg soil	5.1	6.5	5.8
Mango kernels @ 2g /kg soil	0.5	4.4	2.5
Mango kernels @ 4g /kg soil	0.5	5.1	2.8
Unamended check	1.2	2.7	1.9
Mean	2.2	5.6	

CD $_{0.05}$

Amendment = 1.3

Nematodes = 0.74

Amendment x Nematodes = 1.96

The papaya seed powder was the most effective in controlling the root-knot nematode. The root-knot indices were 2.9, 4.0, 4.3, 3.6 in P4, C4, P2 and C2 treatments, respectively. Mango kernel amendment was ineffective in controlling nematode as root-knot indices in both the levels (4.9, 5.0) were comparable to unamended check (5.0). Presence of volatile oils, flavonoids, fatty acids and glucosides in these plant materials may have the inhibitory effect on larval mobility, hatching or penetration.

In similar observations, two species of *Chenopodium* have been found quite effective against both *M. javanica* and *M. incognita* as nursery soil treatments (Anonymous, 1998; Tabil and Walia, 1997). Both manurial as well as nematicidal effects were observed.

In the present studies, the plant materials used have not been investigated earlier and it could be inferred that papaya seed powder can be used as nursery soil treatment effectively against root-knot nematode in tomato.

REFERENCES

Anonymous. (1998). *Studies on the Integrated Management of Root Knot Nematode (Meloidogyne spp.) in Vegetable Crops.* Final Technical Report of PL 480 Project, 231 p.

Dastur, J. F. (1977). *Medicinal Plants of India and Pakistan.* Taraporevala Sons & Co. Pvt. Ltd. Bombay, 212 p.

Johnson, A.W. (1985). In: *Advanced Treatise on Meloidogyne* Vol. 1. *Biology and Control* (Eds. Sasser, J.N. and Carter, C.C.), N.C. State Univ. Graphics, Raleigh, N.C., USA, pp. 283-301.

Tabil, M.A. and Walia, K.K. (1997). Control of root-knot nematode, *Meloidogyne incognita* by amendment of nursery soil with *Chenopodium* species. *Ann. Plant Prot. Sci.* 5: 61-65.

Integrated Plant Disease Management (2005)
Eds. Sharma, R.C. and Sharma, J.N.
Scientific Publishers (India), Jodhpur, pp. 147-149.

23

Integrated management of *Meloidogyne incognita* (Kofoid and White) Chitwood in tomato (*Lycopersicon esculentum* Mill.)

ANJU S. KHANNA and SUNIL KUMAR

Department of Entomology and Apiculture, Dr. Y.S. Parmar University of Horticulture and Forestry, Nauni, Solan – 173 230 (H.P.)

ABSTRACT

Meloidogyne incognita constitutes one of the major pest problems on tomato in mid hill region of Himachal Pradesh. The damage caused by this pathogen is increasing year after year. An experiment conducted to emphasize the minimum use of expensive and hazardous nematicides revealed that cow dung cakes burnt over the nematode infested nursery land produced similar results to that of pre-sowing soil application of carbofuran. Neem cake amendment in field checked the population of *M. incognita* and increased the fruit yield significantly. Also, the crop transplanted in March suffered less as compared to June transplanted crop.

INTRODUCTION

Tomato (*Lycopersicon esculentum* Mill.) is an important solanaceous crop covering the global area of 3.64 million ha. In India, tomato is cultivated as a leading vegetable crop covering an area of 0.36 million ha under its cultivation with an annual production of 5.50 million tonnes (Anonymous, 2001). Mid-hill region of Himachal Pradesh is most suitable for its cultivation and the crop grown in this region ripens at the time when it is a lean period in the plains of northern India, thus fetching highly remunerative prices to the tomato growers of this region.

Among the various biotic factors that cause qualitative and quantitative damage to the tomato crop, plant parasitic nematodes play a significant role. Of these most prevalent is the root knot nematode, *Meloidogyne incognita* (Kofoid and White) Chitwood that has been found to incur yield losses to the tune of 33.3 per cent on tomato in the state (Anonymous, 1998). With the ever increasing emphasis to minimize the use of chemicals/nematicides as a management tactic in recent times, it was thought desirable to conduct an experiment on the management of *M. incognita* in tomato using an integrated approach.

MATERIALS AND METHODS

The nematode susceptible tomato cultivar Pride was used for the experiment. Seeds were sown in well prepared nematode infested nursery beds showing an average nematode population of 246 juveniles of *M. incognita* per 200 cc of soil. Nursery beds were treated separately in three different ways (T_1 to T_3). These nursery seedlings were transplanted in well ploughed field plots of 2x2 m size separately as per following treatments:

Treatments

T_1 = Nursery bed treatment with carbofuran @ 0.3 g a.i./m^2

T_2 = Nursery bed treatment with dazomet @ 0.3 g ai/m^2

T_3 = Dry cow dung cakes burnt slowly over the well prepared nursery bed @ 5 kg/m^2

T_4 = T_1 + neem cake application in field @ 400 g/4m^2

T_5 = T_2 + neem cake application in field @ 400 g/4m^2

T_6 = T_3 + neem cake application in field @ 400 g/4m^2

T_7 = Untreated control

T_8 = Neem cake application in field @ 400 g/4 m^2

Each treatment was replicated four times. The experimental design used was randomized block design.

Observations recorded

- Initial root knot nematode population in nursery, 246/200 cc of soil
- Initial root knot nematode population in field 260/200 cc of soil
- Root gall index of seedling at transplanting (1-5 scale)*

Carbofuran treated nursery - 2.4

Dazomet treated nursery - 2.2

Cow dung cake treated nursery - 1.8

Untreated control - 2.8

* Root gall index 1-5 scale - 1 = No gall, 2 = 1-10 galls, 3 = 11-30 galls, 4 = 31-100 galls and 5 = > 100 galls

RESULTS AND DISCUSSION

M. incognita (Kofoid and White) Chitwood population reduced significantly in all the treatments as compared to untreated control. The reduction in galling was significantly similar in all the treatments except T_8 where untreated nursery seedlings were transplanted in the field amended with neem cake. Though an increase in fruit yield was recorded in all the treatments, this increase was non-significant in T_1 (carbofuran treated nursery) as compared to control (T_7). Nurseries treated with dazomet or cow dung cake produced significantly higher yields which were statistically similar to the fruit yields produced in the plots (T_4, T_5 and T_6) where seedlings from treated nurseries were transplanted in neem cake amended plots. Highest fruit yield of 12.90 kg/4 m^2 bed was recorded in T_5 having dazomet treatment in nursery and neem cake amendment in the field. This was closely followed by T_6 (12.77 kg/4m^2) in which cow dung cakes were burnt over infested nursery bed before sowing and seedlings from this bed were transplanted in neem cake amended plots (Table 1). With an increase in root knot nematode population, root galling increases correspondingly. Higher the root gall index and root-knot nematode population, lower are the fruit yields recorded.

Table 1. Effect of various treatments on fruit yield of tomato and nematode population of *M. incognita*

Treatment	Yield (kg)	Per cent increase in yield over UTC	Root gall index	Final nematode population/200 cc soil
T_1 (Carbofuran)	9.31	30.94	2.55	364.0
T_2 (Dazomet)	11.27	58.50	2.20	241.2
T_3 (Cow dung)	10.75	51.19	2.15	239.5
T_4 (T_1 + NC)	10.77	51.47	2.00	228.7
T_5 (T_2 + NC)	12.90	81.43	1.95	199.2
T_6 (T_3 + NC)	12.77	79.60	1.90	184.0
T_7 UTC	7.11	0	3.95	710.0
T_8 (NC only)	7.85	10.40	3.35	522.5
$CD_{0.05}$	3.24		0.88	162.34

NC = Neem cake; UTC = Untreated control

REFERENCES

Anonymous. (1998). Annual Report. Department of Entomology and Apiculture, Dr. Y.S. Parmar University of Horticulture and Forestry, Nauni, Solan (HP).

Anonymous. (2001). *Quarterly Bulletin of Statistics*. FAO, Rome.

Integrated Plant Disease Management (2005)
Eds. Sharma, R.C. and Sharma, J.N.
Scientific Publishers (India), Jodhpur, pp. 151-157.

24

Management of root-knot nematode (*Meloidogyne incognita*) problem in tomato on farmers' fields

G.C. SHARMA and A.S. KASHYAP

Dr. Y. S. Parmar University of Horticulture and Forestry, Horticultural Research Station, Kandaghat (Solan)- 173 215 (H.P.)

ABSTRACT

Among the various biotic factors responsible for the low productivity in *kharif* vegetables being grown in the mid hill region of Himachal Pradesh, the role of plant parasitic nematodes (especially root-knot nematode) is of very high significance. Tomato being a leading cash and highly nematode prone crop (mainly to *Meloidogyne incognita*), in the present study the nematode problem was tried to be managed through eco-friendly means in the farmers' fields. Experiments were conducted for the four consecutive years (w.e.f. year 2000-2003), in the nematode infested fields of the farmers, belonging to Solan, Kandaghat and Kunihar blocks of the district. In the first instance before sowing the seeds in the nursery, its soil was thoroughly treated/mixed with granular nematicidal formulations like phorate (Thimet 10 G)/ carbofuran (Furadan 3 G) @ 0.3 g a.i./m^2 after watering it to the saturation level. Before transplanting the seedlings in the field, they were given root dip treatment in carbosulfan (Marshal 25 EC) @ 0.05 per cent for 20- 30 minutes in the shade. The fields selected for the crop were given two deep ploughings at an interval of 15-20 days. Results showed 15-31 per cent reduction in the root gall index of transplanting seedlings by treating the nursery soil. Nursery soil treatment in combination with the root dip treatment of the transplanting seedlings reduced the root gall index to the tune of 34-49 per cent in the fields. During different years of experimentation yield in the treated plots improved by 27-64 per cent over control with a net profit of 33-193 per cent.

INTRODUCTION

The mid hill region of Himachal Pradesh (altitude, between 4000-6000 ft amsl) is blessed with agro-climatic conditions which are highly congenial for the cultivation of vegetable crops and is a pioneer producer of vegetables like tomato, capsicum, ginger, beans, cauliflower, cabbage, peas etc. In this region, Solan and Kandaghat blocks of the district Solan are the major vegetable growing areas with tomato cultivation on 60-80 per cent of total cultivable area (Anonymous, 1979-1997). Among the various biotic factors responsible for low productivity in tomato, the role of phytoparasitic nematodes (mainly root-knot nematode; *Meloidogyne incognita)* is of very high significance (Bhatti and Jain, 1971).

It was in view of severity of the nematode problem and on growers' demand that in the present investigation experiments were laid out in their fields so as to demonstrate them the management strategies against the nematode and their impact on the growth, vigour and yield of the crop.

MATERIALS AND METHODS

The experiment was conducted for four consecutive years w.e.f. 2000-2003 in the farmers' fields of Solan, Kunihar and Kandaghat blocks of Solan district. Every year experiments were laid out in four to six farmers of three-four localities. During every year of the experimentation, before the onset of cropping season (February-March), the following treatments were given:

Pre-sowing nursery treatments

1. After preparing the nursery beds, they were thoroughly covered with waste material like stubble, hedges, rubbish, straw, sawdust etc. which was put on fire with the objective to increase the soil temperature and thereby to destroy the nematodes and other pathogenic microorganisms in the rhizosphere.
2. After two- three days of this operation, the nursery soil was watered up to saturation level. It was thoroughly applied/ mixed with granular formulations of the nematicides, carbofuran (3 G) or phorate (10 G) @ 0.3-0.5 g a.i./m^2 (i.e. 10-15 g or 3-5 g per m^2, respectively) along with some inert material like sand, ash etc. Nursery beds were again thoroughly hoed and planked.

Pre-transplanting field treatment

1. Before the final preparation of the earmarked fields, they were given two deep ploughings at an interval of 15-20 days that too without clodding and planking. It was done with the objective to expose the deeper soil to the open sun/ air.
2. For heavily infested fields an irrigation to the saturation level was given. Thereafter, fields were thoroughly applied/mixed with nematicide phorate (10 G) @ 2.5-3.0 kg/ bigha (800 sq. m) by mixing it with some inert material. The fields were then again ploughed,

clodded and planked thoroughly so as to make the fumigant action of the chemical an effective one.

Treatment of transplanting seedlings

Uprooted seedlings for transplanting were thoroughly observed for the presence of nematode galls on the root system. Seedlings with galls were discarded and destroyed. Uprooting was done in the evening hours and before transplanting them in the fields, they were given root dip treatment in carbosulfan (Marshal 25 EC) @ 500 ppm in a shallow container enabling only to dip the root system of the seedlings. The seedlings were given dip treatment for half an hour in shade, so as to avoid drooping effect. Immediately after the treatment, the seedlings were transplanted in the fields.

Post-planting treatments

Farmers were asked to perform frequent weeding and hoeing practices (at an interval of 20-25 days) in their treated plots. It was done with the objective to disturb the multiplying populations of the nematode and to hinder/restrict the entry and establishment of the nematodes (2nd stage juveniles) into the root system of their host plants (Sharma and Khan, 1995).

To make the farmers familiar with the impact of management strategies on the growth, vigour and yield of the crop, untreated/unmanaged nurseries and field plots were also maintained side by side.

During the present course of experimentation/ demonstrations, the tomato cultivars/varieties being grown by the peasants in their fields were; Rakshita, Nutun, Sun Seed-7711 (Avtar), Tolstoy, Manisha- 99, Pride, Yash, Him Sona, and 2000+. The plot size maintained by the peasants for the treated and untreated crop ranged between 60- 1600 and 70- 1700 sq. m, respectively.

Initial status of the nematode was assessed for both the nursery and field plots by recording the population of 2nd stage juveniles before soil preparation/treatments. Observations were recorded on the root gall index of the seedlings and the field plants (n= 25) before transplanting and ten days prior to the termination of the crop, respectively.

At the termination of the crop, besides collecting the yield data, observations were also recorded for the final count of the 2nd stage juveniles of the nematode. Information was also gathered from the individual farmer regarding their total input costs and sale price of the crop.

The root gall indexing was done on 1-5 scale:

Index	No. of galls
1	Nil
2	1-10
3	11-30
4	31-100
5	>100

RESULTS AND DISCUSSION

Root gall-index

(i) At transplanting : By treating the nursery soil, farmers could reduce the nematode infestation in their transplanting seedlings to varying levels during different years of experimentation/ demonstrations. In the year 2000, it was reduced to the tune of 24 per cent (with its range 14-38 %) over control. Per cent reduction recorded for the other three years were 20 (10-32), 15 (9-23) and 31 (20-41), respectively, in the ascending order. Overall results of the four years showed 22 per cent decrease in the root gall index on the transplanting seedlings (Table 1).

Table 1. Impact of management strategies on the infestation level of root knot nematode (*M. incognita*) and yield of tomato, at farmers field

Demo No.	Initial J2 count (per 200 cc soil)		Root gall index (1-5 scale)				J2 population at harvest (per 200 cc soil)		Yield (q/ ha)		Out put (Rs in lakh/ ha)		Cost benefit ratio		Net profit (Rs. In lakh/ ha)	
	Nur-sery	Main field	At transp-lanting		At harvesting											
			Treat.	Contr	Treat.	Contr	Treat.	Contr	Treat.	Contr	Treat.	Contr	Treat.	Contr	Treat.	Contr
A. Year 2000																
1	1005	190	1.00	1.16	2.20	2.80	1500	3600	711	613	3. 55	3. 06	1: 5.7	1:4.9	2. 93	2. 44
			(14)		(21)		(58)		(16)*						(20)*	
2	1120	320	1.28	1.72	2.60	4.60	2100	9000	537	411	2. 68	2. 05	1: 4.3	1: 3.3	2. 06	1. 43
			(26)		(43)		(77)		(31)*						(44)*	
3	130	250	1.00	1.24	2.40	3.20	1200	2000	271	209	1. 04	1. 35	1: 2.2	1: 1.7	0. 73	0. 42
			(19)		(25)		(40)		(30)*						(74)*	
4	280	165	1.00	1.32	2.00	3.00	800	3800	706	625	3. 53	3. 12	1: 5.6	1: 5.0	2. 90	2. 50
			(24)		(33)		(79)		(13)*						(16)*	
5	325	250	1.40	2.24	2.20	4.00	2100	5000	1156	790	5. 78	3. 95	1: 9.2	1: 6.3	5. 15	3. 32
			(37)		(45)		(58)		(46)*						(55)*	
6	160	300	1.00	1.16	1.40	2.80	500	3500	320	240	1. 60	1. 20	1: 2.6	1: 1.9	0. 97	0. 57
			(14)		(50)		(86)		(33)*						(70)*	
B. Year 2001																
1	100	120	1.04	1.16	1.00	2.40	120	400	833	757	4. 17	3. 78	1: 5.7	1: 5.6	3. 54	3. 16
			(11)		(58)		(70)		(11)*						(12)*	
2	100	120	1.00	1.24	2.40	4.60	50	170	520	400	2. 60	2. 00	1: 3.2	1: 2.2	1. 97	1. 37
			(20)		(48)		(70)		(30)*						(44)*	
3	120	150	1.00	1.20	1.40	3.60	nil	20	857	600	4. 28	3. 00	1: 5.9	1: 3.8	3. 66	2. 37
			(17)		(61)		(100)		(43)*						(54)*	

4	180	200	1.04 (32)	1.52	1.40 (42)	2.40	70 (75)	280	1000 (25)*	800	5. 00	4. 00	1: 7.0	1: 5.4	4. 37 (30)*	3. 37
5	100	200	1.04 (32)	1.52	1.60 (38)	2.60	nil (100)	40	1000 (33)*	750	5. 00	3. 75	1: 7.0	1: 5.0	4. 37 (40)*	3. 12
6	200	150	1.20 (10)	1.33	1.60 (43)	2.80	nil (100)	90	437 (28)*	341	2. 19	1. 87	1: 2.5	1: 2.0	1. 56 (25)*	1. 25

C. Year 2002

1	150	200	1.00 (19)	1.24	1.60 (53)	3.40	170 (80)	880	361 (86)*	175	2. 53	1. 36	1: 4.0	1: 2.2	1. 90 (160)	0. 73
2	150	300	1.08 (23)	1.28	2.00 (33)	3.00	450 (67)	1350	483 (32)*	367	3. 38	0. 57	1: 5.4	1: 4.1	2. 76 (42)*	1. 94
3	100	150	1.28 (9)	1.40	2.40 (33)	3.60	360 (72)	1280	416 (66)*	250	2. 91	1. 75	1: 4.6	1: 2.8	2. 29 (105)*	1. 12
4	120	120	1.08 (18)	1.32	2.60 (28)	3.60	360 (75)	1455	417 (43)*	292	2. 92	2. 04	1: 4.6	1: 3.2	2. 29 (61)*	1. 42

D. Year 2003

1	250	100	1.20 (33)	1.80	2.80 (39)	4.60	1500 (62)	4000	703 (87)*	375	4.92	2.62	1: 3.9	1: 1.6	3. 92 (142)*	1.62
2	350	180	1.60 (20)	2.00	2.40 (37)	3.80	1000 (71)	3500	656 (59)*	412	4.59	2.88	1: 3.6	1: 1.9	3.59 (91)*	1.88
3	280	250	1.30 (28)	1.00	1.80 (44)	3.20	800 (75)	3200	625 (67)*	375	4.37	2.62	1: 3.1	1: 1.6	3.37 (108)*	1.62
4	150	200	1.00 (37)	1.60	2.40 (40	4.00	600 (76)	2500	781 (150)*	312	5.47	2.18	1: 4.5	1: 1.2	4. 47 (279)*	1.18
5	150	100	1.20 (25)	1.60	2.60 (28)	3.60	1200 (81)	6200	796 (33)*	600	5.57	4.20	1: 4.6	1: 3.2	4. 57 (43)*	3. 20
6	200	250	1.00 (41)	1.70	2.40 (25)	3.20	900 (59)	2200	656 (31)*	500	4.59	3.50	1: 3.6	1: 2.5	3.59 (44)*	2.50

(ii) At harvesting : The crop raised through the treatment of nursery soil, transplanting seedlings, field soil and post-planting management practices was found loaded with reduced number of nematode galls. Four years results showed 40 per cent decrease in the root gall index, while as per the year-wise observations the figures showed 37 (22-50), 49 (38-61), 37 (33-53) and 36 (25-44) per cent reduced gall index, respectively in the ascending order (Table 2).

Juvenile nematode population

During the year 2000, at harvesting time there was recorded an average decrease of 69 per cent (with its range 50-86%) in the population of 2nd stage juveniles of the nematode in the treated plots over control. For the remaining three years the reduction recorded in the nematode population was 76 (70-

100), 73 (67-80) and 72 (59-81) per cent, respectively in the ascending order. Overall results revealed 72 per cent reduction in the populations of nematode (juveniles) in the rhizosphere of the treated plots at harvesting time (Tables 1,2).

Table 2. Overall impact of management strategies on the infestation level of root-knot nematode (*M. incognita*) and yield of tomato, at farmers field

Year	Initial J2 count (per 200 cc soil)		Root gall index (1-5 scale)				J2 population at harvest (per 200 cc soil)		Yield (q/ ha)		Out put (Rs in lakh/ ha)		Cost benefit ratio		Net profit (Rs. In lakh/ ha)	
	Nursery	Main field	At transplanting		At harvesting		Treat	Cont	Treat	Cont.	Treat	Cont.	Treat	Cont.	Treat	Cont.
			Treat	Contr	Treat	Contr										
2000	503	246	1.11 (24)	1.47	2.13 (37)	3.40	1366 (69)	4483	617 (28)*	481	3.03	2.23	1: 4.9	1: 3.8	2.46 (38)*	1.78
2001	133	157	1.05 (20)	1.32	1.56 (49)	3.06	40 (76)	166	774 (27)*	608	3.87	3.06	1: 5.2	1: 4.0	3.24 (33)*	2.44
2002	130	192	1.11 (15)	1.31	2.15 (37)	3.40	335 (73)	1241	419 (52)*	276	2.93	1.43	1: 4.6	1: 3.1	2.31 (78)*	1.30
2003	230	180	1.21 (31)	1.75	2.40 (36)	3.73	1000 (72)	3600	703 (64)*	429	4.92	3.00	1: 3.9	1: 2.0	3.92 (96)*	2.00
CD 0.05	_		0. 21		0. 52		1572. 9		135.57		_		_		_	

Figures in the parenthesis represents the per cent decrease over control.

* Figures in the parenthesis represents the per cent increase over control.

Inputs per hectare:	Years 2000- 2002	= Rs. 62, 500
	Year 2003	= Rs. 1.00, 000
Average Sale Rate:	Years 2000 & 2001	=Rs. 5 per kg.
	Years 2002 & 2003	= Rs. 7 per kg

Yield

Yield increase in demonstrations over control ranged between 16-46 per cent during the year 2000. The treated plots yielded the minimum produce of 271 q/ha, while maximum of 1156 q/ha was exceptionally high. In untreated plots, crop yield varied between 209-790 q/ ha. For the year 2001 yield increased to the tune of 11-43 per cent in treated plots over untreated ones with maximum yield of 1000 and 800 q/ha in treated and untreated plots, respectively. For the other two years (2002 and 2003), the yield harvested from the treated plots ranged between 361-483 q and 625-796 q/ha, respectively with the per cent increase over their control ranging from 43-86 and 59-150, respectively. The yield increase of 86 and 150 per cent recorded in the two demonstrations (one each during 2002 and 2003, respectively) were exceptionally high. Overall results of the four years demonstrations showed 28-64 per cent improvement in the crop yield by implementing the above mentioned management strategies against the nematode problem (Table 2).

Cost benefit ratio/net profit

Since, there were variations in the inputs being used by the peasants during different years of experimentation, cost benefit ratio also varied accordingly. Net profit earned by the peasants over control varied between 16-74 per cent during the year 2000 while for the other three consecutive years the per cent increase remained between 12-54, 61-160 and 43-142, respectively. Overall results of the individual years in the ascending order showed 38, 33, 78 and 96 per cent increase in the net profit over their control, respectively.

REFERENCES

Anonymous. (1979-1997). *Two Decades of Nematology*. All India Coordinated Research Project. Department of Entomology, Dr. Y S Parmar University of Horticulture and Forestry, Solan (HP).

Bhatti, D.S. and Jain, R.K. (1971). Estimation of losses in okra, tomato and brinjal yield due to *Meloidogyne incognita. Indian J. Nematol.* 7: 37- 41.

Sharma, G.C. and Khan, M.L. (1995). Efficacy of different chemicals against root-knot nematode (*Meloidogyne incognita*) infesting tomato. *Ann. Agric. Res.* 16: 351- 353.

Integrated Plant Disease Management (2005)
Eds. Sharma, R.C. and Sharma, J.N.
Scientific Publishers (India), Jodhpur, pp. 159-161.

25

Prevention of tomato leaf curl virus spread by soil mulching

DHARMESH GUPTA[1], P.D. THAKUR, P.K. SHARMA, ANIL HANDA and USHA SHARMA

[1]Apple Scab Monitoring and Research Laboratory, YSPUHF, Kotkhai, Shimla 171 202 (H.P.)
Department of Mycology and Plant Pathology, Dr.Y.S. Parmar University of Horticulture and Forestry, Nauni, Solan-173 230 (H.P.)

ABSTRACT

Management of tomato (*Lycopersicon esculentum* Mill.) leaf curl disease caused by virus belonging to *geminivirus* group in Dhaulakuan area in Himachal Pradesh was attempted by using polythene mulches. Using yellow polythene as mulch reduced the incidence of the disease besides increasing the yield of tomato.

INTRODUCTION

Tomato (*Lycopersicon esculentum* Mill.) is one of the most important vegetables grown throughout the world and is known for its outstanding nutritional value. It is a rich source of vitamins A and C and has a pleasing taste. It has now become an important component of the vegetable market enterprises and earns the price patronage world-over. Due to versatility of uses, commercial expediency and a lucrative business enterprise; there is a great upsurge in its cultivation within the farming community. It is quite exhilarative that the agro-climatic conditions of Himachal Pradesh are such that this crop can be cultivated appositely as an off- season crop and this enables it to fetch premium prices during June to October, which usually is an off-season for its cultivation in the northern plains. Tomato crop is infected by about 36 viruses and their strains (Chenulu and Giri, 1985). Among these, tomato leaf curl virus is most important and causes serious losses in yield. The virus is transmitted by a vector, whitefly (*Bemisia tabaci*).

MATERIALS AND METHODS

Experiments to examine the effect of polythene mulches of different colours (blue, yellow, white, green and black) were conducted during the years 1997 and 1998 at the experimental field of Horticultural Research Station, Dhaulakuan in Himachal Pradesh. The trials were laid out in a randomized block design with four replications with plot size of 3 x 1 m. The planting was done at a spacing of 90 x 30 cm. After a week of transplanting, polythene mulch was spread by making holes in it for the plants. All other agronomic operations needed for the crop were kept same for all treatments. Data on disease incidence and yield in different treatments were recorded and analyzed statistically.

RESULTS AND DISCUSSION

The data presented in Table 1 show that all the mulch treatments significantly reduced the disease incidence in tomato during both the years of experimentation. However, the lowest disease incidence of 43.75 and 45.00 per cent was recorded in both the years in the plots mulched with yellow polythene. *Bemisia tabaci* was found to be attracted more to yellow polythene than to white, blue or other polythene mulch, where they are killed by the high temperature reflected by the mulch. It should be noted that the untreated plots served as a reservoir of vast population of virus carrying *B. tabaci* and thus increased the chances of the neighbouring plants in the treated plots becoming infected. These conditions are quite different from those in commercial fields. Cohen and Melamed- Madjar (1978) also observed that soil mulching with yellow polythene sheets delayed the spread of tomato yellow leaf curl virus on tomato for at least 20 days. Similar reduction in disease incidence has been reported by Keren *et al.* (1991) and Zamir *et al.* (1991). The hypothesis put forward previously (Cohen *et al.*, 1974) as to the connection between the attractiveness of mulch for *B. tabaci* and the prevention of TYLCV spread was proved under field conditions. Thus, more flies were attracted to yellow mulch than to other colours, and gave better results in preventing the spread of leaf curl virus.

Yellow polythene mulching differed significantly in showing superiority over the white polythene mulch. The black mulch which recorded the highest disease incidence of 73.75 and 71.25 per cent during 1997 and 1998, respectively, was also found to be superior over control and was statistically at par with green polythene mulch treatment during both the cropping seasons.

Also, the effect of mulches on the yield was significant over control in both the years. The yellow polythene mulch recorded highest yield of 16.42 and 15.32 kg per plot and was significantly superior over green, black, blue and white mulch treatments during both the years. However, during 1997, blue and white polythene mulch were found statistically at par with each other. Mound (1962) stated that *B. tabaci* is more attracted by blue and ultraviolet part of spectrum. Similarly, green and black polythene mulches were found statistically at par with each other.

Table 1. Effect of polythene mulch on leaf curl disease incidence and yield of tomato

Colour of polythene	Disease incidence (%)		Average yield (Kg/plot)		Increase in yield over control (%)		Average yield (q/ha)	
	1997	1998	1997	1998	1997	1998	1997	1998
Blue	57.50 (49.31)	62.50 (52.24)	14.10 (22.06)	13.05 (21.13)	37.58	36.01	313.33	290.00
Yellow	43.75 (41.38)	45.00 (42.13)	16.42 (23.89)	15.32 (23.03)	46.40	45.49	364.88	340.44
White	53.75 (47.12)	55.00 (47.87)	15.10 (22.87)	14.37 (22.22)	41.72	41.89	335.55	319.33
Green	68.75 (55.98)	66.25 (54.45)	12.15 (20.36)	11.47 (19.73)	27.57	27.20	270.00	254.88
Black	73.75 (59.15)	71.25 (57.54)	11.15 (19.46)	10.22 (18.63)	21.07	18.29	247.77	227.11
Control	80.00 (63.44)	75.00 (60.00)	8.80 (17.26)	8.35 (16.74)	—	—	195.55	185.55
$CD_{0.05}$	(5.21)	(4.25)	(1.073)	(1.012)				

Figures in parentheses are arcsine transformed values

REFERENCES

Chenulu, V.V. and Giri, B.K. (1985). Viral and mycoplasma diseases of tomato and their management. In: *Perspectives in Plant Virology*. (Eds. Gupta, B. M., Singh, B.P., Verma, H.N. and Srivastava, K.M.). Printing House (India) Lucknow, 322 p.

Cohen, S. and Melamed- Madjar, V. (1978). Prevention by soil mulching of the spread of tomato yellow leaf curl virus transmitted by *Bemisia tabaci* in Israel. *Bull. Entomol. Res.* 68: 465-470.

Cohen, S., Melamed- Madjar, V. and Hameiri, J. (1974). Prevention of the spread of tomato yellow leaf curl virus transmitted by *Bemisia tabaci* in Israel. *Bull. Entomol. Res.* 64: 193-197.

Keren, J., Decco, Z., Cohen, S. and Bar Yossef, R. (1991). Reduction of damage from yellow top virus and tobacco whitefly in tomatoes by yellow plastic mulch. *Hassadeh* 71: 864-868.

Mound, L.A. (1962). Studies on the olfaction and colour sensitivity of *Bemisia tabaci*. *Entomol. Exp. Appl.* 5: 99-104.

Zamir, D., Zakay, Y., Zeidan, M. and Czosnek, H. (1991). Combating the tomato leaf curl virus in Israel: The agrotechnical and genetic approaches. In: *Resistance of Tomato to TYCLV* (Eds. Laterott, H. and Trousse, C.), INRA, Avignon, France, p. 9-14.

Integrated Plant Disease Management (2005)
Eds. Sharma, R.C. and Sharma, J.N.
Scientific Publishers (India), Jodhpur, pp. 163-168.

26

Partial characterization of a *potyvirus* causing mosaic disease in tomato (*Lycopersicon esculentum* Mill.) in Himachal Pradesh

DHARMESH GUPTA[1], P.D. THAKUR[3], S.C. CHOWFLA[3], I.D. GARG[2] and USHA SHARMA[3]

[1]Apple Scab Monitoring and Research Laboratory, YSPUHF, Kotkhai, Shimla – 171 202 (H.P.)
[2]Division of Plant Pathology, Central Potato Research Institute, Shimla- 171 001 (H.P.)
[3]Department of Mycology and Plant Pathology, Dr. Y.S. Parmar University of Horticulture and Forestry, Nauni, Solan-173 230 (H.P.)

ABSTRACT

Tomato is an important cash crop of Himachal Pradesh and also an important economic horticultural crop of many other tropical and subtropical countries. The chracterization of a *potyvirus* causing mosaic disease in tomato (*Lycopersicon esculentum* Mill.) in Himachal Pradesh, was carried out by studying symptomatology, transmission, host range, biophysical properties, particle morphology, cytopathology and immuno electron microscopy. The symptoms produced by the virus were vein- clearing, mild mosaic, downward rolling of leaves and reduction in size of the leaves. The virus isolate was found to be transmissible through sap and aphid vectors viz., *Myzus persicae, Aphis craccivora* and *Aphis gossypii.* The virus has a narrow host range restricted to a few members of the family solanaceae. The virus remained infectious in dilution between 10^{-3} to 10^{-4}, upon storage for 72 h at room temperature (24-29°C), 120 h at 5±1°C and when heated for ten minutes between 50-55°C. Electron microscopic studies of virus revealed presence of flexuous filamentous particles measuring 740 x 12 nm. IEM studies using DLDV and PTV resulted in enhanced trapping and clumping indicated the virus to be a definite member of the genus *potyvirus*.

INTRODUCTION

Tomato is an important cash crop of Himachal Pradesh cultivated in about 3,000 hectares (Chhabra, 1992). It is also an important economic horticultural crop of many other tropical and sub-tropical countries. It is host to many plant viruses some of which are ubiquitous in occurrence (Brunt *et al.*, 1996). Previous investigators in Himachal Pradesh have reported the occurrence of PVY^N (Behl *et al.*, 1987), TMV and a *tospovirus* (Sharma, 1998). Present investigation dealt with characterization of another *potyvirus* causing mosaic in tomato in Solan district of Himachal Pradesh. The virus was found to be distinct from many potyviruses including PVY^N (Behl *et al.*, 1987), PTV (Raymer *et al.*, 1972) and tomato mild mottle virus (TMMV) (Walkey *et al.*, 1994) reported earlier on tomato.

MATERIALS AND METHODS

Naturally infected tomato leaves with mosaic symptoms confirmed with electron microscope to carry only the flexuous virus particles measuring 40 x 12 nm were the source of the virus. The standard inoculum (SI) of the virus was prepared from tomato leaves (cv. Naveen) to which virus had been transmitted by *Myzus persicae* Sulz. non-persistently by grinding 5 g infected leaves in 5 ml of 0.2 M phosphate buffer pH 7.2. For host range study, ten plants of each species belonging to different families were sap inoculated with SI of the virus using carborundum powder as an abrasive. The inoculated test plants were observed for 3-5 weeks for symptoms. Stock culture of the virus was maintained in *Datura metel* plants. Biophysical properties were determined as described by Noordam (1973).

Aphid species, *Aphis craccivora, Rhopalosiphum maidis* and *Brevicoryne brassicae* maintained, respectively, on tobacco, maize and cauliflower were used to test their transmission efficacy. The aphids were given 45 minutes pre-acquisition fasting in the dark and acquisition access of 5 min on young systemically infected leaves. Ten aphids per plant were given inoculation access for 30 min.

To check seed transmission of the virus, 500 seeds from the diseased tomato plants (cv. Naveen) inoculated artificially under glasshouse conditions were selected and sown in pots having sterilized potting mixture. Two hundred such seedlings were transplanted in pots for further observation on the development of symptoms.

To study the soil transmission nature of virus, fifty seedlings of tomato were transplanted in soil collected from rhizosphere of naturally infected plants under insect proof conditions. These plants were observed for two months for development of symptoms.

Electron microscopic studies comprised (i) the examination of clarified virus extract, after negative staining with 2 per cent aqueous uranyl acetate to determine morphology of virions by employing the method described by Hitchborn and Hill (1965); (ii) cytopathology of infected tomato leaf cells by employing the procedure described by Roland (1978) and (iii) immune electron microscopy (IEM) by using procedure described by Garg and Khurana (1992, 1993). IEM studies included immuno- sorption or trapping of the virions on the antiserum coated grid and clumping i.e. virus-antiserum interaction in liquid phase followed by examination of the clumps, if any formed under electron microscope. Trapped virions were counted per 20 cm^2 of screen area at 21,000 X and then trapping index was calculated as follows:

$$\text{Trapping index} = \frac{\text{Number of virions trapped with antiserum}}{\text{Number of virions trapped with normal serum}} \times 100$$

The count of virions used was the average of ten counts at random from two different holes of grid per antiserum. Various antisera employed and their sources were as follows:

Potato virus Y (PVY^0) from CPRI, Shimla; Peruvian tomato virus (PTV) from Dr. L.F. Salazar, International Potato Centre, Lima-Peru; datura leaf distortion *potyvirus* (DLDV) from Dr. P. Srinivasulu, SVU, Tirupathi; potato virus A (PVA) from Dr. Kowalska, Poland; tobacco etch *potyvirus* (TEV), Arizona pepper *potyvirus* (AzPV) and pepper veinal mottle *potyvirus* (PeVMV) from Ivory Coast; bean common mosaic *potyvirus* (BYMV) from Dr. Z. Maat, The Netherlands and peanut mottle *potyvirus* (PMV) from Dr. Ashok Mishra, GAU Anand. All the antisera were diluted before use with 0.86 per cent NaCl solution to a final micro- precipitin titre of 1:20.

RESULTS AND DISCUSSION

Symptomatology

Field symptoms in tomato comprised vein clearing, mild mosaic, downward rolling of leaves and reduction in size of leaves along with stunting of infected plants. The sequence of symptoms upon sap inoculation was as follows: vein clearing after 14-15 days, mild mosaic expressed after 15-20 days and downward rolling of leaves after 30 days. In advanced stage of infection, leaf size was reduced and plants had stunted growth with fewer numbers of small fruits. Similar symptoms were observed by aphid transmission.

Transmission

The virus was readily sap and aphid transmissible. Aphids viz., *Myzus persicae, Aphis craccivora* and *Aphis gossypii* transmitted the virus in non-persistent manner but *Rhopalosiphum maidis* and *Brevicoryne brassicae* failed to transmit the virus. The virus was neither soil nor seed transmitted.

Host range

Virus was found to infect *Capsicum annuum, Datura metel* and *Datura metel* var. *fastuosa*. However, the virus did not infect other plant species tested such as *Gomphrena globosa, Chenopodium amaranticolor, C. quinoa, C. murale, C. album, Zinnia elegans, Brassica oleracea* var. *capitata, B.o.* var.

sativus, Cucumis sativus, Phaseolus vulgaris, Pisum sativum, Vigna sinensis, Datura stramonium, Nicotiana glutinosa, N. rustica, N. tabacum var. White Burley, *Solanum tuberosum, S. khasianum, Physalis floridana* and *Nicandra physaloides*.

Biophysical properties

In *Lycopersicon esculentum* sap, virus remained infectious in dilution between 10^{-3}-10^{-4}, upon storage for 72 h at room temperature (24-29°C), 120 h at 5±1°C and when heated for ten minutes between 50-55°C.

Particle morphology

Dip preparations from clarified virus extract (CVE) when examined under electron microscope revealed presence of flexuous filamentous virus particles having modal dimensions 740 x 12 nm.

Cytopathology

Infected cells showed presence of cytoplasmic cylindrical inclusions comprising pinwheels, scrolls, lamellar aggregates and tubes.

IEM studies

Results of trapping studies given in Table 1 revealed that antisera to only DLDV and PTV caused marked increase in the trapping index. There were fewer enhancements in trapping index with other antisera. Like trapping, clumping of virions, in liquid phase was also observed only with antisera to DLDV and PTV. Clumped virions also showed binding of antibodies.

Table 1. Trapping index of *potyvirus* infecting tomato

Antiserum	Number of virions trapped	Trapping index
PVY	1.5	3
PTV	3.0	6
BCMV	1.0	2
MYMV	1.3	2.6
PVA	1.2	2.4
PMV	1.2	2.4
AzPV	1.0	2.0
PVMV	1.3	2.6
DLDV	3.5	7.0
Normal serum	0.5	—

Virion morphology, cytoplasmic cylindrical inclusions comprising pinwheels, scrolls, lamellar aggregates and tubes in the infected cells and reactivity of the virus with antisera to DLDV and PTV as revealed by enhanced trapping and clumping indicated the virus to be a definite member of the genus *potyvirus*. The virus did not show any appreciable relationship to many other potyviruses namely, PMV, PVMV, PVA, BCMV, BYMV, and

AzPV in IEM studies. The virus is distinct from PVY^N reported on tomato by Behl *et al.* (1987) since it did not show any seroaffinity for PVY in IEM studies and also differed in host range. Though serologically related to DLDV and PTV potyviruses, yet it differed from both in host range, the host range of present virus being highly restricted as compared to these viruses (Prasanna *et al.*, 1996; Fribourg and Fernandez-Northcote, 1982).

The present virus, like TMV (Walkey *et al.*, 1994) showed seroaffinity for PTV which in turn is very closely related to PVY (Fernandez Northcote, personal communication). However, it differed from TMMV in host range and in the cytoplasmic cylindrical inclusions which in case of TMMV lacked lamellar aggregates but possessed dichotomously branched pinwheels.

Thus, this virus appears to be distinct from potyviruses reported to infect tomato in India and abroad. However, classical taxonomic parameters, namely biological properties and conventional serology may no longer be considered appropriate in distinction of potyviruses. N-terminal of the coat protein of potyviruses has been shown to account for much of the diversity in potyviruses where sequence homology is only 31-71 per cent between distinct potyviruses and more than 90 per cent among strains of same virus (Shukla and Ward, 1988, 1989). HPLC profiles of digested coat protein have been used to differentiate BCMV strains into two distinct potyviruses (McKern *et al.*, 1992a,b).

Therefore, further characterization of virus by way of nucleotide sequencing of the viral genome is required to ascertain its exact phylogenic position as a distinct *potyvirus.*

REFERENCES

Behl, M.K., Chowfla, S.C. and Khurana, S.M. Paul. (1987). Tomato shoe-stringing disease incited by necrotic strains of potato virus Y in Himachal Pradesh – A new record. *Curr. Sci.* 56: 27–29.

Brunt, A., Crabtree, K., Dallwitz, M., Gibbs, A. and Watson, L. (1996). *Viruses of Plants: Descriptions and Lists from the VIDE Database*, CAB International, Wallingford. 1484 p.

Chhabra, N. (1992). The tomato research, production and processing. *Indian Hortic.* 37: 13.

Christie, S. R., Purcifull, D.E., Crawford, W.E. and Ahmed, N.A. (1987). *Electron Microscopy of Negatively Stained Clarified Viral Concentrates Obtained from Small Tissue Samples with Appendices on Negative Staining Techniques.* Technical Bulletin 872, Agricultural Experimental Station, Institute of Food and Agricultural Sciences, University of Florida, 45 p.

Fribourg, C.E. and Fernandez – Northcote, E.N. (1982). Peru tomato virus. CMI/ABB. *Descriptions of Plant Viruses.* No. 255.

Garg, I.D. and Khurana, S.M. Paul. (1992). Factors influencing immune electron microscopy of flexuous potato viruses. *Acta Virol.* 36 : 435-442.

Garg, I.D. and Khurana, S.M. Paul. (1993). Morphological changes in the flexuous potato viruses upon decoration in immunosorbent electron microscopy. *Acta Virol.* 37 : 407–411.

Hitchborn, J.H. and Hill, G.J. (1965). The use of negative staining in the electron microscope examination of plant viruses in crude extracts. *Virology* 27: 528–540.

McKern, N.M., Mink, G.I., Barnett, O.W., Mishra, A., Whittaker, L.A, Silbernagel, M.J., Ward, C.W. and Shukla, D.D. (1992a). Isolates of bean common mosaic virus comprising two distinct potyviruses. *Phytopathology* 82: 923–929.

McKern, N.M., Ward, C.W. and Shukla, D.D. (1992b). Strains of bean common mosaic virus consist of at least two distinct potyviruses. *Arch. Virol.* 5 (Suppl.): 407-414.

Noordam, D. (1973). *Identification of Plant Viruses. Methods and Experiments.* Centre for Agricultural Publications and Documentation (Pudoc), Wageningen, 207 p.

Prassana, G.B., Rajeswari, A., Satyanarayana, T., Kiranmai, G., Gopinath, K., Sirinivasulu, P. and Naidu, M.V. (1996). Characterization of a virus causing leaf distortion disease of *Datura metel* in India. *Indian J. Virol.* 12: 129–135.

Raymer, W.B., Kahn, R.P., Hikida, H.R. and Waterworth, H.E. (1972). A new tomato virus from Peru. *Phytopathology* 62: 784 (Abstr.).

Roland, J.C. (1978). General preparation and staining of thin sections. In: *Electron Microscopy and Cytochemistry of Plant Cells* (Ed. Hall, J.L.), Elsevier/North Holland Biomedical Press, Amsterdam, p.1–62.

Sharma, P.K. (1998). *Identification and Management of Viruses Infecting Tomato in Himachal Pradesh.* Ph.D. Thesis, Dr. Y.S. Parmar University of Horticulture and Forestry, Nauni, Solan (HP).

Shukla, D.D. and Ward, C.W. (1988). Amino acid sequence homology of coat proteins as a basis for identification of the potyvirus groups. *J. Gen. Virol.* 69: 2709-2710.

Shukla, D.D. and Ward, C.W. (1989). Identification and classification of potyviruses on the basis of protein sequence data and serology. *Arch. Virol.* 106: 171-200.

Walkey, D.G.A., Spence, N.J., Clay, C.M. and Miller, A. (1994). A potyvirus isolated from solanaceous hosts. *Plant Pathol.* 43: 931– 937.

Integrated Plant Disease Management (2005)
Eds. Sharma, R.C. and Sharma, J.N.
Scientific Publishers (India), Jodhpur, pp. 169-172.

27

Epidemiology of a mosaic disease of tomato in Himachal Pradesh

P.K. SHARMA, ANIL HANDA[1], P.D.THAKUR[1], USHA SHARMA[1] and DHARMESH GUPTA[2]

Sher-e-Kashmir University of Agricultural Sciences & Technology (Jammu), Regional Horticultural Research Station, Bhadarwah, District Doda -182 202 (J&K)
[1]Department of Mycology and Plant Pathology, Dr. Y.S. Parmar University of Horticulture and Forestry, Nauni Solan – 173 230 (H.P.)
[2]Apple Scab Monitoring and Research laboratory, YSPUHF, Kotkhai, Shimla – 171 202 (H.P.)

ABSTRACT

Epidemiology of mosaic disease of tomato was studied with respect to weather parameters. The simple correlation for temperature, relative humidity, rainfall and vector population showed positive correlation with disease incidence. The partial correlation coefficient of disease incidence with vector population, sunshine hours and relative humidity was positive. Multiple correlation coefficients during the course of investigations were 0.9381, indicating thereby that the interaction between plants and virus was greatly influenced by these factors.

INTRODUCTION

A viral disease appeared in severe form in tomato crop in Solan district of Himachal Pradesh. The plants showed mild mosaic followed by vein clearing and malformations such as downward curling and filiformy of leaves. The infected plants were stunted. The virus was identified as a *potyvirus* resembling leaf distortion virus of datura. The filiformy in tomato has been reported to be caused by necrotic strain of potato virus Y (PVY). It is very destructive and could cause complete loss of crop (Behl *et al.*, 1987). Therefore, investigations were conducted to find out the role played by the environmental factors and vector population in the development of disease.

MATERIALS AND METHODS

Field trials were conducted to find out the correlation between the disease incidence and prevailing meteorological conditions and insect vector population. Tomato plants of highly susceptible variety Naveen were raised in field trials.

Data were recorded on disease incidence at seven days interval following first appearance of disease symptoms till the maturity of the crop. Data on weather variables viz., temperature, relative humidity, rainfall, wind velocity and sunshine hours were recorded daily and an average of seven days was taken for analysis of data. For taking vector counts, yellow metallic pan traps (30x30x10 cm) and sticky traps were positioned about one metre above the crop in an inclined position. The aphid count and disease incidence were recorded at weekly intervals and expressed as per cent increase in every seven days. Simple, partial and multiple correlations of dependent variable (disease incidence) with independent variables (weather parameters) were worked out (Gomez and Gomez, 1983).

RESULTS AND DISCUSSION

The disease appeared in the month of May in both the cropping seasons when mean air temperature was 25.5^{0}C and relative humidity 32.00 per cent. The disease became severe by the end of June and was static after mid July. The simple correlation coefficient (Table 1) determined for temperature, relative humidity, cumulative rainfall, and vector population showed strong positive correlation with disease incidence.

Table 1. Simple correlation coefficients between epidemiological parameters and disease incidence

Correlation pair	Correlation coefficient
Disease incidence x Temperature	+ 0.1764
Disease incidence x Relative humidity	+ 0.9298 **
Disease incidence x Total sunshine hours	– 0.5857*
Disease incidence x Wind velocity	– 0.8342**
Disease incidence x Cumulative rainfall	+ 0.3581
Disease incidence x Vector population	+ 0.8051**
Vector population x Temperature	+ 0.3028
Vector population x Relative humidity	+ 0.7535**
Vector population x Total sunshine hours	+ 0.4190
Vector population x Wind velocity	– 0.7174**
Vector population x Cumulative rainfall	+ 0.1101

** Significant at 1 per cent level of significance

* Significant at 5 per cent level of significance

The vector population was positively and significantly correlated with relative humidity, whereas it had negative and significant correlation with wind velocity. The partial correlation coefficient (Table 2) of disease incidence with vector population, total sunshine hours and relative humidity was positive.

Table 2. Partial correlation coefficients between epidemiological parameters and disease incidence

Correlation pair	Correlation coefficient
Disease incidence x Temperature	– 0.2744
Disease incidence x Relative humidity	+ 0.8193*
Disease incidence x Total sunshine hours	+ 0.5785
Disease incidence x Wind velocity	+ 0.3948
Disease incidence x Cumulative rainfall	– 0.0076
Disease incidence x Vector population	+ 0.1263
Vector population x Temperature	+ 0. 2519
Vector population x Relative humidity	+ 0. 1927
Vector population x Total sunshine hours	– 0.0192
Vector population x Wind velocity	– 0.1338
Vector population x Cumulative rainfall	–0.3675

* Significant at 5 per cent level of significance

Similarly, vector population had a positive partial correlation with temperature and relative humidity and negative with wind velocity and cumulative rainfall. The multiple correlation coefficients between the disease incidence and group of independent variables were 0.9381 during the course of experimentation (Table 3). This meant that epidemiological parameters play a significant role in the development and spread of the disease.

Table 3. Multiple correlation and regression coefficients between disease incidence, weather parameters and vector population

Coefficient of determination (R^2)	Regression equation $Y = b_0 + b_1 X_1 + b_2 X_2 + b_3 X_3 + b_4 X_4 + b_5 X_5 + b_6 X_6$	R^2 (%)
0.9381	$Y = -31.038 - 1.5419 X_1 + 1.4765 X_2 + 0.6406 X_3 - 4.9654 X_4 - 0.0032 X_5 + 0.1376 X_6$	93.81

Y = Disease incidence (%); X_1 =Mean air temperature (°C); X_2 =Relative humidity (%) X_3 =Total sunshine hours; X_4 =Wind velocity (Km/hr.); X_5 =Cumulative rainfall (mm); X_6 = Vector population.

Temperature was observed to be positively correlated with the disease incidence in the present studies. Nagaich *et al.* (1969) and Bindra *et al.* (1974) found that the incidence of PVY and potato leaf roll virus increased with an optimum temperature of 26°C which was also congenial for aphid population. With the increase in temperature, there is possible buildup of aphid population (Berg, 1984). Chowfla *et al.* (1989) reported significant positive correlation between temperature and PVY incidence in Himachal Pradesh.

The significant positive correlation of relative humidity with disease incidence during the present studies indicated that relative humidity itself is not independent but is controlled by some other environmental factors. The effect of humidity every year might be on the activity of vectors rather than on the virus. Nitzany (1975) in Israel noticed that tomato leaf curl virus outbreaks were always in months with a mean relative humidity of more than 60 per cent. The rainfall in general, has a negative effect on the vector population and disease incidence (Singh *et al.*, 1984; Androtra, 1990; Parmar, 1991). The correlation of cumulative rainfall with disease incidence was negative. The vector population had positive correlation with total sunshine hours in the present investigation.

REFERENCES

Andotra, P.S. (1990). *Studies on Virus Diseases of Cucumber* (*Cucumis sativus* L.) *in Himachal Pradesh*. Ph.D. Thesis, Dr. Y.S. Parmar University of Horticulture and Forestry, Nauni, Solan (H.P.).

Behl, M.K., Chowfla, S.C. and Khurana, S.M.P. (1987). Tomato shoe- string disease incited by necrotic strain of potato virus Y in Himachal Pradesh. *Curr. Sci.* 56: 27-29.

Berg, G.N. (1984). Non-persistent viruses: Epidemiology and control. *Adv. Virus Res.* 31: 387-429.

Bindra, O.S., Singh, M. and Sikka, L.C. (1974). Prevalence of virus spread in seed potato. *Prog. Farming* 11: 9-10.

Chowfla, S.C., Behl, M.K. and Baruah, B.P. (1989). Correlation of meteorological parameters with tomato shoestring disease in Himachal Pradesh. *Indian J. Plant Pathol.* 3: 252 - 254.

Gomez, K.A. and Gomez, A.A. (1983). *Statistical Procedures for Agricultural Research.* 2nd ed. John Willey and Sons, New York, 680 p.

Nagaich, B.B., Pushkarnath, Bhardwaj, V.P., Giri, B.K., Anand, S.K. and Upreti, G.C. (1969). Production of disease free seed potatoes in the Indo Gangetic plains. *Indian J. Agric. Sci.* 39: 238-243.

Nitzany, E.F. (1975). Tomato yellow leaf curl virus. *Phytopathol. Mediterr.* 14: 127-129.

Parmar, G.S. (1991). *Epidemiology and Management of Virus Induced Diseases of Bell Pepper in Himachal Pradesh.* Ph.D. Thesis, Dr. Y.S. Parmar University of Horticulture and Forestry, Solan (H.P.).

Singh, M.N., Nagaich, B.B. and Agrawal, H.O. (1984). Spread of potato virus Y and leaf roll by aphids in potato fields. *Indian Phytopathol.* 37: 241-261.

Integrated Plant Disease Management (2005)
Eds. Sharma, R.C. and Sharma, J.N.
Scientific Publishers (India), Jodhpur, pp. 173-177.

28

Post-harvest losses of bell pepper and tomato fruits in Himachal Pradesh

R.L. SHARMA, MANJU CHOUDHARY and ARTI SHUKLA

Department of Mycology and Plant Pathology, Dr. Y.S. Parmar University of Horticulture and Forestry, Nauni, Solan - 173 230 (H.P.)

ABSTRACT

The quality of bell pepper and tomato fruits in the markets of nine districts of Himachal Pradesh was assessed and the losses were determined. The losses varied from 5.0 to 26.1 per cent in case of tomato and 6.1 and 19.1 per cent in bell pepper. Maximum losses of tomato fruits were recorded during the rainy season (July to September) and that of bell pepper during the month of August.

INTRODUCTION

Both bell pepper and tomato fruits are vulnerable to post-harvest losses because of their perishable nature. In Himachal Pradesh, market surveys have revealed fourteen fungal species responsible for different diseases of tomato (Sharma, 1994) and thirteen fungi for eight types of rots of bell pepper (Sharma, 1998) at post-harvest stage. Little evidence is available about post-harvest losses in India, although some studies have reported 7-33 per cent loss of tomato due to fungal rots during storage and marketing (Thakur and Chenulu, 1970), entailing a colossal wastage of labour, time, land and money in terms of millions of rupees. No work has been done to determine the quantity of losses in local market trade. This paper reports the losses of bell pepper and tomato fruits caused by various storage rots during the year 1998-1999 in different markets of the state.

MATERIALS AND METHODS

Samples of tomato and bell pepper fruits were obtained from different vegetable markets in each of the nine districts of the state from July to June during 1998-1999 in case of tomato and July to December during 1998 in case of bell pepper. Each market was visited once in a month. From each market samples were brought from three randomly selected sellers. They were sorted into healthy and unwholesome groups, the former showed no external blemishes while the later contained fruits that showed symptoms of physical, physiological or pathological disorders. Physical disorders included sun scorch, mechanical damage etc. Physiological disorders included only those symptoms which were evident such as growth cracks and non-pathogenic blossom end rots. Pathological disorders included dry or soft rots from which fungal causal organisms were isolated. All the fruits of unwholesome group were counted and total loss (incidence) was calculated at each location (market). Average loss for each district was calculated for bell pepper and tomato fruits separately after adding the loss data of each market. Similarly average loss data were computed for each month in case of bell pepper and for each season in case of tomato.

RESULTS AND DISCUSSION

Incidence of post-harvest rots of tomato and bell pepper was recorded in every vegetable market of each district surveyed. The incidence varied from 5.0 to 26.1 per cent in case of tomato and 6.1 to 19.1 per cent in bell pepper. The average figures being 14.9 and 10.6 per cent, respectively. Maximum post-harvest loss of tomato (28.1%) and bell pepper (21.8%) was recorded in Paonta vegetable market in Sirmour district. Losses of tomato fruits were quite alarming in Dhaulakuan (27.4%), Una (24.1%), Theog (24.0%), Mehatpur (23.7%), Shimla (23.3%), Nahan (22.2%) and Solan (22.0%) vegetable markets. The bell pepper losses were comparatively less and almost similar in Nahan (20.2%), Shimla (19.3%), Dhaulakuan (16.5%), Solan (15.9%), Theog (15.7%), Una (17.4%), Kullu (15.0%) and Chambaghat (14.0%). Post-harvest losses of tomato and bell pepper were comparatively less in other vegetable markets surveyed but least in Berthin, Baijnath, Jawalamukhi, Joginder Nagar where the losses never exceeded single digit.

(a) Bell pepper

District-wise losses: Data on district-wise losses of bell pepper from July to December (Table 1) revealed that there was not much variation. However, maximum losses were recorded in Shimla (16.2%) and Sirmour (15.8%) districts. In rest of the seven districts, post-harvest losses were more or less the same varying from 9.1 to 11.8 per cent, with minimum losses being in Bilaspur district.

Table 1. Post-harvest losses of bell pepper in different districts of Himachal Pradesh during 1998

District	Disease incidence (%) during the months						Mean
	July	August	Sept.	Oct.	Nov.	Dec.	
Bilaspur	10.5	12.5	10.0	9.1	6.4	6.1	9.1
Hamirpur	12.5	14.5	12.5	11.1	8.5	7.5	11.1
Kangra	14.5	15.0	14.7	7.3	6.5	5.0	10.5
Kullu	12.9	17.9	13.5	10.8	7.9	6.6	11.6
Mandi	10.3	15.2	13.5	12.0	6.5	5.5	10.5
Solan	14.5	15.7	15.2	11.0	8.3	6.1	11.8
Shimla	20.4	22.0	20.1	16.1	10.5	8.0	16.2
Sirmour	17.9	25.9	18.9	13.7	9.9	9.1	15.8
Una	11.5	16.0	13.5	10.9	8.1	6.6	11.1
Mean	13.9	17.1	14.6	11.3	8.1	6.7	12.0

Month-wise losses: Survey for post-harvest rots of bell pepper in various markets was conducted from July to December, as this vegetable is available in markets only during these months. The losses were maximum (17.1%) in August as perhaps the post-harvest pathogens were most active during the warm and humid conditions prevalent during that period. Losses were almost equal in September (14.6%) and July (13.9%) months but slightly less in October (11.3%). With the onset of winters, the incidence decreased to half of the maximum in November (8.1%) and December (6.7%) months, perhaps due to low temperature and less humidity prevalent during this part of the year.

(b) Tomato

District-wise losses: The total post-harvest/market losses of tomato fruits in the nine districts of the state reveal (Table 2) that maximum losses occur in Sirmour (22.0%) and Shimla (21.8%) followed by Una (19.8%) and Hamirpur (17.2%) districts. With 16 per cent wastage, the losses of tomato fruits were similar in Kullu and Solan district and less in Mandi (13.7%), Bilaspur (12.3%) and Kangra (11.8%) districts.

Seasonal losses: Since there is a continuous and uninterrupted flow of tomatoes in the markets round the year, so the losses were also calculated on the basis of seasons in each district. The whole year was divided into four seasons, having three months per season starting from January to March and so on. The mean values (Table 2) indicated that there was significant variation of losses in summer and winter months. Seasonal losses were maximum (21.4%) during rainy season (July to September) and summer months of April to June (20.0%). Losses markedly decreased during October to December (13.6%) and January to March (11.9%) seasons.

Table 2. Post-harvest losses of tomato in different districts of Himachal Pradesh during 1998-99

District	Disease incidence (%) during the seasons				Mean
	July-Sept. 1998	Oct.-Dec. 1998	Jan.-March 1999	April-June 1999	
Bilaspur	15.9	9.0	8.9	15.4	12.3
Hamirpur	22.8	12.0	12.0	21.8	17.2
Kangra	14.2	10.6	10.1	12.1	11.8
Kullu	21.9	11.6	11.0	19.5	16.0
Mandi	16.9	11.5	10.6	15.8	13.7
Solan	19.9	13.0	13.2	17.9	16.0
Shimla	27.9	17.3	15.1	26.9	21.8
Sirmour	28.5	17.1	14.5	27.9	22.0
Una	24.5	20.5	11.5	22.7	19.8
Mean	21.4	13.6	11.9	20.0	16.7

In Himachal Pradesh, post-harvest losses of tomato and bell pepper up to 26 and 19 per cent, respectively is a severe financial burden to growers, middleman and retailers. Most losses are caused by preventable post-harvest factors such as rough handling, inadequate transportation and inappropriate storage methods (Snowdon, 1991). Mechanical injury can arise at any stage of these vegetable crops. To maintain good quality, mechanical injury should be avoided since it produces structural and physiological changes in tissues that facilitate entry of pathogens.

It is difficult to assess accurately, qualitatively and quantitatively, the proportion of tomato and bell pepper fruits lost in vegetable markets. The losses of tomatoes reported here confirm those from 19.3 per cent from Delhi by Thakur and Chenulu (1970) and are less than 50 per cent as reported by Sohi (1984) in Chandigarh vegetable markets. Those investigations, however, focussed only on losses due to pathogens. Some estimates observed in literature reflect variation in loss not only among different vegetable markets and among commodities but also between post-harvest stages and seasons of the year (Dasgupta and Mandal, 1989). A countrywide survey covering the major zones of India and at different seasons would reveal variation in the importance of different categories of losses encountered during the survey.

ACKNOWLEDGEMENT

The authors are grateful to Indian Council of Agricultural Research for providing funds for research on post-harvest diseases of bell pepper and tomato.

REFERENCES

Dasgupta, M.K. and Mandal, N.C. (1989). *Postharvest Pathology of Perishables*. Oxford and IBH Publ. Co. Pvt. Ltd., New Delhi, 623 p.

Sharma, R.L. (1994). Prevalence of post-harvest diseases of tomato in Himachal Pradesh. *Plant Dis. Res.* 9: 195-197.

Sharma, R.L. (1998). Prevalence of post-harvest diseases of bell pepper in Himachal Pradesh. *Plant Dis. Res.* 13: 193-194.

Snowdon, A.L. (1991). *A Colour Atlas of Post-harvest Diseases and Disorders of Fruits and Vegetables.* Vol. 2 *Vegetables.* Wolfe Scientific Ltd., London, 416 p.

Sohi, H.S. (1984). Present status of our knowledge of important fungal diseases of selected vegetables in India and future needs. *Indian J. Mycol. Plant Pathol.* 14: 1-23.

Thakur, D.P. and Chenulu, V.V. (1970). Survey of Delhi market for fungal diseases of fruits and vegetables. In: *Plant Disease Problems.* (Eds. S.P. Raychaudhuri *et al.*). IPS New Delhi, pp. 301-308.

Integrated Plant Disease Management (2005)
Eds. Sharma, R.C. and Sharma, J.N.
Scientific Publishers (India), Jodhpur, pp. 179-181.

29

Detection of seed-borne infection of bacterial wilt in capsicum

PREETI AGGARWAL[1] and A.K. SOOD[2]

[1]Hill Agriculture Research and Extension Centre,CSK HPKV, Dhaulakuan - 173 001
[2]Department of Plant Pathology, CSK HPKV, Palampur - 176 062 (H.P.)

ABSTRACT

Ralstonia solanacearum is primarily a soil-borne pathogen but there are some reports, which have indicated that pathogen may be carried through infected seed. The seeds were surface sterilized by dipping in 0.1 per cent $HgCl_2$ for 30 seconds followed by three washings in sterilized distilled water. Fifteen seeds of each variety were placed in a plate of tetrazolium chloride medium. Another set of 25 surface sterilized seeds of each variety was crushed in 1 ml sterile distilled water individually and 0.1 ml of the extract was spread on TZC agar plates separately. The plates were incubated at $28\pm1^{0}C$ for 48 h and observed for typical virulent colonies/ growth of *R. solanacearum* on TZC medium. The seeds plated from wilted plants showed 100 per cent infection whereas those from apparently healthy plants collected from resistant varieties were either found free from infection or seed infection ranged between 20.0 to 66.7 per cent. On spreading of 0.1 ml extract, no growth of bacterium was observed in resistant cv. CE 469905, whereas whole Petri plates were full of typical bacterial growth and individual colonies could not be counted in case of seed from the wilted plants. In other cultivars, the number of colonies varied between 13-116. This study reveals the seed-borne infection of *R. solanacearum*. Secondly, the seeds of varieties showing resistant reaction to bacterial wilt under field conditions may carry the latent infection of the bacterium.

INTRODUCTION

Bacterial wilt incited by *Ralstonia solanacearum* has been reported as a serious disease of many plants in India. The occurrence of bacterial wilt of chilli in India was first reported by Khan *et al.* (1979) from Karnataka state. Bacterial wilt is now one of the most important constraints for chilli production in various parts of the country. The typical wilt symptoms along with vascular discolouration can be observed just after two to three weeks after transplanting when there is high soil moisture and temperature. Copious bacterial exudation can be noticed from infected plant parts on ooze test. The bacterium is primarily a soil-borne pathogen but there are some reports, which have indicated that pathogen, may be carried through infected seed (Serbakoff, 1917; Weber and Ramsey, 1926; Van Haltern, 1935). Singh (1994) has reported seed transmission of bacteria in tomato and brinjal. The present investigation was thus planned with the objective to detect the presence of bacterium in the seeds of capsicum.

MATERIALS AND METHODS

Seed samples of eight varieties of capsicum were collected from healthy and wilted plants from farmers' fields and experimental fields of the Department of Plant Pathology, CSK HPKV, Palampur. The seeds were surface sterilized by dipping in 0.1 per cent $HgCl_2$ for 30 seconds followed by three washings in sterilized distilled water. Fifteen seeds of each variety were picked up with sterilized forceps and placed in a Petri plate of tetrazolium chloride medium. Another set of twenty five surface sterilized seeds of each variety was crushed in 1 ml sterilized distilled water individually and 0.1 ml of this extract was spread on TZC agar plates separately. The plates were incubated at $28 \pm 1^{0}C$ for 48 h and observed for typical virulent colonies/growth of *R. solanacearum* on TZC medium. Gram reaction of bacteria from each variety was also studied.

RESULTS AND DISCUSSION

The seeds plated from wilted plants showed 100 per cent infection whereas those from apparently healthy plants collected from resistant varieties were either found free (EC 469905) from infection or seed infection ranged between 20.0 to 66.7 per cent (EC 469898, EC 469899, EC469900, S-2530). On spreading of 0.1 ml extract, no growth of bacterium was observed in resistant cultivar EC 469905 whereas whole Petri plate was full of typical bacterial growth and individual colonies could not be counted in case of seed from wilted plants viz., Jhiri (Chilli), Jhiri (Bell Pepper) and Shapur (Chilli). In other cultivars EC 469898, EC 469899 and EC 469900 the number of colonies ranged between 13-116. Gram reaction of bacterial isolates from each variety revealed the presence of plant pathogenic Gram negative rods, which confirms the bacterium associated to be *Ralstonia solanacearum*. The study reveals the seed-borne infection of the pathogen, which corroborate with the findings of Singh (1994) who reported seed-borne infection of bacterial wilt in tomato and brinjal. The results are also in accordance with Shakya (1993) who confirmed the seed-borne nature of the pathogen. Secondly, the seeds of varieties showing resistant reaction to bacterial wilt under field conditions may carry the latent infection of the bacterium. This might be one of the reasons of breakdown of resistance in Surajmukhi cultivar of chilli (Sood and Aggarwal, 2004) in Himachal Pradesh.

REFERENCES

Khan, A.N.A., Shetty, K.S. and Patil, R.B. (1979). Occurrence of bacterial wilt in chilli in Karnataka and its relationship to the wilts of other solanaceous crops. *Indian Phytopathol.* 32: 507-512.

Serbakoff, C.D. (1917). some important diseases of truck crops in Florida. *Fla. Agric. Exp. Stn. Bull.* 139: 193-277.

Shakya, D.D. (1993). Seed transmission of *Pseudomonas solanacearum* in tomato. In: *Bacterial* Wilt (Eds. Hartman, G.L. and Hayward, A.C.). Proceedings of an International Conference held at Kaohsiung, Taiwan, October 28-31, 1992. ACIAR Proceedings No. 45, pp. 371-372.

Singh, Rupali. (1994). Seed transmission studies with *Pseudomonas solanacearum* in tomato and eggplant. *ACIAR Bacterial Wilt Newsletter*. 11: 12-13.

Sood A.K. and Aggarwal, Preeti. (2004). Breakdown of resistance in Surajmukhi cultivar of chilli in Himachal Pradesh. *ACIAR Bacterial Wilt Newsletter* (Communicated).

Van Haltern, F. 1935. Control of tomato seed bed diseases of southern states. *Ga. Agric. Exp. Stn. Bull.* 187: 39.

Weber, G.F. and Ramsey, G.B. (1926). Tomato diseases in Florida. *Fla. Agric. Exp. Stn. Bull.* 185: 138.

Integrated Plant Disease Management (2005)
Eds. Sharma, R.C. and Sharma, J.N.
Scientific Publishers (India), Jodhpur, pp. 183-186.

30

Diseases of bell pepper under protected cultivation conditions and their management

S.K. GUPTA, R. BHARDWAJ[1], M.K. SHARMA[1] and Y.C. GUPTA[2]

Department of Mycology and Plant Pathology,
[1]Department of Vegetable Crops,
[2]Department of Floriculture and Landscaping,
Dr. Y.S. Parmar University of Horticulture and Forestry, Nauni, Solan-173 230 (H.P.)

ABSTRACT

Bell pepper (*Capsicum annuum* L.) is an important vegetable crop, which is also grown during off-season under protected cultivation. During the cultivation the crop was affected by powdery mildew (*Leveillula taurica* (Lev.) Arnaud, Fusarium wilt (*Fusarium oxysporum* Schlecht. ex Fr. f.sp. *capsici*) and Cercospora leaf spot (*Cercospora capsici* Heald & Wolf). Among these diseases powdery mildew was most severe with 84.17 per cent disease severity while the incidence of Fusarium wilt and Cercospora leaf spot was 8.33 and 2.0 per cent, respectively. In the management of Fusarium wilt and powdery mildew, soil application of *T. harzianum* multiplied on neem cake (500g/m^2) and hexaconazole (0.05%) sprays, respectively, were found highly effective in reducing these diseases.

INTRODUCTION

Bell pepper (*Capsicum annuum* L.) is an important vegetable crop also grown under protected cultivation conditions throughout the world. Recently, the protected cultivation of this crop has been initiated in Himachal Pradesh.

During cultivation, the crop is affected by several diseases, which reduced the fruit yields substantially. Studies were, therefore, conducted to record the incidence/severity of various diseases and their management through integrated practices including chemical and biological agents.

MATERIALS AND METHODS

Bell pepper hybrid Indira was planted in polyhouse for two consecutive crop seasons i.e. February 2003 and July 2003. The data on the incidence/ severity of various diseases were recorded.

A trial for the management of Fusarium wilt and powdery mildew was also laid out with eight treatments including control in completely randomized block design in the polyhouse during July 2003 and each treatment was replicated thrice. The treatments were drenching planting bed with copper oxychloride (0.4%) followed by need based sprays of carbendazim (0.1%)/ mancozeb (0.25%); drenching with copper oxychloride (0.4%) followed by sprays with neem oil (1.5%); drenching with copper oxychloride (0.4%) followed by sprays with neem seed kernel extract (4%); drenching with copper oxychloride (0.4%) followed by sprays with hexaconazole (Contaf 0.05%); application of *Trichoderma viride* multiplied on neem cake (500g/m^2) followed by spray with *T. viride*; application of *T. harzianum* multiplied on neem cake (500g/m^2) followed by spray with *T. harzianum* and drenching planting bed with copper oxychloride (0.4%) followed by sprays of *Pseudomonas fluorescens*. Drenching with copper oxychloride was done before transplanting. In all four sprays of fungicides, bioagents and botanical pesticides were applied at 10 days interval. The data on diseases severity were recorded on 0-4 scale and per cent disease severity was calculated (McKinney, 1923).

Mass culture of *Trichoderma* spp.

Trichoderma spp. for soil applications were grown on moistened neem cake containing 20 per cent moisture. The polypropylene bags containing neem cake were inoculated with mycelial discs of *Trichoderma viride* and *T. harzianum* taken from actively growing 3 days old culture. The cultures were incubated at 25 $\pm$ 1° C for 14 days.

For spray application of *Trichoderma viride* and *T. harzianum*, cultures were raised in potato dextrose broth and flasks were incubated at 25$\pm$1°C for 15 days. After 15 days, 10 ml broth per litre of water was used for spray.

Mass culture of *Pseudomonas fluorescens*

For spray application of *P. fluorescens*, nutrient broth was prepared. Flasks containing nutrient broth were inoculated with *P. fluorescens* culture. The inoculated flasks were incubated at 25$\pm$1°C for 5 days. After 5 days, 10 ml broth per litre of water was used for spray.

RESULTS AND DISCUSSION

In February 2003 planted crop, only powdery mildew appeared in the month of April and continued till the final harvest of the crop. Samples were collected and the pathogen was identified as *Leveillula taurica* (Lev.) Arnaud while in July 2003 planted crop, powdery mildew and incidence of Fusarium wilt (*Fusarium oxysporum* Schlecht. ex Fr. f.sp. *capsici*) and Cercospora leaf spot (*Cercospora capsici* Heald & Wolf) diseases were recorded. Among these diseases, powdery mildew was most severe and the severity was 84.17 per cent followed by Fusarium wilt and Cercospora leaf spot and the incidence was 8.33 and 2.0 per cent, respectively.

Integrated disease management

The data (Table 1) revealed that soil application of *T. harzianum* multiplied on neem cake was found effective in reducing the incidence of Fusarium wilt to 1.66 per cent as compared to 8.33 per cent in control followed by soil application of *T. viride* while soil drenching with copper oxychloride followed by neem oil and *Pseudomonas fluorescens* sprays were least effective with 4.66 per cent incidence.

Table 1. Effect of integrated disease management practices on the diseases of bell pepper under protected cultivation

Treatment	Fusarium wilt incidence (%)	Powdery mildew severity (%)
D + need based sprays of carbendazim/mancozeb	3.66(1.91)^{c}	38.33 (38.24)^{b}
D+ - sprays with neem oil (1.5%)	4.66(2.15)^{e}	54.17 (57.40)^{c}
D+ - sprays with neem seed kernel extract (4%)	4.33(2.07)^{d}	50.00 (45.00)^{c}
D+ - sprays with hexaconazole (0.05%)	3.66(1.91)^{c}	14.17 (21.98)^{a}
B + -sprays of *T. viride*	2.33(1.52)^{b}	64 17 (55.25)^{c}
C + -sprays of *T. harzianum*	1.66(1.27)^{a}	63.33 (52.78)^{c}
D+- sprays of *Pseudomonas fluorescens*	4.66(2.15)^{e}	81.67 (64.81)^{d}
Control	8.33 (2.88)^{f}	84.17 (66.65)^{d}
$CD_{0.05}$	0.24	9.00

Figures in parentheses for powdery mildew are arcsine transformed values

Figures in parentheses for Fusarium wilt are square root transformation values

D = Drenching with copper oxychloride (0.4%)

B = Soil application of *T. viride* (500g/m^2)

C = Soil application of *T. harzianum* (500g/m^2)

In case of powdery mildew, drenching of planting beds with copper oxychloride (0.4%) followed by sprays with hexaconazole (0.05%), was significantly superior and reduced the disease severity to 14.17 per cent followed by drenching the planting beds with copper oxychloride (0.4%) followed by alternate sprays of mancozeb/carbendazim resulting in 38.33 per cent disease severity. However, drenching planting beds with copper oxychloride (0.4%) followed by sprays of *Pseudomonas fluorescens* was least effective with 81.67 per cent disease severity.

Present studies clearly indicated the efficacy of soil application of *Trichoderma harzianum* and *T. viride* multiplied on neem cake in the management of Fusarium wilt. Yan *et al.* (1999) also advocated the use of

T. harzianum for the control of *F. oxysporum* on *Capsicum annuum*. The decomposition of organic matter like neem cake helps in alteration of the physical, chemical and biological conditions of the soil thus reducing the inoculum potential of soil-borne diseases. Neem cake can also be mixed with other bio-products like species of *Trichoderma*, *Gliocladium* and *Bacillus* to have an effective control of various diseases (Dohroo and Gupta, 1995). However, there is no report in literature about the growing of *Trichoderma* spp. on neem cake and its efficacy in the management of wilt in *C. annuum* and cannot be compared with. Powdery mildew was best contained by hexaconazole sprays which is in agreement with Negron *et al.* (1991) and Desai *et al.* (2002) who reported the efficacy of various triazoles like penconazole, hexaconazole and triadimefon against this disease. However, other treatments including bioagents were not found effective against powdery mildew under protected conditions.

ACKNOWLEDGEMENTS

Authors are grateful to National Agricultural Technology Project for financial assistance in the form of Mission Mode Project on Protected Cultivation of Vegetables and Flowers in Plains and Hills.

REFERENCES

Desai, S.A., Nagaraj, M.S. and Naik, K.S. (2002). A note on penconazole – A new triazole molecule in the control of powdery mildew of chilli and grapevine. *Karnataka J. Agric. Sci.* 15: 386.

Dohroo, N.P. and Gupta, S.K. (1995). Neem in plant disease control. *Agric. Rev.* 16: 133-140.

McKinney, H.H. (1923). Influence of soil temperature and moisture on infection of wheat seedlings by *Helminthosporium sativum*. *J. Agric. Res.* 26: 195-217.

Negron, J.A, Rodrigues, R. and Aviles, L. (1991). Fungicide evaluation for control of powdery mildew, *Leveillula taurica* (Lev.) Arn. of peppers in Puerto Rico. *J. Agric. Univ. Puerto Rico* 75: 119-123.

Yan, S.H., Lu, D.Q. and Yang, Y.H. (1999). Control of hot pepper wilt with *Trichoderma harzianum* rice bran. *Chinese J. Biol. Cont.* 15: 97.

Integrated Plant Disease Management (2005)
Eds. Sharma, R.C. and Sharma, J.N.
Scientific Publishers (India), Jodhpur, pp. 187-194.

31

Integrated management of cucurbit mosaic

RAVINDER KAUR and S.S. KANG

Department of Plant Pathology, P.A.U. Ludhiana – 141 004 (Punjab)

ABSTRACT

The effect of cultural practices viz., date of sowing, depth of sowing and barrier crops (sunflower, *gobhi sarson* and *raya*) alone and in combination with sprays of two insecticides namely Malathion 50 EC and Thiodan 35 EC either on barrier crop or on main crop was studied for the management of cucurbit mosaic on summer squash and muskmelon. All the treatments reduced the incidence of mosaic in both the crops. Crop sown in the second fortnight of March registered least disease incidence (52.1 and 54.0%) and severity (54.3 and 57.9%) in summer squash and muskmelon, respectively, without reduction in yield. Further delay in sowing up to end April reduced the disease incidence but yield was also reduced. Increase in sowing depth up to 9.00 cm reduced the disease incidence by 23.34 per cent in summer squash and 24.61per cent in muskmelon but adversely affected the germination as well as crop stand. The disease was also lowest in these crops when sunflower was grown as barrier crop. However, sowing of sunflower as barrier crop and spray of thiodan on main crop (cucurbit) further reduced the disease incidence (37.24%) and severity (40.34%). The sowing of *gobhi sarson* and *raya* alone or the sprays of insecticides on these crops had no positive effect in disease reduction. However, the sprays of insecticides on main crop reduced the disease incidence and severity even when *gobhi sarson* and *raya* were sown as barrier.

INTRODUCTION

Cucurbits are one of the most important groups of summer vegetables. These vegetables are prone to number of bacterial, fungal and viral pathogens but the viruses (cucumber mosaic virus, watermelon mosaic virus, bottle gourd mosaic virus and their strains) are the main pathogens known to cause considerable losses in yield and fruit quality of cucurbitaceous vegetables. They cause systemic infection, therefore, very difficult to control by means available for the management of fungal and bacterial diseases. So, the present investigations have been carried out for the management of the mosaic (caused by watermelon mosaic virus and potato virus Y) in cucurbits by the integration of various control methods.

MATERIALS AND METHODS

All the experiments were conducted using summer squash var. Pb *Chappan kadoo*-1 and muskmelon var. Pb *Sunehri.*

Date of sowing

The seeds of both the crops were sown on eleven different dates starting from end January to end June at 15 days interval. The observations were recorded for disease incidence and disease severity.

Sowing depth

This experiment was conducted in polythene bags by sowing seeds of both the crops at different depths (1.5, 3.0, 4.5, 6.0, 7.5 and 9.0 cm). The observations were recorded for seed germination, disease incidence, disease severity and aphid population.

Barrier crops

Three barrier crops viz., sunflower var. SH-3322, *raya* var. RLM-619 and *gobhi sarson* var. GSL-1 were sown in three rows around the main crops of summer squash and muskmelon one month before the sowing of main crops. The experiment was laid out in randomized block design (RBD) with the three replications. The observations on disease incidence, disease severity and aphid population were recorded.

Barrier crops sprayed with insecticides

Barrier crops were sown around main crops as described earlier and were sprayed with two insecticides viz., Malathion 50 EC and Thiodan 35 EC @ 300ml/acre in 150 litres water at 15 day interval starting at disease initiation after one month of seed germination. Total three sprays of insecticides were given and subsequently observations were recorded on disease incidence, disease severity and aphid population.

Barrier crops and insecticides sprayed on main crop

In this experiment, main crops of summer squash and muskmelon were sprayed with Malathion 50 EC and Thiodan 35 EC @ 300ml/acre in 150 litres water at 15 day interval starting at disease initiation. Observations on disease incidence, disease severity and aphid population were recorded.

RESULTS AND DISCUSSION

Effect of date of sowing

It is evident from Fig. 1 and 2 that late sowing reduced the disease incidence and severity. The crop sown in second fortnight of March, registered less disease incidence without reduction in yield as compared to normal sowing in February. Further delay in sowing reduced the disease incidence but yield was also reduced. The disease incidence was positively correlated with aphid population. Aphid population was maximum in February and March resulting in maximum disease incidence in these months due to secondary spread through vectors. In summer squash maximum disease incidence (89.3%) was recorded in crop sown on March 2 followed by 79.3 per cent in February 15 sown crop. Late sowing of the crop resulted into lower disease incidence where per cent incidence was 52.1, 32.9, 28.4 and 25.7 in March 17, April 1, April 16 and May 1 sown crop, respectively. However the germination of the crop was significantly reduced after first April.

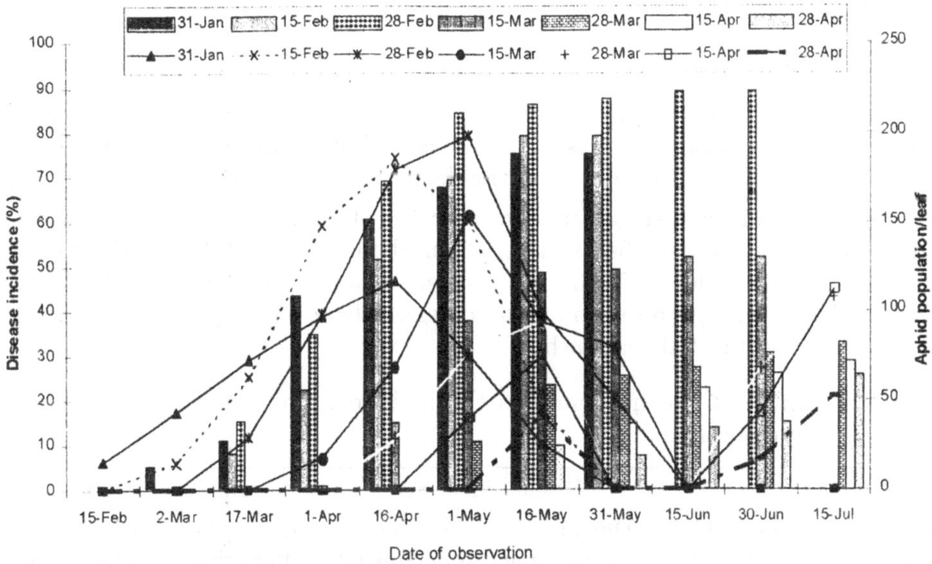

Fig. 1. Effect of date of sowing on mosaic of summer squash

Similar trend in disease incidence and aphid population buildup was observed in muskmelon crop. Maximum disease incidence (83.3%) was recorded in March 2 sown crop followed by 81.9 per cent in February 15 sown crop (Fig. 2). Least disease incidence (34.2%) was recorded in May 1 sown but with reduction in germination percentage. This decrease in disease incidence and severity in the late sown crop was due to decrease in aphid population in the later months with increase in temperature and decrease in relative humidity.

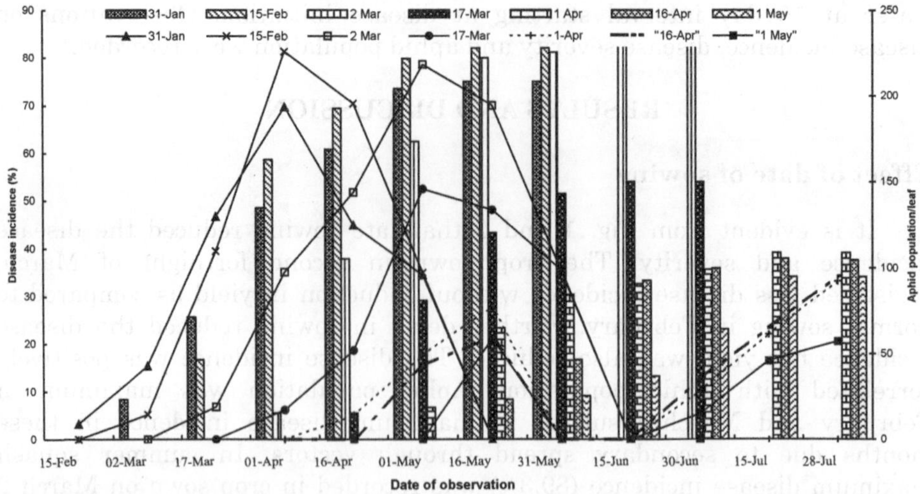

Fig. 2. Effect of date of sowing on incidence of muskmelon mosaic

A similar observation on the effect of date of sowing on the mosaic disease in cucurbits was also reported by Singh (1984) and Singh *et al.* (1999).

Sowing depth

Disease incidence was found to be negatively correlated with depth of sowing. Germination of seeds also reduced significantly at deeper depths, fewer plants become established at the deeper seedling depths.

In summer squash after two months of sowing, maximum disease incidence (63.83%) was observed at 1.5 cm depth (Table 1) followed by 63.09 per cent at 3.0 cm depth. Minimum disease incidence of 48.93 per cent was at 9.0 cm depth. The germination also decreased from 96.02 per cent at 1.5 cm to 68.13 per cent at 9.0 cm depth.

Table 1. Effect of sowing depth on the incidence and severity of mosaic in summer squash

Depth (cm)	Per cent germination	Disease incidence		Disease severity	
		One month after sowing	Two months after sowing	One month after sowing	Two months after sowing
1.5	96.02 (9.84)	4.47 (2.33)	63.83 (8.05)	5.32 (2.51)	65.89 (8.18)
3.0	96.04 (9.85)	4.57 (2.36)	63.09 (8.00)	5.26 (2.50)	64.96 (8.12)
4.5	80.98 (9.05)	3.98 (2.23)	58.14 (7.69)	4.93 (2.43)	60.28 (7.83)
6.0	74.27 (8.67)	3.54 (2.13)	53.92 (7.41)	4.80 (2.41)	57.41 (7.64)
7.5	72.75 (8.59)	1.79 (1.67)	50.76 (7.19)	3.38 (2.09)	54.82 (7.47)
9.0	68.13 (8.31)	1.68 (1.64)	48.93 (7.06)	3.23 (2.05)	53.19 (7.36)
$CD_{0.05}$	0.62	0.25	0.17	0.18	0.21

Figures in parentheses are the square root transformed values

Similarly, in muskmelon (Table 2), sowing at 1.5 cm depth exhibited maximum disease incidence of 67.69 per cent after two months of sowing and decreased to 51.03 per cent at 9.0 cm depth. Germination of seeds decreased from 97.94 per cent at 1.5 cm to 70.38 per cent at 9.0 cm depth.

Table 2. Effect of sowing depth on the incidence and severity of mosaic on muskmelon

Depth (cm)	Per cent germination	Disease incidence		Disease severity	
		One month after sowing	Two months after sowing	One month after sowing	Two months after sowing
1.5	97.94 (9.95)	5.53 (2.55)	67.69 (8.29)	6.48 (2.73)	70.04 (8.43)
3.0	97.05 (9.90)	5.55 (2.56)	65.18 (8.13)	5.80 (2.61)	69.95 (8.42)
4.5	84.90 (9.27)	4.70 (2.39)	64.22b (8.07)	5.41(2.53)	67.47 (8.27)
6.0	79.33 (8.96)	3.71 (2.17)	59.34 (7.77)	4.88 (2.42)	63.18 (8.01)
7.5	76.15 (8.78)	1.97 (1.72)	54.47 (7.45)	3.77 (2.18)	60.53 (7.84)
9.0	70.38 (8.45)	1.79 (1.67)	51.03 (7.21)	3.69 (2.16)	58.29 (7.70)
$CD_{0.05}$	0.44	0.25	0.66	0.13	0.11

Figures in parentheses are the square root transformed values

Jones and Proudlove (1991) reported the lower incidence of CMV on narrow - leafed lupin at 8 and 11 cm depths as compared to sowing at 5 cm depth due to inactivation of virus while travelling from deeper depths. But during present studies, most of the infected seed did not germinate and hence reduced the source of primary inoculum.

Effect of barrier crops and insecticides

Data presented in Table 3 indicate the effect of different treatments with barrier crops on aphid population, disease incidence and severity of mosaic on summer squash. Among the three-barrier crops, minimum disease incidence and severity was 55.00 and 59.10 per cent, respectively, when sunflower was grown around the main crop, which was significantly lower than the control. However, sowing of *gobhi sarson* and *raya* alone had no effect on the disease incidence and severity of summer squash mosaic and it was at par with the control.

Observation on aphid population showed that the aphid-landing rate was similar in protected and unprotected plots. This indicates that the disease control occurred as a result of decrease in the infectivity of vectors that landed on the barrier crop before reaching the main crop.

Spraying of insecticides on *gobhi sarson* and *raya* as barrier crops had no significant effect on disease incidence and severity as compared to control and barrier crop alone. Minimum disease incidence and severity, 78.12 and 80.96 per cent recorded in *raya* sprayed with thiodan was at par with control that registered the disease incidence and severity of 83.32 and 86.80 per cent, respectively. Although the insecticides spray on sunflower significantly reduced the disease incidence and severity as compared to control, but this difference was non-significant within the treatments.

Table 3. Effect of barrier crops and insecticides spray on aphid population, disease incidence and severity of summer squash mosaic

Treatment	Aphid population /leaf	Disease incidence (%)	Disease severity (%)
Sunflower alone*	104.17 (10.25)	55.00 (7.48)	59.10 (7.75)
Sunflower + Malathion	83.08 (9.17)	54.93 (7.48)	57.26 (7.63)
Sunflower + Thiodan	80.10 (9.00)	54.20 (7.43)	56.46 (7.58)
Sunflower + Malathion on main crop	74.92 (8.71)	53.33 (7.37)	52.12 (7.29)
Sunflower + Thiodan on main crop	70.52 (8.46)	52.29 (7.29)	51.78 (7.26)
Gobhi sarson alone*	112.30(10.64)	80.20 (9.01)	83.48 (9.19)
Gobhi sarson + Malathion	91.78 (9.63)	78.54 (8.92)	80.07 (9.00)
Gobhi sarson + Thiodan	87.89 (9.43)	74.27 (8.67)	77.39 (8.85)
Gobhi sarson+ Malathion on main crop	85.07 (9.28)	72.72 (8.59)	74.59 (8.69)
Gobhi sarson + Thiodan on main crop	79.45 (8.97)	68.32 (8.32)	66.27 (8.20)
Raya alone*	114.32(10.74)	83.00 (9.16)	86.39 (9.35)
Raya + Malathion	93.44 (9.72)	80.48 (9.03)	82.77 (9.15)
Raya + Thiodan	89.59 (9.52)	78.12 (8.89)	80.96 (9.05)
Raya + Malathion on main crop	84.18 (9.23)	72.44 (8.57)	73.59 (8.63)
Raya + Thiodan on main crop	82.97 (9.16)	68.16 (8.32)	69.91 (8.42)
Control **	112.37(10.65)	83.32(9.18)	86.80 (9.37)
$CD_{0.05}$	0.46	0.52	0.53

Figures in parentheses are the square root transformed values

* - Barrier crop without the spray of insecticide

** - Main crop without barrier crop

Insecticidal sprays of malathion and thiodan on main crop reduced the disease incidence, disease severity and aphid population as compared to control in all the treatments or combinations. Minimum disease incidence (52.29%) and severity (51.78%) was recorded in sunflower and thiodan on main crop and maximum 72.72 and 74.59 per cent in *gobhi sarson* + malathion on main crop (Table 3). The differences when compared with barrier crop alone or barrier crop + spray on barrier crop were significant in *gobhi sarson* and *raya* whereas these were non- significant in sunflower. The two insecticides namely, malathion and thiodan did not differ in their efficacy in all the treatments. However, thiodan was slightly better than malathion.

Similar observations were recorded on the effect of barrier crop alone or in combination with insecticides on the incidence and severity of muskmelon mosaic (Table 4).

Table 4. Effect of barrier crops and insecticides spray on aphid population, disease incidence and severity of muskmelon mosaic

Treatment	Aphid population/ leaf	Disease incidence (%)	Disease severity (%)
Sunflower alone*	106.38 (10.87)	57.84 (7.67)	58.11 (7.64)
Sunflower + Malathion	89.81 (9.53)	57.19 (7.63)	58.11 (7.69)
Sunflower + Thiodan	83.08 (9.17)	55.77 (7.53)	56.84 (7.60)
Sunflower + Malathion on main crop	76.96 (8.83)	53.94 (7.41)	53.50 (7.38)
Sunflower + Thiodan on main crop	70.83 (8.47)	52.73 (7.33)	53.50 (7.38)
Gobhi sarson alone*	120.88(11.04)	87.42 (9.40)	93.18 (9.70)
Gobhi sarson + Malathion	105.75(10.33)	84.18 (9.23)	89.45 (9.51)
Gobhi sarson + Thiodan	96.13 (9.85)	80.67 (9.03)	85.45 (9.30)
Gobhi sarson + Malathion on main crop	90.61 (9.57)	79.59 (8.98)	80.19 (9.01)
Gobhi sarson + Thiodan on main crop	89.14 (9.49)	75.81 (8.76)	78.19 (8.90)
Raya alone*	117.32(10.88)	88.96 (9.48)	91.87 (9.64)
Raya + Malathion	108.15(10.45)	85.13 (9.28)	89.87 (9.53)
Raya + Thiodan	97.88 (9.94)	82.87 (9.16)	85.87 (9.32)
Raya + Malathion on main crop	93.78 (9.73)	78.73 (8.93)	79.83 (8.99)
Raya + Thiodan on main crop	94.64 (9.78)	76.44 (8.80)	77.91 (8.88)
Control**	117.73(10.89)	89.03 (9.49)	94.62 (9.78)
$CD_{0.05}$	0.56	0.68	0.54

Figures in parentheses are the square root transformed values

* Barrier crop without the spray of insecticide

** Main crop without barrier crop

Treatments with sunflower as barrier crop were significantly better than other treatments and control. In sunflower, minimum disease incidence (52.73%) and severity (53.50%) was recorded in sunflower + thiodan on main crop and maximum 57.84 and 58.11 per cent in sunflower alone. Effect of barrier crops and insecticides on the virus diseases has been studied by Deol and Rataul (1978), Cheema *et al.* (1993) and Collar *et al.* (1997).

REFERENCES

Cheema, S.S., Kapila, S. and Kumar, A. (1993). Efficacy of various bio-products and chemicals against tobacco mosaic virus in tomato and cucumber mosaic virus in bottle gourd. *Plant Dis. Res.* 8: 110-114.

Collar, J.L., Avilla, C., Dugue, M. and Fereres, A. (1997). Behavioural response and virus vector ability of *Myzus persicae* (Homoptera: Aphididae) probing on pepper plants treated with aphicides. *J. Econ. Entomol.* 90: 1628-1634.

Deol, G.S. and Rataul, H.S. (1978). Role of various barrier crops in reducing the incidence of cucumber mosaic virus in chilli, (*Capsicum annuum* L). *Indian J. Entomol.* 40: 261-264.

Jones, R.A.C. and Proudlove, W. (1991). Further studies on cucumber mosaic virus infection of narrow leafed lupin (*Lupinus augustifolius*): Seed borne infection, aphid transmission, spread and effects on grain yield. *Ann. Appl. Biol.* 118: 319-329.

Singh, P. (1984). *Epidemiology of Virus Disease of Vegetable Marrow* (*Cucurbita pepo* L.) *in Punjab.* M.Sc. Thesis, Punjab Agricultural University, Ludhiana, India.

Singh, R.P., Mohan, J. and Singh, D.P. (1999). Assessment of losses and management of ridge gourd mosaic virus. *Plant Dis. Res.* 14: 134-138.

Integrated Plant Disease Management (2005)
Eds. Sharma, R.C. and Sharma, J.N.
Scientific Publishers (India), Jodhpur, pp. 195-198.

32

Management of radish mosaic disease in Himachal Pradesh

USHA SHARMA, S.C CHOWFLA, P.D. THAKUR, ANIL HANDA and BHUPESH GUPTA

Department of Mycology and Plant Pathology, Dr.Y.S. Parmar University of Horticulture and Forestry, Nauni, Solan-173 230 (H.P.)

ABSTRACT

Management of radish (*Raphanus sativus* L.) mosaic disease caused by virus(es) belonging to *potyvirus* group in Solan area of Himachal Pradesh was attempted by planting barrier crops alone or in combination with insecticide (Rogor). Wheat along with Rogor was found to be more effective in reducing the incidence and increasing the yield. Rogor was also found effective in reducing the disease incidence.

INTRODUCTION

Radish is one of the important vegetables cultivated for its enlarged edible roots. However, its successful cultivation has been hampered due to infection with viruses causing a serious mosaic disease resulting in reduced yield. Based on host range, particle morphology and serological relationships; these isolates were identified as turnip mosaic *potyvirus*. An attempt was made to save the crop from the ravages of mosaic disease by planting barrier crops alone and in combination with insecticide (Rogor) and spraying or applying insecticides and plant extract. These results are described in this article.

MATERIALS AND METHODS

Barrier crops alone and in combination with insecticide sprays

A field trial in randomized block design (RBD) with replication in a plot size of 2.5 x 2.5 m with main crop and outer 0.5 m with trap crop, each plot accommodating four rows with plant to plant and row to row distance of 30 x 60 cm, respectively laid out in cropping seasons of 1998-99 and 1999- 2000. Wheat, oat and barley were used as barrier crops alone and in combination with insecticide (0.03% Rogor 30 EC).

Insecticide sprays

To study the effect of insecticides and plant extracts, a field trial was laid out during 1998-1999 and 1999-2000 in randomized block design using five Insecticides (0.05% Malathion, 0.01% Monocrotophos, 0.0025% Karate, 0.03% Rogor, 0.025% Metasystox) and an organic compound (0.3% Virex-H). The plot size was kept 2 x 2 m and plant to plant and row to row distance was 30 x 60 cm, respectively. The insecticidal sprays were done at 10 days interval.

RESULTS AND DISCUSSION

The effect of barrier crops alone and in combination with insecticide in reducing disease incidence indicated encouraging results. It was observed (Table 1) that all the three crops alone and in combination with insecticides were significantly effective in reducing the incidence of disease and increasing the yield as compared to control. Wheat along with Rogor sprays was found to be more effective than other treatments in both the cropping seasons. There were no significant differences in disease incidence amongst all other treatments. Similarly, the yield was also significantly higher in the treatment where wheat was grown as barrier crop in combination with insecticidal spray. Similarly, other workers have also reported that the planting of barrier crops reduced the incidence of virus disease (Danju, 1993; Yang *et al.*, 1998).

Table 1. Effect of planting barrier crops alone and in combination with insecticide on mosaic disease incidence and yield of radish

Treatment	Disease incidence (%)		Average yield (kg/plot)		Per cent increase in yield over control		Average yield (q/ha)	
	1998- 99	1999-2000	1998- 99	1999-2000	1998- 99	1999-2000	1998- 99	1999-2000
Wheat	3.12 (1.69)	3.40 (1.84)	5.86	5.70	11.77	2.11	146.50	142.50
Oat	6.26 (2.48)	7.52 (2.73)	5.45	6.33	5.14	11.94	136.25	158.33
Barley	1.32 (1.12)	5.39 (2.32)	5.48	6.28	5.65	9.39	137.00	157.08

Continue....

Treatment	Disease incidence (%)		Average yield (kg/plot)		Per cent increase in yield over control		Average yield (q/ha)	
	1998- 99	1999-2000	1998- 99	1999-2000	1998- 99	1999-2000	1998- 99	1999-2000
Wheat + Rogor	0.22 (0.41)	2.33 (1.49)	7.33	7.52	29.44	34.79	183.18	187.93
Oat + Rogor	0.33 (0.46)	3.39 (1.84)	6.44	6.33	19.72	11.94	161.00	158.33
Barley + Rogor	0.22 (0.46)	2.63 (1.56)	6.37	6.25	18.84	10.73	159.25	156.18
Control	21.76 (4.66)	23.73 4.87)	5.17	5.58	-	-	129.25	139.43
$CD_{0.05}$	(0.6067)	(0.5735)	(0.9652)	(2.4750)				

Figures in parentheses are arcsine transformed values

Effect of insecticidal sprays also proved effective in increasing the yield of crop. It was found (Table 2) that among all insecticides and plant extract, Rogor 30EC (0.05%) was most effective in reducing disease incidence. It was followed by Metasystox, Karate, Monocrotophos and Malathion. However, Virex- H was found to be the least effective.

Table 2. Effect of botanical extract and insecticides on mosaic disease incidence and yield of radish

Treatment	Disease incidence (%)		Average yield (kg/plot)		Per cent increase in yield over control		Average yield (q/ha)	
	1998-99	1999-2000	1998-99	1999-2000	1998-99	1999-2000	1998-99	1999-2000
Malathion (0.05%)	1.49 (1.20)	4.26 (2.05)	6.34	7.45	18.94	23.96	158.425	186.33
Monocrotophos (0.01%)	4.38 (2.09)	7.49 (2.73)	8.24	6.31	37.68	10.23	206.075	157.83
Karate (0.0025%)	5.32 (2.26)	12.29 (3.50)	6.66	6.22	22.83	8.85	166.425	155.43
Rogor (0.03%)	2.30 (1.27)	1.68 (1.21)	7.12	7.22	27.85	21.52	178.00	180.53
Metasystox (0.025%)	3.15 (1.46)	2.29 (1.27)	7.16	5.71	28.28	0.89	179.08	142.83
Virex-H (0.3%)	8.51 (2.91)	12.56 (3.54)	6.26	4.31	21.92	-31.39	156.58	107.83
Control	21.36 (4.62)	23.52 (4.85)	5.14	5.67	-	-	128.43	141.68
$CD_{0.05}$	(1.26)	(0.89)	(4.08)	(4.53)				

Figures in parentheses are arcsine transformed values

Szalay- Marzso and Solymosy (1962) were able to reduce infection of CMV in *Capsicum annuum* in Hungary by 33 per cent after spraying Metasystox at 10 days interval. Rathore and Agnihotri (1985) from Jobner, Rajasthan (India) reported 39 per cent reduction in incidence of yellow mosaic by spraying crop with Rogor (0.02%) at 15 days interval.

REFERENCES

Dhanju, K.S. (1993). *Epidemiology and Management of Virus Disease(s) of French Bean (Phaseolus vulgaris* L.). Ph.D. Thesis, Dr. Y.S. Parmar University of Horticulture and Forestry, Nauni, Solan (HP), 111 p.

Rathore, G.S. and Agnihotri, J.P. (1985). Effect of insecticides on yellow mosaic of moth bean (*Vigna acontifolia*). *Indian J. Virol.* 1: 92-94.

Szalay- Marzso, L. and Solymossy, F. (1962). Investigations on the aphid vectors of the cucumber mosaic virus on *Capsicum* in Hungary. *Acta Agron.* 11: 329-343.

Yang, C.S., Zhao, S.M., Wang, W.L. and Zhu, H.Y. (1998). Studies on ecological control of virus diseases on Weiqing radish. *Acta Phytophylacica Sinica* 25: 111-114.

Integrated Plant Disease Management (2005)
Eds. Sharma, R.C. and Sharma, J.N.
Scientific Publishers (India), Jodhpur, pp. 199-201.

33

Occurrence and management of siliqua blight of cabbage seed crop in dry temperate conditions of Himachal Pradesh

NARENDER K. BHARAT and B.S. DOGRA

Dr. Y.S.Parmar University of Horticulture and Forestry, Regional Horticultural Research Station, Sharbo, Kinnaur – 172 107 (H.P.)

ABSTRACT

Siliqua blight has emerged as an important disease causing huge losses in cabbage seed crop in dry temperate conditions of Himachal Pradesh. The disease caused by a fungus *Alternaria brassicicola* appears in the month of July as black spots on the siliquae (pods). Two sprays of Kavach (0.2%) or Indofil M-45 (0.25%) with first spray just after the disease appearance and the second one after 20 days were found effective in checking the disease as well as yield losses.

INTRODUCTION

Cabbage (*Brassica oleracia* var. *capitata* L.) is an important vegetable crop of Himachal Pradesh. Its seed production is successfully done in Kinnaur district which falls under dry temperate zone. Previously, there was hardly any attack of diseases of cabbage seed crop in this area due to dry weather conditions. But for the last 2-3 years, a disease causing blight of siliquae (pods) because of rains has been reported to cause heavy losses during summer. As the disease is new to the area, no work has been done on its symptomatology, etiology and management. Present investigation was, therefore, carried out to record various symptoms, morphological characteristics of the pathogen associated and the management of the disease.

MATERIALS AND METHODS

This study was conducted at Dr. Y.S. Parmar University of Horticulture and Forestry, Vegetable Research Station, Kalpa, district Kinnaur, located at 3,000 m above msl in dry temperate zone of Himachal Pradesh. Symptoms of the disease were recorded since the time of its first appearance on pods. The isolation of the pathogen was done on potato dextrose agar medium with standard methods. Morphological characters of the pathogen were recorded under light microscope. The detail of the management trials is as under:

Cultivar	: Pride of India
Design of experiment	: Randomized block design
Number of treatments	: Seven
Number of replications	: Three
Plot size	: 2 x 1 m
Date of transplanting	: 06.04.2002
Time of sprays	: 20.07.2002 (1st) and 09.08.2002 (2nd)

Date of harvesting and final observation: 28.08.2002

Twenty five pods were collected randomly from each plot and based on the area infected, pods were graded on 0-5 scale and disease index was calculated based on the formula of McKinney (1923).

RESULTS AND DISCUSSION

Symptomatology

The disease appeared in second week of July. Small black coloured spots appeared on siliquae (pods) and rachises. Spots enlarged, coalesced and formed larger necrotic area. Some of the severely infected pods turned dark and dried up prematurely. Similar symptoms of the disease have also been described on cabbage and other members of brassicaceae family by Babadoost *et al.* (1993).

Etiology

The infected pods were brought to the laboratory and were examined under microscope. The observation revealed the association of dark coloured conidia. Conidia were septate, pale to dark brown in colour with 11-15 transverse and 2-5 longitudinal septa. The conidia were having rounded basal cell and short (sometimes absent) beak. The conidia measured 80-225 µm . On the basis of the morphological characters and their analogy with that given by Wiltshire (1947), it was identified as *Alternaria brassicicola* (Schw.) Wiltshire. The fungus is also known to cause leaf spot disease on the same crop in association with another fungus *Alternaria brassicae* (Singh, 1987).

Management

Perusal of the data presented in Table 1 indicated that all the test fungicides reduced the disease and increased seed yield significantly over control. However, Kavach (chlorothalonil) was adjudged as the best fungicide followed by Indofil M-45 (mancozeb) amongst the six fungicides tested. Two sprays of Kavach resulted 83.36 per cent disease control. Bavistin was found to be the least effective fungicide against the disease. The seed yield per plot was statistically at par in all the fungicide treatments except for Bavistin and was significantly higher as compared to the control plot. Chlorothalonil and mancozeb fungicides have also been found effective in checking Alternaria blight of cabbage, cauliflower and carrot by Babadoost *et al.* (1993), Mande *et al.* (1984) and Sandhu and Dhiman (1985).

Table 1. Effect of different fungicides on siliqua blight and yield of cabbage seed crop

Fungicide	Conc. (%)	Disease index (%)	Disease control (%)	Seed yield (g/2m^2 plot)
Indofil M-45 (mancozeb)	0.25	12.83 (19.98)	75.26	148.16
Bavistin (carbendazim)	0.05	26.57 (32.72)	48.77	137.83
Kavach (chlorothalonil)	0.20	8.63 (15.84)	83.36	148.53
Kocide (copper hydroxidce)	0.30	14.20 (21.02)	72.62	148.00
Blue copper(copper oxychloride)	0.30	15.63 (23.18)	67.99	146.50
Companion (mancozeb + carbendazim)	0.30	16.60 (23.18)	67.99	146.50
Control	—	51.86 (50.89)	—	98.77
$CD_{0.05}$	—	3.96	—	8.42

Figures in parentheses are arcsine transformed values

REFERENCES

Babadoost, M., Gabrielson, R.L., Olson, S.A. and Mulanax, M.W. (1993). Control of *Alternaria* diseases of *Brassica* seed crops caused by *Alternaria brassicae* and *Alternaria brassicicola* with ground and aerial fungicide applications. *Seed Sci. Technol.* 21: 1-7.

Mande, R.B., Humpherson Jones, F.M. and Shuring, C.G. (1984). Treatments to control *Phoma* and *Alternaria* infection of *Brassica* seed. *Plant Pathol.* 33: 525-535.

McKinney, H.H. (1923). Influence of soil temperature and moisture on infection of wheat seedlings by *Helminthosporium sativum*. *J. Agric. Res.* 26: 195-217.

Sindhu, T.S. and Dhiman, J.S. (1985). Evaluation of different varieties and fungitoxicants against *Alternaria* and *Cercospora* blight of carrot. *J. Res. Punjab Agric. Univ.* 22; 675-678.

Singh, R. (1987). *Diseases of Vegetable Crops*. Oxford & IBH Publishers, New Delhi, 362 p.

Wiltshire, S.P. (1947). *Species of Alternaria* on *Brassicae*. Mycol. Paper 12, CMI, Kew England.

Integrated Plant Disease Management (2005)
Eds. Sharma, R.C. and Sharma, J.N.
Scientific Publishers (India), Jodhpur, pp. 203-211.

34

Variations in the physiological and antagonistic behaviour of wild and mutant strains of *Trichoderma viride*

S.K. GANDHI and M.K. SAIKIA

Department of Plant Pathology, CCS Haryana Agricultural University,
Hisar - 125 004 (Haryana)

ABSTRACT

Wild and mutant strains of *Trichoderma viride* developed through exposure to UV irradiation were compared for their growth at different pH levels, temperature and production of toxic metabolites inhibitory to *Rhizoctonia solani*. The wild strain was quite sensitive to higher pH, while mutant strains showed considerable tolerance when evaluated both on liquid and solid media. Similarly, at higher temperature, mutant strains performed better than wild strain. The mutant strains produced both heat labile and heat stable toxic metabolites while wild strain produced only heat labile toxic metabolites. Further, mutant strains were capable of producing volatile compounds that were more inhibitory to the growth of *R. solani* as compared to those produced by the wild isolate.

INTRODUCTION

Trichoderma species have been used successfully against several soil-borne diseases owing to various antagonistic mechanisms operating in different isolates (Hjeljord and Tronsmo, 1998). Wide range of environmental conditions tolerated by various *Trichderma* spp. make it possible to select isolates suited for conditions under which plant pathogens can cause disease. It has been amply demonstrated that antagonistic potential of bio-agents is largely governed by physical, chemical and biological characteristics of the soil. That's why, an antagonist effective in one crop at one location behaves differently on the same crop in another location. Hence, there is need to develop better antagonist that must be ecologically fit to survive, become established and function with in particular conditions of the eco-system. Since natural populations of *T. viride* are reported to be highly sensitive to MBC fungicides thus limiting the integration of benzimidazoles with them for the control of soil-borne plant pathogens. Hence, few strains of this antagonist tolerant to carbendazim were developed by repeated and long exposure of wild strains to UV irradiation. In present investigations, variation in terms of physiological and antagonistic behaviour of wild and mutant strains of *T. viride* against *Rhizoctonia solani* causing stem rot in cauliflower was studied.

MATERIALS AND METHODS

The wild strain of *T. viride* was designated as TV-W and new isolates raised after exposure to UV irradiation were designated as TV-M-1, TV-M-2 and TV-M-3.

Effect of pH

Both wild and UV irradiated strains of the antagonist were grown on solid (PDA) and liquid (PDB) medium adjusted to pH 5,6,7,8, and 9 with 1.0 N NaOH or HCl. The Petri plates with PDA and flasks containing PDB were inoculated separately with each strain of *T. viride* by transferring 5 mm mycelial discs cut from 5 day old culture of wild and UV irradiated strains of the antagonist. The inoculated plates and flasks were incubated at 25±1°C. Colony diameter on PDA was recorded after 4th day of inoculation. Mycelial mats on PDB were harvested after 10 days of inoculation and dried in oven at 60°C by keeping them between folds of filter papers and mycelial dry weight was taken.

Effect of temperature

The growth performance of all the strains of the antagonist was evaluated by inoculating PDA and PDB with each strain separately and incubating at temperature ranging from 15-30°C with 5°C intervals. Colony diameter and mycelial dry weight were recorded as in the preceding experiment.

Inhibition of *R. solani*

Five mm mycelial discs cut from margin of 5 day old culture of each strain of the antagonist (grown on PDA) were placed individually at the periphery of Petri plates containing 15 ml of sterilized PDA. Similarly a disc of the pathogen *R. solani* was placed on the opposite side of the antagonist. The inoculated plates were incubated at 25±1°C along with the control (inoculated with pathogen only). Radial growth of the pathogen (paired with antagonist) was recorded after 3 days of inoculation and per cent inhibition of growth over control was calculated.

Production of fungitoxic metabolites

All the strains (UV induced and wild) of the antagonist were grown separately on PDB in Erlenmeyer flasks. After 10 days of incubation at 25°C culture filtrates were separated from mycelial mats and filtrates of each strain divided into two lots. One lot was sterilized by passing through millipore filter (size 0.45 µ) and the other lot was autoclaved at 15lb pressure/ sq. inch for 15 minutes. Sterilized culture filtrates from both the lots of each strain were added separately @10, 25 and 50 per cent (v/v) to cooled (45°C) sterile PDA and 8 ml of this amended medium was dispensed in each Petri plate (5 cm dia). After solidification, 5 mm mycelial discs cut from margin of actively growing culture of the pathogen were transferred to centre of each plate and incubated as earlier. The colony diameter of the inoculated pathogen was recorded on 3rd day of inoculation and per cent inhibition over control was calculated using standard formula.

Production of volatile compound : The method suggested by Dennis and Webster (1971) was adopted for this study. Petri plates containing sterile PDA were inoculated separately with 5 mm mycelial discs of each strain of the antagonist and incubated at 25±1°C for 5 hours. After incubation, the lid of plate was replaced by a bottom of Petri dish pair inoculated with the pathogen and sealed together with adhesive tape and incubated as earlier. The colony diameter of the pathogen was measured on 4th day of inoculation and the per cent reduction of the pathogen's growth was calculated.

RESULTS AND DISCUSSION

Effect of pH on the growth

Maximum linear and mycelial growth of all the strains of *T. viride* occurred at pH 6 (Tables 1,2). However, the growth (both linear and mycelial) of UV irradiated strains TV-M-2 and TV-M-3 is consistently higher than other two strains at all pH levels. At higher pH (9) these two strains produced 67.6 and 70.3 mg dried mycelium respectively, where wild mother strain produced 32.6 mg only. Highest growth of the wild as well as the UV irradiated strains at pH 6 is in accordance with the work of Chet and Baker (1980). Moreover, Okada (1976) was of the opinion that better growth of *Trichoderma* spp. in acidic condition is due to increased activities of enzymes. The enhanced growth of TV-M-2 and TV-M-3 (irrespective of pH) may be attributed to the production of higher amount of enzymes which enable them to utilize the medium more efficiently than their wild mother strain. Enhanced growth of UV induced strains of *Trichoderma* species was also reported by Papavizas and Lewis (1983) and Rajappan (1997).

Table 1. Linear growth of UV irradiated and wild strains of *Trichoderma viride* at different pH levels

pH	Colony diameter (mm) after 3 days of incubation				Mean
	TV-W	TV-M-1	TV-M-2	TV-M-3	
5.0	60.66	62.00	71.66	69.00	65.83
6.0	71.00	73.66	82.00	84.33	71.14
7.0	50.00	52.33	61.00	63.33	56.66
8.0	43.33	47.66	54.00	52.33	49.33
9.0	31.00	38.66	44.00	47.66	40.33
Mean	51.98	54.86	62.53	63.33	—

$CD_{0.05}$

Antagonist = 2.85
pH = 3.07
Antagonist x pH = 6.15

Table 2. Mycelial growth of UV irradiated and wild strains of *Trichoderma viride* at different pH levels

pH	Mycelial dry weight (mg) after 10 days of incubation				Mean
	TV-W	TV-M-1	TV-M-2	TV-M-3	
5.0	82.66	85.33	118.33	121.66	101.99
6.0	96.33	94.00	130.33	134.33	113.66
7.0	69.33	67.00	108.00	104.33	87.16
8.0	57.66	62.00	96.33	97.66	78.41
9.0	32.66	54.00	67.00	70.33	55.99
Mean	67.72	72.44	103.99	105.59	—

$CD_{0.05}$

Antagonist = 4.78
pH = 5.23
Antagonist x pH = 10.45

Effect of temperature on the growth

The growth of wild and UV irradiated strains of *T. viride* under different incubation temperatures (Tables 3, 4) indicated that a temperature range of 25-30°C is optimum for growth of *T. viride* irrespective of strain of the antagonist.

Table 3. Effect of incubation temperature on the linear growth of UV irradiated and wild strains of *Trichoderma viride*

Temperature (°C)	Colony diameter (mm)				Mean
	TV-W	TV-M-1	TV-M-2	TV-M-3	
15	44.66	45.00	47.00	45.66	45.58
20	50.66	52.00	56.66	55.33	53.66
25	66.00	64.66	74.66	77.00	70.58
30	64.66	65.00	75.33	74.66	69.91
35	47.00	48.33	61.00	62.33	54.66
Mean	54.59	54.99	62.93	62.99	—

$CD_{0.05}$

Antagonist = 4.34
Temperature = 4.86
Antagonist x Temperature = 9.72

Table 4. Effect of incubation temperature on the mycelial dry weight of UV irradiated and wild strains of *Trichoderma viride*

Temperature (°C)	Mycelial dry weight (mg)				Mean
	TV-W	TV-M-1	TV-M-2	TV-M-3	
15	65.00	64.33	67.33	66.00	65.66
20	74.66	76.33	82.00	84.33	79.33
25	99.33	97.66	137.66	135.33	117.49
30	101.00	98.06	134.66	136.66	117.59
35	71.33	73.33	86.00	89.33	79.91
Mean	82.26	81.94	101.53	102.33	—

$CD_{0.05}$

Antagonist = 5.56
Temperature = 6.22
Antagonist x Temperature = 12.44

However, growth of all the strains declined sharply with the increase or decrease in temperature beyond this range. Similar observations have been made by Sesan (1984). No strain of the antagonist grew well at 15°C. At higher temperatures, growth performance of TV-M-2 and TV-M-3 is better than the other two strains (TV-W and TV-M-I). Tronsmo and Dennis (1977) however, reported that *Trichoderma* spp. can grow well even at temperature as low as 2°C. Probably *Trichoderma* spp. used by them were isolated from cooler region which may have high degree of adaptability to low temperature. This indicates the rationale of using ecologically well adapted native isolates for achieving successful biocontrol of soil-borne pathogens.

Since natural population of *Trichoderma* spp. are sensitive to high pH, raising of new isolates (with tolerance to high pH) of *T. viride* by UV irradiation may be an effective alternative for successful use of this antagonist in alkaline soil conditions.

Inhibition of *R. solani*

Though all the strains of *T. viride* evaluated reduced the growth of *R. solani* but strains TV-M-2 and TV-M-3 were most effective and reduced the growth to the extent of 68 per cent whereas, the other two strains (TV-W and TV-M-I) inhibited growth between 47-48 per cent only (Table 5).

Table 5. *In vitro* screening of wild and UV irradiated strains of *Trichoderma viride* against *Rhizoctonia solani*

T. viride strains	Growth inhibition (%) of *R. solani*
TV-W	47.3
TV-M-1	48.4
TV-M-2	67.7
TV-M-3	68.0
$CD_{0.05}$	2.02

Enhanced biocontrol ability of UV induced mutants of *Trichoderma* spp. has been reported by earlier workers also (Papavizas and Lewis 1983; Baby, 1998). Production of higher amount of β-I, 4 glucosidase by mutants of *Trichocerma* spp. (raised through UV irradiation) was observed by Ahmad and Baker (1987). Probably in the present investigation also, TV-M-2 and TV-M-3 strains might have produced higher amount of this enzyme which enabled them to grow at a faster rate by utilizing the medium more efficiently than the other two strains and being more inhibitory to the pathogen in comparison to TV-W and TV-M-I.

Production of fungitoxic metabolites

All the stains of the antagonist in the present study were capable of producing fungitoxic metabolites (Table 6).

Table 6. Effect of millipore filtered culture filtrates of UV irradiated and wild strains of *Trichoderma viride* on the radial growth (mm) of *Rhizoctonia solani*

T. viride strain	Concentration (%) of culture filtrate				Per cent inhibition of growth at concentration (%)		
	10	20	50	Mean	10	20	50
TV-W	40.00	16.33	5.33	20.55	0.00	59.17	86.67
TV-M-1	39.33	17.33	5.00	20.55	0.01	56.67	87.50
TV-M-2	13.33	0.00	0.00	4.44	66.67	100.00	100.00
TV-M-3	13.66	0.00	0.00	4.55	65.85	100.00	100.00
Mean	26.58	8.41	2.58	—	—	—	—

$CD_{0.05}$

Antagonist = 1.95

Concentration = 1.38

Antagonist x Concentration = 3.37

However, millipore filtered culture filtrate of TV-W and TV-M-I at lower concentration (10%), was unable to check growth of the pathogen while culture filtrates of the other two strains (TV-M-2 and TV-M-3) at the same concentration inhibited growth of the pathogen by 66 per cent. Again, in media amended with 25 per cent culture filtrate of these two strains, the pathogen did not grow at all and even lost its survivability but the medium amended with culture filtrate of TV-W at the same rate supported the growth and development to a considerable extent. Autoclaved culture filtrate of either TV-W or TV-M-I had no effect on growth of *R. solani* irrespective of concentration (Table 7) thereby indicating clearly that these strains are unable to produce heat stable fungitoxic metabolites. Similarly culture filtrates of TV-M-2 and TV-M-3 at lower concentration (10%), were not much effective to check the growth of the pathogen but at higher concentration (50%), these strains inhibited the growth to the extent of 46 per cent. From these findings, it may be assumed that these two strains produced not only heat labile, but heat stable fungitoxic metabolites as well.

Table 7. Effect of culture filtrates (autoclaved) of UV irradiated and wild strains of *Trichoderma viride* on the radial growth (mm) of *Rhizoctonia solani*

T. viride strain	Concentration (%) of culture filtrate				Per cent inhibition of growth at concentration (%)		
	10	20	50	Mean	10	20	50
TV-W	40.00	40.00	40.00	40.00	0.00	0.00	0.00
TV-M-1	40.00	40.00	40.00	40.00	0.00	0.00	0.00
TV-M-2	38.66	31.33	21.33	30.44	3.35	21.67	46.67
TV-M-3	39.33	31.00	21.66	30.66	1.67	22.50	45.85
Mean	39.49	35.58	30.74	—	—	—	—

$CD_{0.05}$

Antagonist = 0.51
Concentration = 0.45
Antagonist x Concentration = 0.89

The presence of enzymes like chitinase and β I- 3 glucanase in the culture filtrates of *Trichoderma*. spp. was reported earlier by Jones and Watson (1969) which are responsible for degradation of cell wall chitin and glucan of many pathogenic fungi (Stefanova *et al.,* 1999). The greater inhibition of growth of *R. solani* by TV-M-2 and TV-M-3 may be due to production of higher amount of antifungal metabolites in comparison to the other strains. This is in agreement with work of Mandels *et al.* (1971), who reported that the mutant strain of *T. viride* secrete twice as much enzyme as its wild mother strain.

Growth inhibition through production of volatile compound

Inhibition of growth of *R. solani* was observed by volatile compound released by both UV irradiated and wild strain of *T. viride.* However, it was found that UV irradiated strains (TV-M-2 and TV-M-3) inhibited the growth more effectively than the other two strains (Table 8). Inhibitory effect of volatile substance produced by *Trichoderma* spp. on growth of various soil-borne plant pathogens has been reported earlier also (Padmadaya and Reddy, 1996). In soil, the inhibitory volatile substance may diffuse through air filled pores and thus antagonist may inhibit the pathogen without physical contact.

The present investigation concludes that besides well-known mycoparasitic nature of *T. viride,* the antagonist also produces toxic metabolite and volatile compounds inhibitory to growth of *R. solani.* The enhanced potential of the mutants of *T. viride* (raised through UV irradiation) in controlling *R. solani* might be due to their increased capability of producing enzymes and antibiotics that indicates the possibility of exploiting the UV induced mutants of *T. viride* for efficient management of *R. solani.*

Table 8. Inhibition of growth of *Rhizoctonia solani* by volatile compounds produced by UV irradiated and wild strains of *Trichoderma viride*

T. viride strain	Inhibition (%) of radial growth on		
	2nd Day	3rd Day	Mean
TV-W	14.83	21.83	18.33
TV-M-1	14.73	22.56	18.64
TV-M-2	15.93	26.96	21.44
TV-M-3	15.93	28.43	22.18
Mean	15.35	24.94	—

$CD_{0.05}$

Antagonist	= 1.31
Days	= 0.40
Antagonist x Days	= 1.61

REFERENCES

Ahmad, J.S. and Baker, R. (1987). Comparative saprophytic ability and cellulolytic activity of rhizosphere competent mutants of *Trichoderma harzianum*. *Phytopathology* 77: 358-362.

Baby, U.I. (1998). Biocontrol potential of fungicide resistant mutants of *Trichoderma* spp. *Indian J. Microbiol.* 38: 165-166.

Chet, I. and Baker, R. (1980). Induction of suppressiveness to *Rhizoctonia solani* in soil. *Phytopathology* 70 : 994-998.

Dennis, C. and Webster, J. (1971). Antagonstic properties of subspecies group *Trichoderma* II. Production of volatile antibiotics. *Trans. Br. Mycol. Soc.* 57: 41-48.

Hjeljord, L.G. and Tronsmo, A. (1998). *Trichoderma* and *Gliocladium* in biological control: An Overview. In: *Trichoderma and Gliocladium* Vol. 2. (Eds. Harman, G.E. and Kubick, C.P), Marcel Dekker Inc. New York, pp. 129-145.

Jones, D. and Watson, D. (1969). Parasitism and lysis by soil fungi of *Sclerotinia sclerotiorum* (Lib.) de Bary, a phytopathogenic fungus. *Nature* 224: 287-288.

Mandels, M., Weber, J. and Parizek, R. (1971). Enhanced cellulase production by mutant of *Trichoderma viride*. *Appl. Microbiol.* 21: 152-154.

Okada, G. (1976). Enzymatic studies on a cellulose system of *Trichoderma viride*. *J. Biochem.* 80: 913-922.

Padmadaya, B. and Reddy, H.R. (1996). Screening of *Trichoderma* spp. against *Fusarium oxysporum* f. sp. *lycopersici* causing wilt of tomato. *Indian J. Mycol. Plant Pathol.* 26: 266-270.

Papavizas, G.C. and Lewis, J.A. (1983). Physiological and biocontrol characteristics of stable mutants of *Trichoderma viride* resistant to MBC fungicides. *Phytopathology* 73: 407-411.

Rajappan, K. (1997). Induction of mutants in *Trichoderma viride* by UV irradiation. *Plant Dis. Res.* 12: 1-5.

Sesan, T. (1984). Contribution to the study of the biology of some antagonistic fungi. IV. Effect of temperature on the growth and sporulation of *Trichoderma viride* Pers. ex. *Fr. Biol. Vege.* 36: 62-69.

Stefanova, M., Leiva, A., Larringa, L. and Coronado, M.F. (1999). Metabolic activity of *Trichoderma* spp. isolates for control of soil-borne phytopathogenic fungi. *Rev. Fac. Agron. Univ. Zulia* 16: 509-516.

Tronsmo, A. and Dennis, C. (1977). The use of *Trichoderma* sp. to control strawberry fruit rot. *Neth. J. Plant Pathol.* 83: 449-455.

ORNAMENTAL CROPS

Integrated Plant Disease Management (2005)
Eds. Sharma, R.C. and Sharma, J.N.
Scientific Publishers (India), Jodhpur, pp. 215-220.

35

Integration of soil solarization with cruciferous leaf residues for the control of *Fusarium* wilt pathogen of gladiolus

HARENDER RAJ and SACHIN UPMANYU

Department of Mycology and Plant Pathology, Dr. Y.S. Parmar University of Horticulture and Forestry, Nauni, Solan – 173 230 (H.P.)

ABSTRACT

Wilt of gladiolus caused by *Fusarium oxysporum* f.sp. *gladioli* is one of the most serious diseases in the state and other parts of the country. Studies were conducted to find out the efficacy of soil solarization and soil amendment of different crucifer residues on the viability of mycelium and colony-forming units of the wilt pathogen buried in the field under different soil depths. Soil solarization with transparent polyethylene mulch (25 μm) for 40 days during March-April, 2003, resulted in an increase of 9.8°C in average maximum temperature at 5 cm soil depth with average maximum temperature of 41.4° C. Among leaf residues of cabbage, cauliflower, broccoli and sarson tested as soil amendments followed by mulching with polyethylene, leaf residues of cabbage were found most effective against the wilt pathogen. Combination of soil amendment of cabbage leaf residues with soil solarization caused 74.6 per cent reduction in mycelial growth of the pathogen in comparison to unamended and unsolarized control. Soil solarization alone for 40 days reduced the viability of the pathogen propagules by 67.3 per cent in comparison to unsolarized control. However, soil amendment of cabbage leaf residues in combination with soil solarization resulted in complete elimination of the pathogen at 5 cm soil depth.

INTRODUCTION

Gladiolus is one of the important ornamental crops commercially grown in Himachal Pradesh. However, for the past few years, increased attack of various pests and diseases is posing a serious challenge to its cultivation. Among various diseases and pests, *Fusarium* wilt caused by *Fusarium oxysporum* Schlechtend ex Fr. f.sp. *gladioli* (L. Massey) Snyder and Hans. is most important causing huge losses to the growers in the state. The disease assumes serious proportions during the month of July – August with the onset of rains. Tomar (1997) reported 7 to 64 per cent disease incidence in different sub-mountainous regions of the state. Attempts have been made to manage the disease by different chemical pesticides but the disease is not controlled effectively. Moreover, continuous use of chemicals is hazardous to the environment as it leads to residual toxicity in the soil and water. Adoption of eco-friendly techniques would therefore, be advisable for effective disease management. Soil solarization is one such technique in which solar energy is conserved in soil by covering with transparent polyethylene mulch and increase in soil temperature suppresses various pests and pathogens harbouring in the soil. Soil solarization increases soil temperature to the limits, which are lethal/sublethal to the pathogen, thus drastically reducing the viability and potentiality of the pathogen (Katan, 1981). However, by integrating the soil solarization with residues of cabbage and other crucifer crops as well as composted animal manures, an effective disease control is achieved by means of enhanced release of toxic volatile compounds from these amendments in heated soils (Ramirez and Munnecke, 1987; Gamliel and Stapleton, 1993; Smolinska, 2000). Keeping in view the earlier findings, attempts were made to study the effect of soil solarization along with leaf residues of crucifer crops on the growth and propagule count of *Fusarium* wilt pathogen of gladiolus.

MATERIALS AND METHODS

In order to know the combined effect of soil solarization with soil amendments of leaf residues of crucifer crops on the growth and viability of colony forming units (c.f.u.) of gladiolus wilt pathogen (*F.o.* f.sp. *gladioli*), leaf residues of four crucifer crops namely, cabbage, cauliflower, sarson and broccoli were evaluated as soil amendments under semi-controlled conditions to select the most efficacious one for further evaluation under field conditions. Shade dried leaf tissues, 2 g each of cabbage, cauliflower, sarson and broccoli were surface sterilized with ethanol, chopped and mixed with 20 g of sterilized soil contained in the lower lid of Petri plates (90 mm) followed by addition of small quantity of water to just moisten the soil so that the water applied is held by the soil particles and not in loosened condition. Lids of Petri plates of same diameter containing 24 h old culture of *F.o.*f.sp. *gladioli* on PDA (agar-agar in double strength) were inverted over the lids already containing soil + leaf tissues and sealed with cello tape under aseptic conditions. Control plates containing soil but without any leaf tissue were also kept for comparison. Each treatment was replicated thrice in a completely randomized block design. The plates were subsequently buried at 5 cm depth in the soil and half of the Petri plates of each treatment were tarped with transparent polyethylene mulch (25 μm) for 7 days, while other half number of Petri plates were kept as such in the soil. Solarization resulted in increase in soil temperature (42.3°C) in comparison to 34.8°C in unsolarized soil. Data on the diametric growth (mm) of each treatment were recorded after 7 days.

Field experiment was laid out at the farm of the Department of Floriculture and Landscaping, UHF, Nauni, Solan during 2003 in randomized block design. The experiment had two sets of treatments comprising half of the plots with the treatment of solarization and another half without solarization. For the purpose, 10 tonnes per hectare well chopped shade dried leaves of cabbage were incorporated in the plots (1 x 1 m) up to 20 cm soil depth, whereas 10 tonnes per hectare was applied as a 7-10 cm thick layer prior to the solarization. The same treatments were applied to the rest half number of plots without solarization. Culture of wilt pathogen (*F. o.* f.sp. *gladioli*) prepared on sand corn meal medium was taken in nylon bags (15 x 7 cm) @10 g culture in each bag which were buried at 5, 10 and 20 cm depths in the plots amended with FYM, cabbage leaves and unamended plots both under solarized as well as unsolarized conditions and replicated thrice. The bags were retrieved after 10, 20 and 40 days of solarization. The bags containing the pathogen were brought to the laboratory and analysed for the viable number of colony forming pathogen propagules by serial dilution technique (Johnson, 1957) using *Fusarium* specific medium (Nash and Snyder, 1962). Soil temperature was also simultaneously recorded daily at 2.00 pm during the course of solarization to know the comparative effect of solarization on the soil temperature. Solarization raised the temperature of solarized soil by 9.8° C with an average maximum temperature of 41.4°C as compared to unsolarized soil (31.6° C) at 5 cm depth.

Determination of pathogen population

Soil dilution was prepared by mixing 1 g of pathogen culture in 9 ml sterilized water in a test tube and it was further diluted with the help of a cyclomixer to get the final dilution of 1 x 10^{-3}. Out of the final dilution, 0.1 ml was spread in sterilized Petri plates and 25 ml of the medium was poured over it. The plates were given gentle swirl so as to distribute the suspension uniformly in the medium. The plates were incubated at 25±1°C for 5 days with each treatment replicated thrice and number of propagules of *F.o.* f.sp. *gladioli* were counted.

RESULTS AND DISCUSSION

It is evident from the data (Table 1) that soil solarization alone and in combination with leaf residues of different crucifer crops had a marked inhibitory effect on the mycelial growth of the pathogen, which was reduced to 21.40 mm as compared to 47.53 mm growth of the pathogen without solarization i.e. Petri plates not covered with transparent polyethylene mulch. Solarization of soil amended with cabbage leaf tissues proved to be most efficacious and reduced the mycelial growth to 17.50 mm as compared to 27.17 mm in control (soil solarization alone). Soil solarization and soil amendment with leaf tissues of sarson was found next best to cabbage with reduced growth of 19.00 mm closely followed by cauliflower (20.83 mm) and broccoli (22.50 mm) and did not differ significantly from each other.

Table 1. Effect of soil solarization and soil amendment of crucifer crops on the mycelial growth of *F.o.f.* sp. *gladioli*

Crop	Diametric growth of mycelium (mm)	
	Solarized	Unsolarized
Cabbage	17.50^{a}	55.83^{b}
Cauliflower	20.83^{a}	38.50^{a}
Sarson	19.00^{a}	36.33^{a}
Broccoli	22.50^{a}	38.17^{a}
Control	27.17^{b}	68.83^{c}
Mean	21.40^{a}	47.53^{b}

$CD_{(0.05)}$

Leaf residues = 4.02

Solarization = 2.54

Leaf residues x Solarization = 5.68

However, under unsolarized conditions soil in the Petri plates amended with leaf residues of sarson, broccoli and cauliflower proved very effective in reducing the growth of mycelium to 36.33, 38.17 and 38.50 mm, respectively, as compared to 68.83 mm growth of fungus in control. Amendment of soil with cabbage was proved least effective in reducing the growth of the fungus, may be the anti-fungal substances present in the leaf tissues are released at the higher temperatures achieved by way of solarization of the soil.

Soil solarization along with soil amendment with FYM and cabbage leaf residues also resulted in a significant decrease in pathogen population as compared to unamended unsolarized plots (Table 2). After 10 days of solarization, plots amended with cabbage leaf residues and FYM recorded 61.84 x 10^3 and 59.44 x 10^3 cfu of pathogen population, respectively, as compared to the unamended plot (71.33 x 10^3 cfu) which further decreased with the duration of solarization. However, no viable pathogen propagule was observed in the plots amended with cabbage leaf residues after 40 days of solarization followed by FYM (1.33 x 10^3 cfu) as compared to 28.22 x 10^3 cfu g^{-1} soil in unamended control. Moreover, in all the treatments including control, the pathogen population increased with the increase in the soil depth.

Thus, from the data (Table 2) it can be concluded that though solarization alone had a negative effect on the pathogen population but soil amendment with FYM and cabbage leaf residues coupled with solarization had a synergistic effect which reduced the pathogen population to approximately negligible after 40 days. Hence, supplementation of these amendments in the soil, especially cabbage leaves along with solarization is useful against this pathogen.

Table 2. Effect of soil solarization, farmyard manure and soil amendment with cabbage leaf residue on the viability of *Fusarium oxysporum* f.sp. *gladioli*

Treatment		No. of pathogen propagules (x 10^3cfu) /g soil after solarization (days) at soil depths (cm)												Mean
		10 Days			Mean	20 Days			Mean	40 Days				
		5	10	20		5	10	20		5	10	20		
FYM	Solarized	40.00	69.33	76.33	61.89	33.00	44.67	54.67	44.11	0.00	1.67	2.33		1.33
		(1.61)	(1.84)	(1.89)	(1.78)	(1.53)	(1.66)	(1.74)	(1.64)	(0.00)	(0.42)	(0.52)		(0.31)
	Unsolarized	50.67	73.00	84.33	69.33	43.33	68.67	74.00	62.00	31.00	53.67	67.00		50.56
		(1.71)	(1.87)	(1.93)	(1.84)	(1.64)	(1.84)	(1.87)	(1.79)	(1.50)	(1.74)	(1.83)		(1.69)
Mean		45.33	71.17	80.33	-	38.17	56.67	64.33	-	15.50	27.67	34.67		-
		(1.66)	(1.86)	(1.91)		(1.59)	(1.75)	(1.81)		(0.75)	(1.08)	(1.17)		
Cabbage leaf residue	Solarized	38.33	66.67	73.33	59.44	29.67	36.67	52.00	39.44	0.00	0.00	0.00		0.00
		(1.59)	(1.83)	(1.87)	(1.76)	(1.49)	(1.57)	(1.72)	(1.59)	(0.00)	(0.00)	(0.00)		(0.00)
	Unsolarized	46.33	76.33	81.67	68.11	37.67	67.67	73.67	59.67	23.67	49.67	61.33		44.89
		(1.67)	(1.89)	(1.92)	(1.82)	(1.59)	(1.83)	(1.87)	(1.76)	(1.39)	(1.70)	(1.79)		(1.63)
Mean		42.33	71.50	77.50	-	33.67	52.17	62.83	-	11.83	24.83	30.67		-
		(1.63)	(1.86)	(1.89)		(1.54)	(1.70)	(1.80)		(0.70)	(0.85)	(0.89)		
Control	Solarized	53.33	76.00	84.67	71.33	50.00	67.67	70.67	62.78	18.33	28.33	38.00		28.22
		(1.73)	(1.88)	(1.93)	(1.85)	(1.70)	(1.83)	(1.85)	(1.80)	(1.27)	(1.46)	(1.58)		(1.44)
	Unsolarized	57.67	83.67	90.33	77.22	55.00	83.00	85.33	74.44	56.00	69.67	82.67		69.44
		(1.77)	(1.93)	(1.96)	(1.88)	(1.75)	(1.92)	(1.93)	(1.87)	(1.75)	(1.85)	(1.92)		(1.84
Mean		55.50	79.83	87.50	-	52.50	75.33	78.00	-	37.17	49.00	60.33		-
		(1.75)	(1.91)	(1.95)		(1.72)	(1.88)	(1.89)		(1.51)	(1.65)	(1.75)		

Figures in parentheses are log transformed values

$CD_{0.05}$

Treatment	=	0.021
Soil depth	=	0.021
Interval	=	0.021
Solarization	=	0.017
Treatment x Soil depth x Interval x Solarization	=	0.090

Soil solarization with transparent polyethylene mulch has a drastic effect on the viability of the pathogens at different soil depths. However, the effectiveness of soil solarization has been reported to enhance further with the application of soil amendments of dried leaf residues of crucifer crops and farmyard manure (Muchelchen *et al.*, 1990; Angus *et al.*, 1994). During the present investigations, though soil solarization alone was quite effective in reducing the growth of the pathogen, the integration of leaf residues of four crucifer crops (cabbage, cauliflower, sarson and broccoli) as soil amendments further enhanced the effectiveness of soil solarization. Cabbage leaves resulted in the maximum inhibition in the growth of the pathogen when integrated with soil solarization. Further, solarization of soil amended with leaf residues of crucifer crops and farmyard manure also resulted in a marked reduction in viable pathogen propagules at different soil depths. In the present study, plots amended with cabbage leaf residues followed by soil solarization for 40 days resulted in the complete elimination of the pathogen from 5 cm soil depth while farmyard manure amended plots after solarization for 40 days also significantly reduced the pathogen population at the same depth as compared to solarization alone. Similarly, findings at other places also indicate remarkable reduction in the propagule count of *Pythium ultimum*, *Sclerotium rolfsii*, *S. cepivorum*, *F. o.* f.sp. *lycopersici* and *Macrophomina phaseolina* (Gamliel and Stapleton, 1993; Lodha *et al.*, 1997; Stapleton and Duncan, 1998; Smolinska, 2000). Moreover, the residues of crop plants and other amendments have also been reported to enhance the activity of soil microflora which are antagonistic to several soil-borne pathogens by production of biochemical substances during decompo-sition in the soil (Linderman, 1989; Raj and Kapoor, 1997). The information generated in the present investigations and earlier findings suggest, that integration of soil solarization with soil amendment of leaf residues of crucifer crops as well as farmyard manure may serve as an effective tool for the management of *Fusarium* wilt of gladiolus.

REFERENCES

Angus, J.P., Gardner, P.A., Kirkegaard, J.A. and Deshmarchelier, J.K. (1994). Biofumigation : Isothiocyanates released from brassica roots inhibit growth of the take – all fungus. *Plant Soil* 162: 107-112.

Gamliel, A. and Stapleton, J.J. (1993). Characterization of antifungal volatile compounds evolved from solarized soil amended with cabbage residues. *Phytopathology* 83: 899-905.

Johnson, L.Z. (1957). Effect of antibiotics on number of bacteria and fungi isolated from soil by dilution plate method. *Phytopathology* 47: 630.

Katan, J. (1981). Solar heating (solarization) of soil for the control of soil borne pests. *Annu. Rev. Phytopathol.* 19 :211-236.

Linderman, R.G. (1989). Organic amendments and soil borne diseases. *Can. J. Plant Pathol.* 11: 180-183.

Lodha, S., Sharma, S.K. and Aggarwal, R.K. (1997). Solarization and natural heating of irrigated soil amended with cruciferous residues for improved control of *Macrophomina phaseolina. Plant Pathol.* 46: 180-190.

Muchelchen, A.M., Rand, R.E. and Parke, J.L. (1990). Evaluation of crucifer green manure for controlling *Aphanomyces* root rot of peas. *Plant Dis.* 47: 651-654.

Nash, S.M. and Snyder, W.C. (1962). Quantitative estimation of plate counts of propagules of the bean root rot *Fusarium* in field soils. *Phytopathology* 52: 567-572.

Raj, H. and Kapoor, I. J. (1997). Possible management of *Fusarium* wilt of tomato by soil amendments with composts. *Indian Phytopathol.* 50: 387-395.

Ramirez-Villapudua, J. and Munnecke, D.E. (1987). Control of cabbage yellows (*Fusarium oxysporum* f.sp. *conglutinans*) by solar heating of field soil amended with dry cabbage residues. *Plant Dis.* 71: 217-221.

Smolinska, V. (2000). Survival of *Sclerotium cepivorum* and *Fusarium oxysporum* chlamydospores in soil amended with cruciferous residues. *J. Phytopathol.* 148: 343-349.

Stapleton, J.J. and Duncan R.A. (1998). Soil disinfestations with cruciferous amendments and sublethal heating: effects on *Meloidogyne incognita, Sclerotium rolfsii* and *Pythium ultimum. Plant Pathol.* 47: 737-742.

Tomar, M. (1997). *Studies on the Management of Gladiolus Yellows Caused by Fusarium Species.* M.Sc. Thesis, Dr. Y.S. Parmar University of Horticulture & Forestry, Nauni, Solan (HP), 43 p.

Integrated Plant Disease Management (2005)
Eds. Sharma, R.C. and Sharma, J.N.
Scientific Publishers (India), Jodhpur, pp. 221-229.

36

Integrated management of *Fusarium* yellows of gladiolus caused by *Fusarium oxysporum* f. sp. *gladioli* Snyder & Hans. under polyhouse conditions

SHYAMA NAND SHARMA, SUNITA CHANDEL and MANICA TOMAR

Department of Mycology and Plant Pathology, Dr. Y.S. Parmar University of Horticulture and Forestry, Nauni, Solan - 173 230 (H.P)

ABSTRACT

Fusarium yellows of gladiolus was effectively managed by the application of organic amendments and biocontrol agents. Neem cake and *Trichoderma harzianum* emerged as a best treatment in combating the disease up to 72.24 per cent followed by the combination of neem cake plus *T. viride* and *T. harzianum* (alone). Six fungicides, which were compatible with biocontrol agents under *in vitro,* were further tested *in vivo* by integrating them with neem cake and biological control agents. The results of integrated approach revealed a significant superiority of treated plots over untreated plots, however, neem cake+ carbendazim + *T. harzianum* together registered highest disease control vis-a-vis enhanced corm yield and improved the plant health.

INTRODUCTION

Gladiolus (*Gladiolus* spp.) are most fascinating and valuable export earning commercial flowers of the world including India where they can be grown in varied agro-climatic conditions throughout the year. Amongst various pathological problems, diseases caused by fungal pathogens take a heavy toll in terms of plant stand, quality and yield (Protsenko, 1958; Vlasova and Shitan, 1974). *Fusarium* yellows is highly devastating disease entailing into substantial economical losses by assuming considerable increase in disease status in almost all major gladiolus growing areas of our country. It is both seed and soil-borne in nature. Prayal (1909) reported the first occurrence of the disease from California whereas in India, Singh (1969) observed its first evidence from Lucknow. The disease is also reported from Iraq, Taiwan, Czechoslovakia, Pakistan and other countries (Tarabeigh *et al.*, 1981; Hsieh,1985; Vaclavik *et al.*, 1986; Mirza and Shakir, 1991; Chen *et al.*,1994). Recently, the disease was investigated from Himachal Pradesh by Tomar (1997), where the incidence of the disease ranged between 7.12-64.25 per cent. Keeping in view the non-availability of desirable resistance in cultivars and exorbitant cost of chemical strategies for the management of major soil-borne diseases, the introduction of biological and cultural methods is believed to offer an alternative to chemical methods. *Trichoderma* spp. are most potent and widely distributed soil fungi giving continuous and ecofriendly control of many plant pathogens (Elad *et al.*, 1984; Liu and Baker, 1986; Domsch and Gams, 1972; Smith *et al.*, 1990; Mukhopadhyay and Mukerjee, 1996). Incorporation of these methods with minimum chemical use gives better opportunity in managing various soil-borne diseases. Hence, an attempt was made in the present investigations to reduce the effect of gladiolus yellows by integrating cultural, biological and chemical methods.

MATERIALS AND METHODS

Pot culture experiment

Earthen pots measuring 22 cm diameter were filled with 4 kg of formalin sterilized garden soil having 5.5 to 6.0 pH. The soil in the pots was inoculated a week before of corm sowing with 100 g of corn - sand medium containing inoculum of *Fusarium oxysporum* f.sp. *gladioli*. Two biocontrol agents namely, *Trichoderma harzianum* and *T. viride* isolated from rhizosphere of gladiolus plants by serial dilution method (Johnson, 1957). The biocontrol agents were grown on wheat bran and applied @ 2.5 per cent w/w basis per pot at the time of sowing of corms. The mixing of organic amendments such as neem cake, mustard cake and cotton cake was carried out at the same rate (2.5 % w/w) individually and in combinations along with above two biocontrol agents. Four corms of variety American Beauty belonging to medium sized category (B-grade) were sown in each pot with four replications in completely randomized design. The pots were placed on wooden benches in open atmosphere and watered regularly. The observations were recorded on per cent disease incidence at 70 and 90 days after corm sowing.

In vitro evaluation of fungicides against biocontrol agents

To determine the compatibility between fungicides and biocontrol agents, eleven fungicides were evaluated against *Trichoderma* by poisoned food technique (Falck, 1907). Systemic and combifungicides viz., Bavistin (carbendazim), Kri-benomyl (benomyl), Topsin–M (thiophanate methyl),

Triadimefon (bayleton) and Saaf (carbendazim + mancozeb) were used at four different concentrations of 50, 100, 200 and 500 ppm whereas non-systemics viz., Captan (captan), Difolatan (captafol), Diathane–M 45 (mancozeb), Kavach (chlorothalonil), Thiram (thiram) and Blitox (copper oxychloride) were tested at 500, 750, 1000 and 1500 ppm. A double strength potato dextrose agar medium was prepared in distilled water and sterilized in an autoclave at 0.05 kg/cm^2 for 20 minutes. A series of different concentrations of fungicides were prepared separately in sterilized distilled water and added to the cooled medium aseptically. Mycelial discs (4 mm) from actively grown cultures of *Trichoderma* species (5 days old) were placed in the centre of PDA poured Petri plates. The Petri plates were then incubated for a week at 25±1°C. All treatments were replicated four times in a completely randomized design and data pertaining to colony diameter of antagonistic fungi were recorded regularly at 24 h intervals up to 7 days and observations on colony diameter (mm) were recorded. The per cent inhibition in mycelial growth was calculated as per Vincent (1947).

Integrated disease management

The experiment was laid out in a polyhouse in randomized block design. Six fungicides which showed compatible reaction with two biocontrol agents (*T. harzianum* and *T. viride*) under *in vitro* namely, carbendazim, mancozeb, captan, chlorothanonil, Saaf and thiram were incorporated at half of their recommended doses. The best organic amendment i.e., neem cake was kept constant in all the treatments and was mixed thoroughly @ 100 g per row of 1 m length, 20 days prior to sowing of fungicide treated corms. Fungicides were applied to the corms as suspension prior to their sowing by steeping them for 30 minutes @ 0.05 and 0.15 per cent for systemic and non-systemic, respectively. The inoculum of test fungus was applied as corn sand medium @ 50 g per row into the soil one week before sowing of the fungicide treated corms. Untreated corms shown in uninoculated rows were kept as control treatment. The data on per cent disease incidence and growth parameters were recorded at 70 and 90 days after sowing of corms.

RESULTS AND DISCUSSION

Effect of organic amendments and biocontrol agents (BCA) on disease

The data in Table 1 with regard to organic amendments and biocontrol agents (BCA) reveals that all the treatments individually as well as in combination reduced the disease incidence significantly over check. The maximum disease control (72.24%) was registered in case of neem cake + *Trichoderma harzianum* followed by neem cake + *T. viride* (66.68%) and *T. harzianum* (55.67%). Mustard cake was found least effective giving only 11.11 per cent disease control. The results further indicate that overall disease incidence reduced from 50.00 to 12.50 per cent after 70 days and 62.50 to 18.75 per cent after 90 days of BCA application in organic amended soil. In interaction studies, neem cake + *T. harzianum* showed maximum disease reduction after 70 and 90 days. Although, neem cake + *T. viride* also gave good control of the disease up to 70 days but thereafter the effect of this treatment was significantly reduced.

Table 1. Effect of organic amendments and biocontrol agents on the incidence of *Fusarium* yellows under pot culture conditions

Treatment	Disease incidence days after corm sowing (%)		Overall mean	Disease control (%)
	70	0		
Neem cake + *T. harzianum*	12.50 (15.00)	18.75 (22.50)	15.63 (18.75)	72.24
Cotton cake + *T. harzianum*	18.75 (22.50)	37.50 (37.50)	28.13 (30.00)	50.01
Mustard cake + *T. harzianum.*	31.25 (33.75)	43.75 (41.25)	37.50 (37.50)	33.34
Neem cake + *T. viride*	12.50 (15.00)	25.00 (30.00)	18.75 (22.50)	66.68
Cotton cake + *T. viride*	31.25 (33.75)	43.75 (41.25)	37.50 (37.50)	33.34
Mustard cake + *T. viride*	37.50 (37.50)	43.75 (41.25)	40.63 (39.38)	27.78
Neem cake	31.25 (33.75)	43.75 (41.25)	37.50 (37.50)	33.34
Cotton cake	37.50 (37.50)	50.00 (45.00)	43.75 (41.25)	22.23
Mustard cake	43.75 (41.25)	56.25 (48.75)	50.00 (45.00)	11.11
T. harzianum	18.75 (22.50)	31.25 (33.75)	25.00 (28.13)	55.67
T. viride	25.00 (30.00)	37.50 (37.50)	31.25 (33.75)	44.45
Control	50.00 (45.00)	62.50 (52.50)	56.25 (48.75)	—
Overall Mean	29.17 (30.63)	41.15 (39.38)		

Figures in parentheses are angular transformed values

CD $_{(0.05)}$

Treatment = 9.89

Days = 4.04

Treatment x Days = 13.97

It has been reported by various workers that the addition of organic materials or residues usually favours the built up of the antagonistic microbes which decreases the infection caused by potential root pathogens (Singh, 1983; Linderman, 1989). Nagesh *et al.* (1998) reported effectiveness of neem cake in association with biocontrol agents in controlling gladiolus wilt. Similar findings were reported by Sharma and Bedi (1988) in cotton wilt where use of various oil cakes like mustard, groundnut and margo were found promising in lowering down the inoculum of *Fusarium* spp.

In vitro effect of fungicides on biocontrol agents

In chemical sensitivity tests (Table 2), out of eleven fungicides tested, only three namely benomyl, thiophanate methyl and bayleton from systemic category and two of non-systemic type such as foltaf and copper oxychoride did not show compatibility with *T. harzianum* and *T. viride*. However, rest of the fungicides were found compatible at all the tested concentrations except carbendazim at lower concentrations of 50 and 100 ppm. It was thus concluded that carbendazim, Saaf, captan, mancozeb, thiram, and chlorothanonil could be used with BCA for the management of *Fusarium* yellows. The results are in conformity with those reported by Papavizas (1981), Hazarika and Das (1998) and Sharma *et al.* (1999), they obtained the compatibility of Bavistin and Dithane M-45 with *Trichoderma* spp. Ramarethinam *et al.* (2001) also observed compatibility between Dithane M-45 and *T. viride* (Bio-Curf F).

Table 2. Compatibility of fungicides with biocontrol agents

Fungicide	Concentration (ppm)	Compatibility	
		T. harzianum	*T. viride*
Systemic			
Carbendazim	50	+	+
	100	+	+
	250	-	-
	500	-	-
Benomyl	50	-	-
	100	-	-
	250	-	-
	500	-	-
Thiophanate methyl	50	-	-
	100	-	-
	250	-	-
	500	-	-
Bayleton	50	-	-
	100	-	-
	250	-	-
	500	-	-
Saaf	50	+	+
	100	+	+
	250	+	+
	500	+	+
Non –systemic			
Captan	500	+	+
	750	+	+
	1000	+	+
	1500	+	+
Captafol	500	-	-
	750	-	-
	1000	-	-
	1500	-	-
Mancozeb	500	+	+
	750	+	+
	1000	+	+
	1500	+	+
Chlorothalonil	500	+	+
	750	+	+
	1000	+	+

	1500	+	+
Thiram	500	+	+
	750	+	+
	1000	+	+
	1500	+	+
Copper oxychloride	500	-	-
	750	-	-
	1000	-	-
	1500	-	-

+ Compatible; - Incompatible

Integrated disease management

Integrated study revealed a maximum disease control in treatment combinations of neem cake + carbendazim + *T. harzianum* (76.31%) followed by neem cake + *T. viride* (74.51%) and neem cake + Saaf + *T.harzianum* (Table 3). While, lowest disease control (32.24%) was recorded in neem cake + thiram + *T.viride.* The perusal of interaction study between organic amendments, fungicides and BCA's were found highly effective in reducing the incidence of the disease after 70 days of corm sowing. A similar trend was observed after 90 days in same combinations.

Table 3. Combined effect of organic amendment, fungicides and biocontrol agents on the incidence of *Fusarium* yellows of gladiolus

Treatment	Disease incidence days after corm sowing (%)		Overall Mean	Disease control (%)
	70	90		
NC+ Carbendazim+T.h.	9.33 (17.67)	14.83 (22.65)	12.08 (20.16)	76.31
NC+Captan + T.h.	16.67 (24.82)	29.83 (33.10)	23.08 (28.48)	54.75
NC+ Mancozeb+ T.h.	16.33 (24.88)	20.69 (27.05)	16.68 (25.57)	63.37
NC+ Chlorothalonil+ T.h	25.67 (30.43)	36.00 (36.67)	30.83 (33.65)	39.55
NC+Saaf+T.h.	11.33 (19.67)	14.83 (22.65)	13.08 (21.20)	74.35
NC+Thiram+ T.h.	16.20 (24.82)	22.00 (27.97)	19.10 (25.91)	62.54
NC+ Carbendazim +T.v.	10.33 (18.75)	15.67 (22.82)	13.00 (20.83)	74.51
NC+Captan + T.v.	17.67 (24.85)	25.50 (30.31)	21.58 (27.58)	57.69
NC+ Mancozeb + T.v.	14.33 (22.44)	21.00 (27.27)	17.67 (24.76)	65.35
NC+ Chlorothalonil + T.v.	22.83 (31.84)	36.83 (37.36)	29.83 (33.10)	41.50
NC+Saaf+T.v.	12.50 (20.72)	15.67 (23.82)	14.08 (22.04)	72.39
NC+Thiram+ T.v.	28.17 (32.50)	36.33 (37.06)	34.56 (23.08)	32.24
Control	41.67 (40.20)	60.23 (50.97)	51.00 (45.59)	-
Overall Mean	18.69 (24.17)	26.88 (30.67)		

Figures in parentheses are angular transformed values

NC = Neem cake; T.h.= *T. harzianum*; T.v.= *T. viride*

CD $_{(0.05)}$

Treatment	= 1.09
Days	= 0.43
Treatment x Days	= 4.63

Integrated disease management approach increased the germination percentage of gladiolus corms as well as improved the growth characteristics as is evident in Table 4. Maximum germination of corms (86.33%) was recorded with neem cake + carbendazim + *T. harzianum* which was at par with neem cake + carbendazim + *T. viride*, neem cake + thiram + *T. harzianum*, neem cake + thiram + *T. viride* and neem cake + mancozeb + *T. harzianum*. Maximum plant height (97.33 cm), average number of leaves/ plant (10.67), spike length (86.67 cm), average number of florets/spike (12.67), corm weight (57.67 g) and corm size (3.90 cm) were also registered in neem cake + carbendazim + *T. harzianum* as compared to other treatments. Chen *et al.* (1982) and Nagesh *et al.* (1998) reported positive results in integrated management strategies of seed and soil-borne diseases of iris and gladiolus, respectively.

Table 4. Combined effect of organic amendment, fungicides and biocontrol agents on the growth parameters of gladiolus

Treatment	Growth parameters						
	Germination (%)	Plant height (cm)	Av. No. of leaves / plant	Spike length (cm)	No. of florets /spike	Corm weight (g)	Corm size (cm)
NC+Carbendazim+T.h.	86.33	97.33	10.67	86.67	*12.67*	57.67	3.90
NC+Captan + T.h.	78.67	82.67	8.67	80.33	11.00	54.83	3.53
NC+ Mancozeb+ T.h.	80.67	93.67	10.00	81.33	11.67	57.53	3.73
NC+Chlorothalonil +T.h.	72.67	80.00	9.66	74.00	10.67	51.67	3.36
NC+Saaf+T.h	74.33	88.67	10.33	78.33	11.33	56.67	3.60
NC+Thiram+ T.h.	83.00	88.73	9.00	81.33	11.00	57.33	3.73
NC+Carbendazim+T.v.	85.33	94.33	9.00	84.67	11.33	57.53	3.80
NC+Captan + T.v.	78.67	80.33	8.33	76.33	11.00	54.33	3.46
NC+Mancozeb + T.v.	78.00	91.67	10.00	77.67	11.33	57.17	3.66
NC+Cholothanonil+T.v	71.33	81.00	8.67	70.00	9.67	52.83	3.20
NC+Saaf+T.v	71.33	85.33	10.00	76.33	11.00	54.67	3.56
NC+Thiram+ T.v.	80.67	93.00	9.33	81.67	11.67	55.33	3.60
Control	65.33	78.36	8.33	69.67	9.33	49.33	3.06
$CD_{(0.05)}$	6.56	5.48	1.37	6.01	1.61	2.02	0.24

NC = Neem Cake; T.h. = *T. harzianum*; T.v. = *T. viride*

REFERENCES

Chen, L.Z., Gan, X.B., Song, J.X. and Gu, W. (1994). A study on gladiolus root rot. *J. Shanghai Agri. Coll.* 12: 240-246.

Chen, Z., Elad, Y., Kalfen, A., Hadar, Y. and Katan, J. (1982). Integrated control of soil- borne and bulb-borne pathogens in iris. *Phytoparasitica* 10: 229-236.

Domsch, K.H. and Gams, W. (1972). *Fungi in Agricultural Soils.* Longman Group Limited. 290 p.

Elad, Y., Barak, R. and Chet, I. (1984). Parasitisum of sclerotia of *Sclerotium rolfsii* by *Trichoderma harzianum. Soil Biol. Biochem.*16: 301-386.

Falck, R. (1907). Wachstumgesctze, Wachstumataktoren Und temperature Werteder holzerstorenden. *Mycelien* 1: 153-154.

Hazarika, D.K. and Das, K.K. (1998). Biological management of root rot of French bean caused by *Rhizoctonia solani. Plant Dis. Res.* 13: 101-105.

Hsieh, S.P.Y. (1985). Ecology and control of gladiolus *Fusarium* wilt. *Plant Prot. Bull.* Taiwan 27: 247-256.

Johnson, L.F. (1957). Effect of antibiotics on the numbers of bacteria and fungi isolated from soil by dilution-plate method. *Phytopathology* 47: 630-631.

Linderman, R.G. (1989). Organic amendments and soil borne diseases. *Can. J. Plant Pathol.* 11: 180-183.

Liu, S. and Baker, R. (1986). Mechanism of biological control in soil suppressive to *Rhizoctonia solani. Phytopathology* 70: 404-412.

Mirza, J.H. and Shakir, A.S. (1991). First report of fungal pathogen of gladiolus from Pakistan. *Pak. J. Plant Pathol.* 3: 74-76.

Mukhopadhyay, A.N. and Mukerjee, P.K. (1996). Fungi as fungicides. *Int. J. Trop. Plant Dis.* 14: 1-17.

Nagesh, M., Reddy, P.P. and Ramachander, M. (1998). Integrated management of *Meloidogyne incognita* and *Fusarium oxysporum* f. sp. *gladioli* using antagonistic fungi and neem cake. In: *Nematology: Challenges and Opportunities of 21st Century*. Agro- Asian Society of Namatologists, Luton, U.K, pp. 263-266.

Papavizas, G.C. (1981). Survival of *Trichoderma harzianum* in soil and in pea and bean rhizosphere. *Phytopathology* 71: 121-125.

Prayal, W.A. (1909). Diseases of gladioli. *Rural New Yorker*. 68: 1009.

Protsenko, E.P. (1958). Premature yellowing of gladioli. *Bull. Centro. Bot. Gdn.* Moscow 30: 78-84.

Ramarethinam, R., Murugesan, N.V. and Marimuthu, S. (2001). Compatibility studies of fungicides with *Trichoderma viride* used in the commercial formulation- Bio-Curf-F. *Pestology* 25: 2-6.

Sharma, D.D., Gupta, V.P. and Chandrashekher, D.S. (1999). Compatibility of certain biocontrol agents with chemical pesticides and fertilizers. *Indian J. Seri.* 38: 79-82.

Sharma, J.R. and Bedi, P.S. (1988). Effect of soil amendment with oil cakes on the wilt of cotton. *J. Res. Punjab Agric. Univ.* 23: 414-416.

Singh, R.N. (1969). A vascular disease of gladiolus caused by *Fusarium oxysporum* f. sp. *gladioli* in India. *Indian Phytopathol.* 22: 402-403.

Singh, R.S. (1983). Organic amendments for root disease control through management of soil microbiota and the host. *Indian J. Mycol. Plant Pathol.* 13: 1-16.

Smith, V.R., Wilcox, W.F. and Harman, G.E. (1990). Potential for biological control of Phytophthora root and crown rots of apple by *Trichoderma* and *Gliocladium* spp. *Phytopathology* 80: 885.

Tarabeigh, A.M., Michail, S.H., Al- Zarari, A.J. and Sultan, S. (1981). *Fusarium* wilt of gladiolus with reference to varietal response and chemical control in Iraq. *Acta Phytopathol.* 16: 293-297.

Tomar, M. (1997). *Studies on the Management of Gladiolus Yellows Caused by Fusarium Species.* M.Sc. Thesis, Dr. Y.S. Parmar University of Horticulture and Forestry, Nauni, Solan (H.P), 43 p.

Vaclavik, J., Ulrychova, M and Jokes, M. (1986). Gladiolus "grassy top" disease recorded in Czechoslovakia and increased susceptibility of affected corms to *Fusarium* species. *Biol. Plant.* 28: 137-140.

Vincent, J.H. (1947). Distortion of fungal hyphae in the presence of certain inhibitors. *Nature* 15: 580.

Vlasova, V.J. and Shitan, N. (1974). Means of increasing resistance of plants to Fusarium wilt. *Nauchn. Tr.Stravop. S.Kh. Inst.* 37: 127-133.

Integrated Plant Disease Management (2005)
Eds. Sharma, R.C. and Sharma, J.N.
Scientific Publishers (India), Jodhpur, pp. 231-234.

37

Management of *Septoria* leaf spot of chrysanthemum through cultural practices

SUNITA CHANDEL, VANITA CHANDEL and MANICA TOMAR

Department of Mycology and Plant Pathology, Dr Y.S. Parmar University of Horticulture and Forestry, Nauni, Solan -173 230 (H.P.)

ABSTRACT

Septoria leaf spot is an important disease of chrysanthemum. The causal organism *Septoria obesa* was managed through cultural methods by employing mulches and defoliation practices. Amongst mulches, six different mulches, namely polythene sheet, sawdust, wheat straw, eucalyptus leaves, pine needles and grass were included, while five different plant heights were selected for defoliation to see their effect on disease development and spread. Polythene mulch gave maximum (64.28%) disease control with minimum disease severity (18.33%) in comparison to 51.33 per cent disease severity in the untreated control. Leaves plucked up to 25 cm height of the plant gave lowest (15.0%) disease severity and maximum disease control.

INTRODUCTION

Chrysanthemum (*Dendranthema grandiflora*), popularly known as queen of east, is a flower of great commercial and aesthetic value. There are more than 20,000 varieties of chrysanthemum in the world and in India about one thousand varieties are grown (Kher, 1988). About 4,000 hectares of land is under flower cultivation in India out of which 1,500 hectares is under modern cut flowers including chrysanthemum with an earning of Rs. 205 crores in the world market (Gowda, 1996; Dadlani, 1996). In Himachal

Pradesh, its utility and popularity has increased immensely for potted and commercial type flowers. High magnitude of losses have been estimated during rainy season mostly due to fungal diseases, particularly, *Septoria* leaf spot (*Septoria obesa* Syd.) which appears in severe form and is responsible in deteriorating the quality and quantity of the produce. The pathogen usually induce excessive defoliation, thereby, reduced the photosynthetic surface area. Keeping in view the seriousness of the disease, the present study was aimed at working out cultural management practices.

MATERIALS AND METHODS

The experiment pertaining to mulching was laid in the experimental farm of the Department of Mycology and Plant Pathology, Dr. Y.S. Parmar University of Horticulture and Forestry, Nauni, Solan, during the year 2001-2002. The beds of 1x1 m were prepared in the field. The mulching of the beds was done two months prior to transplanting of rooted chrysanthemum cuttings of cultivar, Tata Century with pine needles, polythene sheet (25 μm), wheat straw, sawdust, eucalyptus leaves and grass. The trial was laid out in randomized block design (RBD). Each treatment was replicated thrice and data with respect to disease severity was recorded immediately after the initiation of the disease up to the harvesting stage at 20 days interval.

The effect of defoliation was studied by removing the lower leaves at different heights viz., 5, 10, 15, 20 and 25 cm from the ground level. The defoliation was done during the growth season of the plant. The planting was carried out in 1x1 m beds. The design followed was RBD and each treatment was replicated thrice. The per cent disease severity was observed in each case during the crop season at an interval of 20 days and analyzed statistically (Mathew, 1999).

RESULTS AND DISCUSSION

The data presented in Table 1 revealed that out of six mulches tested, three mulches namely polythene sheet, sawdust and wheat straw performed better and proved effective in reducing the disease severity in comparison to other mulches used in the present study.

Table1: Effect of various mulches on the severity of *Septoria* leaf spot of chrysanthemum

Mulch	Disease severity (%)	Disease control (%)
Eucalyptus leaves	37.00 (37.46)	27.92
Sawdust	35.00 (36.27)	31.81
Polythene sheet	18.33 (25.35)	57.78
Wheat straw	22.67 (28.43)	55.83
Grass	24.00 (29.33)	53.24
Control	51.33 (45.77)	—
$C.D_{0.05}$	2.12 (1.37)	—

Figures in parenthesis are angular transformed values

Minimum severity was recorded in soil covered with polythene mulch (18.33%) followed by sawdust (21.67%) with 64.28 and 57.78 per cent disease control, respectively. Eucalyptus leaves and pine needles gave least 27.92 and 31.81 per cent disease control hence were recorded ineffective in reducing the effect of the diseases. The control plots recorded maximum disease severity of 51.33 per cent. The results of the present experiment are in conformity with the work done by various workers. Dutta (1959) reported brown spot severity three times more when soybean was grown after soybean than after maize, oats, wheat or flax.

The leaves were defoliated at different plant heights and the data recorded on disease severity are presented in Table 2.

Table 2: Defoliation effect on development of *Septoria obesa* in chrysanthemum

Height (cm)	Disease severity (%)	Disease control (%)
5	35.0 (34.21)	16.66
10	32.0 (33.21)	23.80
15	25.0 (24.20)	40.47
20	18.0 (21.25)	57.14
25	15.0 (20.22)	64.28
Control	42.0 (41.20)	—
C.D$_{0.05}$	2.16 (41.20)	—

Figures in parentheses are angular transformed values

Maximum disease control (64.28%) was observed when the plants were defoliated up to 25 cm height. This was followed 20 cm height with a disease control of 57.14 per cent. Maximum (35.0%) severity was observed when the leaves were plucked up to 5 cm plant height. However, it was less as compared to control (42.0%) where no leaves were plucked, thereby proving that plucking of leaves is helpful in reducing the severity of the disease. It can thus be concluded from the study that 20 to 25 cm plant height should be selected for defoliation of the leaves to check the spread of leaf spot disease in chrysanthemum. Kumar and Sugha (2000) reported similar findings in case of *Septoria* leaf spot of tomato. The disease was reduced in plants where defoliation of lower leaves was done. They also advocated the role of spacing, staking and inter-cropping for the management of leaf spot of tomato.

REFERENCES

Dadlani, N.K. (1996). International trade in chrysanthemum cut flowers. *Floric. Today* 3: 15-17.

Dutta, P.K. (1959). Effect of previous crop on growth yield and certain chemical constituents of two soybean varieties. *Dis. Abstr.*19: 277-278.

Gowda, N.J.V. (1996). Potential and marketing of cut flowers. *Floric. Today* 3: 18-20.

Kher, M.A. (1988). *Chrysanthemum in India.* Associating Publishing Company, New Delhi, 79 p.

Kumar, S. and Sugha, S.K. (2000). Role of cultural practices in management of *Septoria* leaf spot of tomato. *Indian Phytopathol.* 53: 105-106.

Mathew, K. Abraham (1999). *Epidemiology, Variability and Management of Angular Leaf Spot of French Bean.* Ph.D Thesis, Dr. Y.S. Parmar University of Horticulture and Forestry, Nauni, Solan (H.P), 107 p.

Integrated Plant Disease Management (2005)
Eds. Sharma, R.C. and Sharma, J.N.
Scientific Publishers (India), Jodhpur, pp. 235-239.

38

Partial characterization of a *tospovirus* and chrysanthemum B *carlavirus* infecting chrysanthemum in Himachal Pradesh

PROMIL KAPOOR

Department of Mycology and Plant Pathology, Dr.Y.S. Parmar University of Horticulture and Forestry, Nauni, Solan -173 230 (H.P.)

ABSTRACT

Chrysanthemum is grown for cut flowers, loose flowers as potted plants and as border plants in garden. Two virus isolates C-I and C-II infecting chrysanthemums under natural conditions were studied which evoked symptoms of prominent green vein banding, mild leaf mottling, chlorotic and/or necrotic spots on leaves and occasional flower distortion. Both isolates were sap transmissible. However, isolate C-II alone was transmissible through an aphid vector. Both isolates differed largely in their biophysical properties, host range, particle morphology and serology. DAS and DAC forms of ELISA tested positive for the presence of tomato spotted wilt *tospovirus* and chrysanthemum B *carlavirus*. Electron microscopic studies confirmed the presence of isometric and flexuous particles for C-I and C-II indicating members of genus *tospovirus* and *carlavirus*, respectively.

INTRODUCTION

Chrysanthemum *(Dendranthema grandiflora)* is important both as cut flower and as potted plant in the international market. Himachal Pradesh situated in the north-western Himalayas is climatically suited for chrysanthemum production. Unfortunately, the crop falls prey to around 10 viruses belonging to six different genera (Horst, 1987; Zaidi, 1992). Two isolates C-I and C-II infecting chrysanthemum under natural conditions were studied which evoked symptoms of prominent green vein banding, mild leaf mottling, chlorotic and/or necrotic spots on leaves and occasional flower distortion.

MATERIALS AND METHODS

Collection and maintenance of virus isolates

The cultures of the present virus isolates were collected from naturally infected chrysanthemum plants on the basis of symptoms. Phosphate buffer of 0.01M strength and pH 7 was used as an extracting medium. The virus cultures were maintained on chrysanthemum cv. Tata Century and tobacco (*Nicotiana tabacum* var. White Burley).

Transmission

(a) Mechanical transmission: Healthy plants of same age with almost uniform leaf size were used for various experimental purposes which involved mechanical inoculation either by rubbing the sap with fore finger or by cotton swab method.

(b) Transmission by aphids: Virus free colonies of commonly encountered aphid species viz., *Myzus persicae* Sulz., *Aphis craccivora* Koch. and *Brevicoryne brassicae* Linn. maintained on chrysanthemum cv. Tata Century and *Nicotiana tabacum* var. White Burley were used to test their transmission efficacy and were observed for 2-3 weeks for symptom development under glasshouse conditions.

Host range

Plants of different species were grown under insect proof glasshouse conditions and tested by mechanical inoculation. The plants which did not show any apparent symptoms were back inoculated on to original host (chrysanthemum cv. Tata Century) to check for symptomless infections.

Biophysical properties

Thermal inactivation point (TIP), dilution end point (DEP) and longevity *in vitro* (LIV) were determined as per standard methods described by Noordam (1973).

Serology

Alkaline phosphatase (ALP) based direct double antibody sandwich (DAS) and direct antigen coating (DAC) indirect forms of ELISA were used for serological detection of the virus as per protocol of the supplier of reagents of ELISA (Agdia Inc., Elkhart, Indiana, U.S.A.)

Electron microscopy

Negatively stained dip preparations were prepared from epidermal strips according to the method of Hitchborn and Hills (1965), and also from the clarified virus concentrates.

RESULTS AND DISCUSSION

Symptomatology

Symptoms produced on the inoculated plants were similar to those observed under natural conditions. The isolate C-I produced chlorotic and/or necrotic spots on leaves and occasionally flower distortion occurs. The incubation period for this isolate was 14 – 18 days. Prominent green vein banding followed by mild leaf mottling were the differentiating symptoms of C- II isolate. The incubation period for this isolate was 15 – 21 days.

Table 1. Host range of chrysanthemum virus isolates

Family/ Host	Virus isolate	
	C – I	C – II
Amaranthaceae		
Amaranthus caudatus L.	-	-
Gomphrena globosa L.	M	-
Chenopodiaceae		
Chenopodium album L.	-	-
C. amaranticolor Coaste & Reyn	CL, NL	CL, NL
C. quinoa Wild	-	-
C. murale L.	-	CL
Compositae		
Zinnia elegans Jacq.	M	MM
Callistephus chinensis (L.) Nees	-	-
Calendula officinalis L.	-	-
Cucurbitacae		
Cucumis sativus L.	CS	-
C. pepo L.	-	-
Luffa actangula L.	-	-
Leguminosae		
Vicia faba L.	M	MM
Vigna sinensis Savi.	M, VB	M, VB
Pisum sativum L.	M	M
Dolichus lablab L.	-	-
Lathyrus odoratus L.	-	-
Cicer arietinum L.	LL	LL
Phaseolus vulgaris L.	GVB	GVB
Crotolaria juncea L.	-	-
Cajanus cajan (L.) Mill	-	-
Melilotus alba Lamk.	-	-
Medicago sativa L.	-	-

LL = Local lesions; NL= Necrotic lesions; GVB = Green vein banding; M = Mosaic; MM = Mild mosaic; VB = Vein banding; CL = Chlorotic lesions; CS = Chlorotic spots

Transmission

Both the isolates were easily sap transmissible. *Myzus persicae* could transmit the C-II isolate but failed to transmit the isolate C-I. *Aphis craccivora* and *Brevicoryne brassicae* could not transmit any of the isolate. *Myzus persicae* has been reported to be the most efficient vector for transmitting chrysanthemum aspermy *cucumovirus* (Gupta and Singh, 1981).

Host range

The isolate C-I infected the plants belonging to the family amaranthaceae, chenopodiaceae, compositae, cucurbitaceae and leguminosae. However, isolate C-II could not infect the plants of amaranthaceae and cucurbitaceae (Table 1). Experimentally infected plants showed mosaic, local lesions, vein banding and chlorotic spots.

Properties of particles in sap

The isolate C-I had lower thermal inactivation point of 40-45°C as compared to C-II which had TIP of 70-75°C. The isolates had similar dilution end point and longevity *in vitro* at room temperature (Table 2).

Table 2. Comparison of biophysical properties of two virus isolates

Isolate	TIP (°C)	DEP	LIV	
			Room temp.	Under refrigeration
C-I	40-45	$10^{-3} - 10^{-4}$	5 hours	3 days
C-II	70-75	$10^{-3} - 10^{-4}$	5 days	7 days

Electron microscopy

Electron microscopic studies of the virus isolate C-I under investigation revealed the presence of isometric virus particles measuring 80 nm. These results are substantiated by findings of Verhoeven *et al.* (1996). Dip preparations of isolate C-II prepared from leaf epidermal strips and from the clarified virus concentrate when examined under electron microscope revealed the presence of flexuous particles with modal dimension of 685 x 13 nm.

Serology

In DAS ELISA the mean optical density (O.D.) values of 1.64 and 1.49 were recorded for the virus isolates C-1 and C–II, respectively (Table 3). As, O.D. values of infected samples were twice the O.D. values of healthy counterpart indicating the presence of virus. Similarly, in DAC ELISA mean O.D. values reveals the presence of tomato spotted wilt *tospovirus* and chrysanthemum B *carlavirus* in isolates C-1 and C-II, respectively (Table 4).

Table 3. DAS ELISA for tomato spotted wilt *tospovirus* and chrysanthemum B *carlavirus*

Isolate	Mean O.D. value (A_{405} nm)	
	Healthy	Infected
C-I	0.09	1.64
C-II	0.10	1.49

Table 4. DAC ELISA for tomato spotted wilt *tospovirus* and chrysanthemum B *carlavirus.*

Isolate	Mean O.D. value (A_{405} nm)	
	Healthy	Infected
C-I	0.16	1.25
C-II	0.08	1.10

Hence, based on parameters studied under the present investigations viz., biophysical properties, host range, particle morphology and serology it can be concluded that the virus isolate C-I is tomato spotted wilt *tospovirus* belonging to the genus *tospovirus* and isolate C-II is chrysanthemum *carlavirus* B belonging to the genus *carlavirus.*

REFERENCES

Gupta, R.P. and Singh, R.P. (1981). Studies on chrysanthemum aspermy virus on chrysanthemum. *Indian J. Mycol. Plant Pathol.* 11: 158-160.

Hitchborn, J.H. and Hills, G.J. (1965). The use of negative staining in the electron microscope examination of plant viruses in crude extracts. *Virology* 27: 528–540.

Horst, R.K. (1987). Chrysanthemum chlorotic mottle. In: *The Viroids.* (Ed. Diener, T.O.), Plenum Press, New York, pp. 291 –295.

Noordam, D. (1973). *Identification of Plant Viruses : Methods and Experiments.* Centre Agricultural Publications Documentation (Pudoc), Wageningen. 207 pp.

Verhoeven, J. Th. J., Roenhorst, J.W., Cortes, I. and Peters, D. (1996). Detection of novel *tospovirus* in chrysanthemum. *Acta Hortic.* 432: 44-51.

Zaidi, A.A. (1992). Ornamental plants, viral diseases and ELISA applications. In: *ELISA as Diagnostic Tool: Prospects and Implications.* (Eds. Visalakshi, S. and Mohan, S.), Wiley East Ltd., New Delhi, pp. 83–103.

Integrated Plant Disease Management (2005)
Eds. Sharma, R.C. and Sharma, J.N.
Scientific Publishers (India), Jodhpur, pp. 241- 245.

39

Effect of VA-mycorrhizae, nitrogen and phosphorus on growth and flowering in tuberose (*Polianthes tuberosa* L.)

SATYAVIR SINGH[1], N.R. GODARA[2] and B.P. SHARMA[1]

[1]Directorate of Extension Education, Dr. Y.S. Parmar University of Horticulture and Forestry, Nauni, Solan – 173 230 (H.P.)
[2]Department of Horticulture, CCS Haryana Agricultural University, Hisar - 125 004 (Haryana)

ABSTRACT

The effect of vesicular arbuscular mycorrhizae, nitrogen (0, 50, 100 and 200 kg/ha) and phosphorus (0, 25, 50 and 100 kg/ha) and their combinations were studied in tuberose (*Polianthes tuberosa* L.) cv. Double during the year 2000 at HAU, Hisar. The increasing levels of nitrogen and phosphorus increased the growth and flowering parameters significantly. The results of phosphorus application @ 50 kg/ha and 100 kg/ha were found statistically at par when applied with VA-mycorrhizae.

INTRODUCTION

Tuberose is an economically important bulbous flower crop. It can successfully be grown in pots, beds, borders and can also be cultivated commercially for its varied uses. The flowers remain fresh for longer time and can fill a useful place in the flower markets. Conventional nutritional applications create threat to the soil health. The application of biofertilizers along with optimum quantity of chemical fertilizers in this crop can be safer and more economic choice. Hence, the present investigation with vesicular arbuscular mycorrhizae (VA-mycorrhizae) in combination with nitrogen and phosphorus in tuberose was taken up.

MATERIALS AND METHODS

The present investigations were carried out on tuberose (*Polianthes tuberosa* L.) cv. Double at research farm, Department of Horticulture, CCS Haryana Agricultural University, Hisar during the year 2000. Half dose of nitrogen (0, 50, 100 and 200 kg/ha) and full amount of phosphorus (0, 25, 50 and 100 kg/ha) were applied at the time of planting while remaining half dose of nitrogen was applied at 3-4 leaves stage of plants. VA-mycorrhizae were applied at the rate of 10 g per bulb by placing it below the bulb at the planting time. The uniform size bulbs of tuberose were planted in April, 2000 in well prepared beds of 1x1 m size at 25x25 cm distance in order to accommodate 16 bulbs per bed. The treatments were replicated thrice in split plot design by taking nitrogen and phosphorus as main plots and VA-mycorrhizae as sub plots. Data were recorded on various parameters of growth and flowering.

RESULTS AND DISCUSSION

Number of leaves per plant

Increasing levels of nitrogen and phosphorus alone and in combination significantly increased the number of leaves per plant (Fig. 1). These findings corroborate with those of Yadav *et al.* (1985) Banker and Mukhopadhayay (1990) in tuberose. Treatments involving VA-mycorrhizae in association with various increasing levels of nitrogen and phosphorus maintained their supermacy over uninoculated ones. In case of VA-mycorrhizae with P_{50} and P_{100} , the results remained at par. Similar findings were also reported by Prabhat Kumar *et al.* (2003) in China aster and Azcon *et al.* (1992) in lettuce.

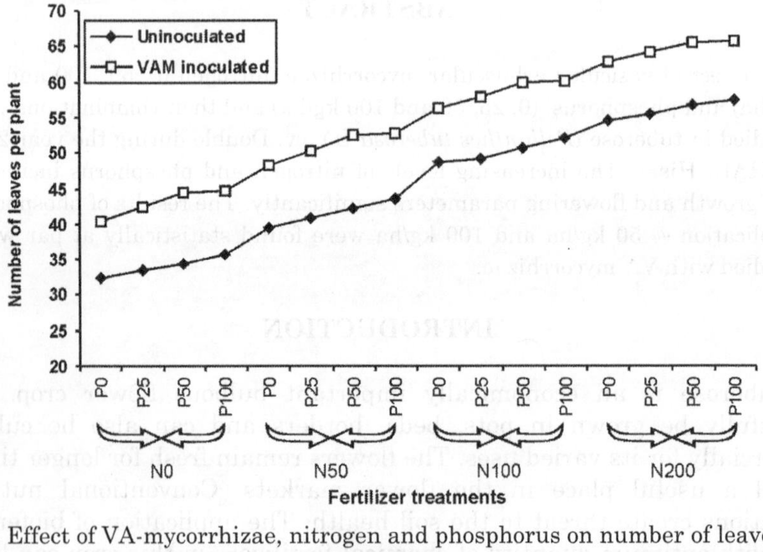

Fig.1. Effect of VA-mycorrhizae, nitrogen and phosphorus on number of leaves per plant

Plant height

The plant height was increased significantly in all the nitrogen and phosphorus treatment combinations (without VA-mycorrhizae) over control (Fig. 2). However maximum plant height (53.00 cm) was observed in N_{200} and P_{100}. The VA-mycorrhizae application with various nitrogen and phosphorus combinations showed significant increment in plant height, however, higher dose of phosphorus (P_{100}) had no significant effects over P_{50}. The encouraging results of VA-mycorrhizae application might have obtained as it explorates more soil volume making the nutrients available for diffusion of phosphate ions and increase the surface area for absorption of nutrients. Similar improvement in plant height was also reported by Crew *et al.* (2003) in three different ornamental crops.

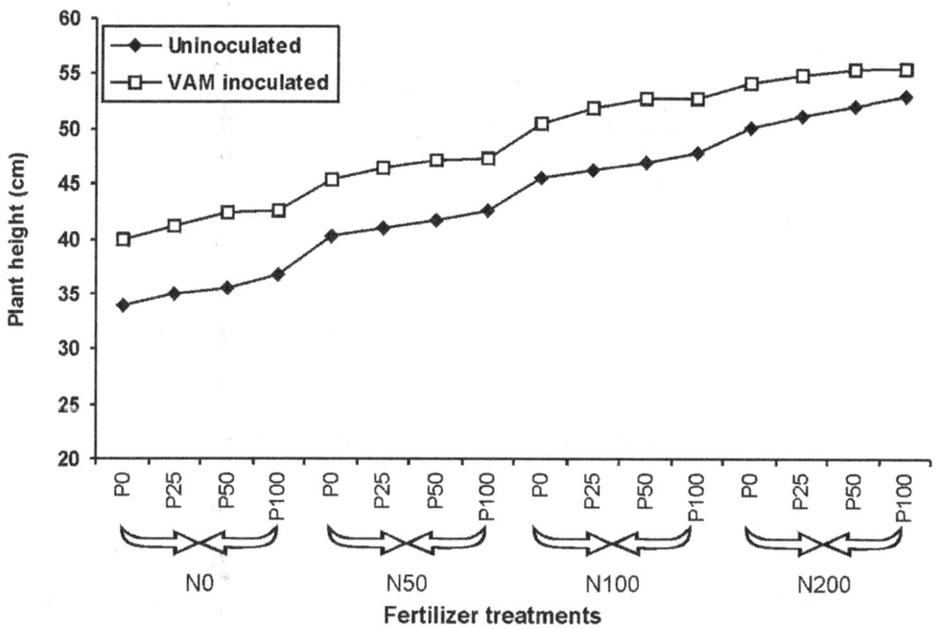

Fig. 2. Effect of VA-mycorrhizae, nitrogen and phosphorus on plant height

Spike length

Spike length (92.18 cm) was maximum under $N_{200}P_{100}$ with VAM which was at par with $N_{200}P_{50}$ with VA-mycorrhizae (92.10 cm). However, minimum spike length (67.28 cm) was observed under N_0P_0 without VA-mycorrhizae inoculation (Fig. 3). All the treatments involving VAM in combination with various levels of nitrogen and phosphorus were effective considerably as compared to other treatments due to enhanced absorption in VA-mycorrhizae inoculated plants leading to increased availability of assimilates that needed for improvement in length of spike. These results were in conformity with those of Bagyaraj and Powell (1985), Mukhopadhyay and Banker (1990).

Number of florets/spike

Number of florets per spike were found significantly different among the treatments (Fig. 4). The increasing levels of nitrogen and phosphorus showed an increase in number of florets per spike. The inoculation of VA-mycorrhizae in combination with nitrogen and phosphorus doses were considerably more effective than uninoculated treatments. Increase in number of florets/spike under VA-mycorrhizae treatments might be due to increased availability of N and P for flower development as a result of better absorption of nutrients. These results were similar to the findings of Maronek *et al.* (1981) in various horticultural crops. Johenson *et al.* (1982) also observed the similar findings in chrysanthemum. It is also observed that the results obtained at P_{50} remained at par with P_{100}. It may be because of fact that increased phosphorus application depressed mycorrhizal root colonization.

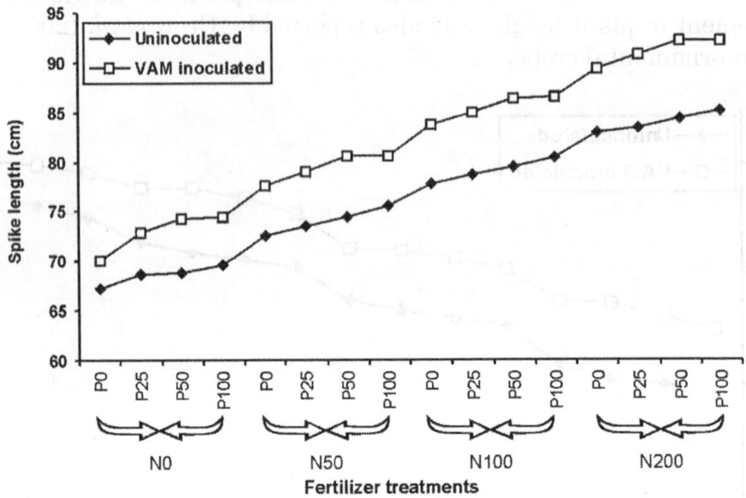

Fig. 3. Effect of VA-mycorrhizae, nitrogen and phosphorus on number of leaves per plant

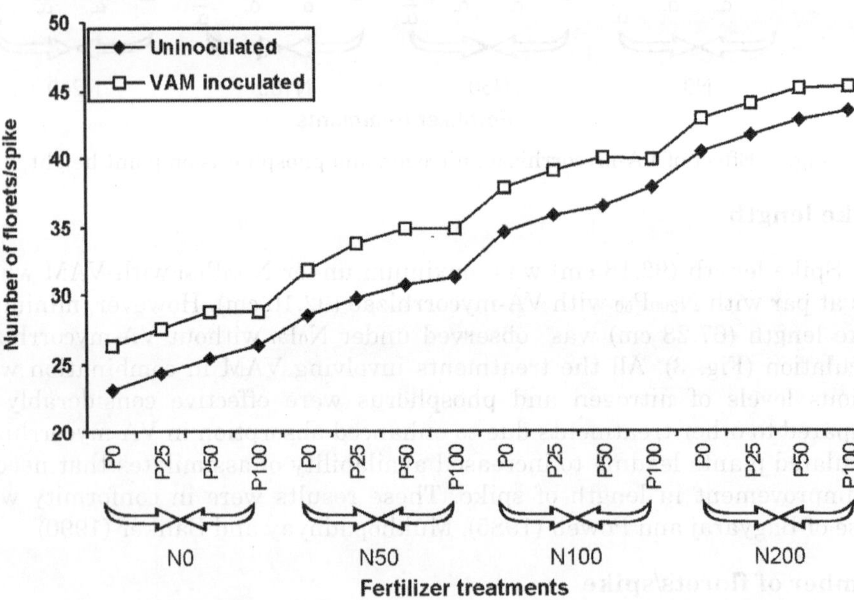

Fig. 4. Effect of VA-mycorrhizae, nitrogen and phosphorus on number of florets per spike

REFERENCES

Azcon, R.M., Gomez, M. and Tober, R. (1992). Effect of nitrogen source on growth, nutrition, photosynthetic rate and nitrogen metabolism of mycorrhizal and phosphorus fertilized plants *Lactuca sativa* L. *New Phytol.* 121: 227-234.

Bagyaraj, D.J. and Powell, C.L. (1985). Effect of VAM inoculation and fertilizer application on growth of marigold. *N.Z.J. Agric. Res.* 28: 165-173.

Banker, G.J. and Mukhopadhyay, A. (1990). Effect of NPK on growth and flowering in tuberose cv. 'Double'. *Indian J. Hortic.* 47: 120-126.

Crew, C.E., Johenson, C.R. and Joiner, J.N. (1978). Benefit of mycorrhizae on growth and development of three woody ornamentals. *Hortic. Sci.* 13: 429-430.

Johenson, C.R., Menge, J.A. and Johenson, E.L.V. (1982). Effect of vesicular – arbuscular mycorrhizae on growth of *Chrysanthemum morifolium* Ramat. *Sci. Hortic.* 17: 265-269.

Maronek, D.M., Handrix, J.W. and Kiernan, J. (1981). Mycorrhizal fungi and their importance in horticultural crop production. *Hortic. Rev.* 3: 172-213.

Mukhopadhyay, A. and Banker, G.J. (1986). Studies on nutritional requirement of tuberose. *S. Indian Hortic.* 34: 167-172.

Prabhat Kumar, Raghava, S.P.S. and Misra, R.L. (2003). Effect of biofertilizer on growth and yield of China aster. *J. Ornamental Hortic.* 6: 85-88.

Yadav, L.P., Bose, T.K. and Maiti, R.G. (1985). Response of tuberose (*Polianthes tuberosa* L.) to nitrogen and phosphorus fertilization. *Prog. Hortic.* 17: 83-86.

Integrated Plant Disease Management (2005)
Eds. Sharma, R.C. and Sharma, J.N.
Scientific Publishers (India), Jodhpur, pp. 247- 251.

40

Production of virus free plants of African marigold (*Tagetes erecta* L.) through shoot meristem culture

S.V. BHARDWAJ, ARPANA MEHRA, ANIL HANDA, SHUBHR JYOTSANA and MADHURIMA BHATNAGAR

Department of Biotechnology, Dr. Y.S. Parmar University of Horticulture and Forestry, Nauni, Solan - 173 230 (H.P.)

ABSTRACT

Viral diseases are responsible for poor quality of flowers of marigold, one of the most remunerative flower crops, in terms of size, shape and discolouration and foliage injury including necrosis, mottling, mosaic etc. Biotechnological approaches like shoot meristem culture have been accepted to be an efficient therapeutic technique capable of eradicating viruses from infected plants. During present investigations, African marigold (*Tagetes erecta* L.) cv. Pusa Narangi Gainda was selected as an experimental material due to the presence of prominent symptoms and wide susceptibility. The infected plants were indexed both serologically and biologically and were found to be infected with potato Y *potyvirus* (PVY). Virus cultures were maintained under *in vitro* conditions by culturing nodal segments from infected plants of African marigold. The most suitable medium for establishment and shoot multiplication was found to be MS medium supplemented with 2.5 mg/l IBA and 0.5 mg/l NAA. Different sizes of meristems ranging from 0.2 to 1.0 mm were cultured on standardized MS medium supplemented with 0.1 mg/l NAA and 4.0 mg/l kinetin. Better size of meristem increases the survival rate also. PVY was not detectable in plants raised from meristems ranging between 0.2-0.4 mm. For root regeneration, half strength MS medium supplemented with 2.5 mg/l IBA and 0.2 per cent charcoal was found to be the best.

INTRODUCTION

Marigold (*Tagetes erecta* L.) belongs to the family asteraceae. It is one of the most popular and commercial flowering annuals. Both leaves and flowers are equally important from medicinal point of view. Although it is an important crop, it falls prey to a number of insect pests as well as fungal, bacterial and viral pathogens resulting in poor quality of flowers, taking a heavy toll of the crop. Amongst these, viruses are becoming more important due to their significant effect on flower quality and quantity. Potato Y *potyvirus* was noticed to infect the crop widely under natural conditions. As the crop is generally produced/multiplied through cuttings, production of virus free elite material has gained high priority. Use of meristem tip culture to produce high quality planting material is being reported in this communication.

MATERIALS AND METHODS

The virus infected plants of marigold cultivar Pusa Narangi Gainda were selected on the basis of visual observations as an experimental material.

Indexing

Extract was prepared for mechanical inoculations by grinding the young leaves of infected plants in a sterilized pestle and mortar with 1 ml of phosphate buffer (0.1M, pH 7.0) per gram of leaf material (1:1v/w). Various possible indicator plants viz., C*henopodium amaranticolor* Coaste & Reyn, *C album* L. and *Nicotiana tabacum* L. were selected for mechanical transmission studies.

Serological indexing using alkaline phosphatase (ALP) based direct antigen coating (DAC) indirect enzyme linked immunosorbant assay (ELISA) was used as per the protocol of the supplier of reagents of ELISA (Agdia Inc., Elkhart Indiana, USA). Scoring was done by ELISA plate reader model MS 5605A (ECIL, Hyderabad) at different intervals.

Micropropagation

Nodal segments were used as explants for establishment of cultures. Surface sterilization was done using Bavistin (0.2%) for 15 min and $HgCl_2$ (0.1%) for 1-2 min, followed by three washings in sterile distilled water. Size of explants used varied between 1 to 2 cm. Surface sterilized explants were inoculated on MS medium supplemented with various combinations of BAP (1.0-2.5 mg/l) and NAA (0.2 mg/l) for establishment and multiplication of cultures.

The auxiliary shoots obtained from shoot tips and lateral buds on the multiplication medium were separated and cultured on rooting medium prepared by adding IBA (0.5-3.0 mg/l) and activated charcoal (0.1-0.2%) in half strength MS medium. After complete development of roots, the plantlets were placed over sterilized soil and FYM mixture and were covered with sterilized sand for hardening. The protocol for micropropagation has been standardized as a prerequisite for obtaining virus free plant using meristem culture.

Meristem tip culture

Meristems were excised using the plants maintained under glasshouse conditions as base material. The shoots from these plants were first surface sterilized using Bavistin (0.2%) for 15 minutes and $HgCl_2$ (0.1%) for two minutes. The meristem consisting of one or two leaf primordia were excised gently and removed from parental tissues and cultured on medium (Murashige, 1974).

The MS medium was supplemented with varying amounts of growth regulators such as NAA (0.0-0.1 mg/l) and kinetin (0-8.0 mg/l) and the meristems excised as above were inoculated on to these modified media. To find out the response of the meristems, different size meristems (0.2-1.0 mm) were excised and cultured on modified MS medium. The plants raised by culturing meristems of different sizes were tested for the presence of the virus through ELISA.

RESULTS AND DISCUSSION

Field crop of Pusa Narangi Gainda cultivar of African marigold was carrying 100 per cent infection of a virus disease exhibiting symptoms like mosaic, mottling and necrosis on leaves. The causal virus was found to be mechanically transmissible to *Chenopodium album* L. and *C amaranticolor* Coaste and Reyn, which produced local lesions in 5 and 10 days, respectively. Similar symptoms have been reported to be caused by marigold mottle virus which has a wide host range and produces local lesions on *C amaranticolor* Coaate & Reyn and mosaic mottling on *Nicotiana tabacum* L. (Naqvi *et al.*, 1981). Large scale application of ELISA appears to be practical and efficient way of testing marigolds for presence of viruses.

Micropropagation

Shoot tips have been used to achieve multiple shoot development under *in-vitro* conditions (Mishra and Datta, 1999) using nodal explants. The most critical factor for establishment of cultures is surface sterilization. Nodal explants were surface sterilized using 0.1 per cent $HgCl_2$ for 1-2 minutes and Bavistin (0.2%) for 15 minutes followed by three washings in sterilized distilled water following (Kothari and Chandra, 1984). Murashige and Skoog (MS medium) proved satisfactory for marigold micropropagation. Amongst the cytokinins tested (Table 1), 2.5 mg/l BAP in combination with 0.5 mg/l NAA was found to be satisfactory for establishment and multiplication of nodal explants.

There is possibility of different requirements during multiplication and rooting by different cultivars. The best rooting medium in this case was half strength with 2.5 mg/l IBA and 0.2 per cent activated charcoal (Mishra and Datta, 1999; Kothari and Chandra, 1984). Once the root system was fully developed, the plantlets were transferred from aseptic conditions to plastic pots containing pre-sterilized mixture of soil and FYM (1:1). Survival after 4 weeks of transplanting was 30.0 per cent.

Table 1. Effect of different combinations of BAP and NAA doses on establishment and multiplication of nodal explants of *Tagetes erecta*

MS medium supplemented with		Uncontaminated cultures (%)	Survival of nodal explant (%)
BAP(mg/l)	NAA(mg/l)		
1.0	0	65.00 (53.75)	55.00 (47.88)
1.5	0	75.00 (60.08)	70.00 (56.84)
2.0	0.1	100.00 (90.00)	80.00 (63.55)
2.5	0.5	100.00 (90.00)	100.00 (90.00)
1.0	2.0	90.00 (71.95)	90.00 (71.95)
1.5	0.5	85.00 (67.41)	75.00 (60.08)
2.0	1.0	75.00 (60.08)	70.00 (56.84)
2.5	2.0	65.00 (53.74)	55.00 (47.88)
$CD_{0.05}$		4.1452 (5.5125)	

Values expressed in parentheses are arcsine transformed values

Meristem tip culture

Meristem tips of African marigold were found to be less totipotent and have shown establishment in combinations i.e. NAA (0.1 mg/l) and kinetin (1.0-8.0 mg/l). The maximum average increase in size of meristem in a period of 75 days was observed on MS medium supplemented with NAA 0.1 mg/l and kinetin 4.0 mg/l (Table 2).

Table 2. Effect of different growth regulators on the survival of meristem cultures of *Tagetes erecta* L. on half strength MS medium

NAA (mg/l)	Half strength MS supplemented with		Survival (%)
	Kn(mg/l)	BAP (mg/l)	
0	0	0.5	0.00 (0.00)
1.0	0	0.5	0.00 (0.00)
1.0	1.0	0.5	0.00 (0.00)
0.1	1.0	0.5	0.00(0.00)
0.1	1.0	0	23.00 (28.29)
0.1	2.0	0	35.00 (36.06)
0.1	4.0	0	55.00 (47.93)
0.1	6.0	0	31.00 (33.79)
0.1	8.0	0	24.00 (28.74)
$CD_{0.05}$			4.737 (3.045)

Values expressed in parentheses are arcsine transformed values

Hundred per cent bud formation in *Chrysanthemum morifolium* when MS medium was supplemented with 2.0 mg/l NAA and 5.0 mg/l kinetin has been reported by Sangwan *et al.* (1987). It is widely accepted that larger the size of meristem excised, greater are the chances of contamination with the virus particles available in the lower region of the plant and our observations are also in consonance (Lankes *et al.*, 1994; Morel and Martin, 1952; Hollings and Stone, 1970; Horst *et al.*, 1976)

In case of African marigold, plants were regenerated from meristem size 0.2-0.4 mm, 0.4-0.6 mm and 0.8- 1.0 mm, however, those regenerated from meristem of size 0.2-0.4 mm only, were found to be free from potato Y *potyvirus*. Only 8.00 per cent establishment of cultures was observed when the size of meristem was between 0.2-0.4 mm (Table 3). Survival percentage increased with increase in size of meristems. Similar results have been obtained with regards to meristem tip culture in other crops (Langhans *et al.*, 1977).

Table 3. Effect of African marigold (*Tagetes erecta* L.) meristem size on their survival

Size of meristem (mm)	Per cent meristem survival (after 75 days)
0.2-0.4	8.00 (0.427)
0.4-0.6	20.00 (0.830)
0.6-0.8	24.00 (1.027)
0.8-1.0	36.00 (1.289)
$CD_{0.05}$	29.067 (0.979)

Values expressed in parentheses are arcsine transformed values

REFERENCES

Hollings, M. and Stone, O.M. (1970). Attempts to eliminate chrysanthemum stunt from chrysanthemum by meristem tip culture after heat treatment. *Ann. Appl. Biol.* 65: 311-315.

Horst, R.K., Smith, S.H. and Ogleene, W.A. (1976). *In vitro* regeneration of shoot and root growth from meristematic tips of *Pelargonium x hortorum* Bailey. *Acta Hortic.* 59: 131-142.

Kothari, S.L. and Chandra, N. (1984). *In vitro* propagation of African marigold. *Horic. Sci.* 19: 703-705.

Langhans, R.W., Horst, R.K. and Earle, E.D. (1977). Disease free plants via tissue culture propagation. *Hortic. Sci.* 12: 149-150.

Lankes, C., Ramsdell, D.C. and Barba, M. (1994). Elimination of raspberry bushy dwarf virus. *Acta Hortic.* 385: 72-75.

Misra, P. and Datta, S.K. (1999). *In vitro* propagation of white marigold (*Tagetes erecta* L.) through shoot tip proliferation. *Curr. Sci.*. 77: 1138-1140.

Morel, G. and Martin, C. (1952). Guerison de dahlias attains duno mlodie a virus. *C. R. Hebd. Seances Acad. Sci.* 235: 1324-1325.

Murashige, T. (1974). Plant regeneration through tissue cultures. *Annu. Rev. Plant Physiol.* 25: 135-166.

Naqvi, Q.A., Hadi, S. and Mahmood, K. (1981). Marigold mottle virus in Aligarh, India. *Plant Dis.* 65: 271-275.

Sangwan, R.S., Detrez, C. and Sangwan-Norreel, B.S. (1987). *In vitro* culture of shoot-tip meristems in some higher plants. *Acta Hortic.* 212: 661-665.

Integrated Plant Disease Management (2005)
Eds. Sharma, R.C. and Sharma, J.N.
Scientific Publishers (India), Jodhpur, pp. 253 - 263.

41

Characterization of a *carlavirus* infecting Oriental lilies

ASHUTOSH BHAIK and ANIL HANDA

Department of Mycology and Plant Pathology, Dr. Y.S. Parmar University of Horticulture and Forestry, Nauni, Solan –173 230 (H. P.)

ABSTRACT

Oriental lilies (*Lilium* sp.) are one of the top ranked cut flowers, grown mostly under greenhouse conditions in India for their fragrance and brilliant coloured flowers. It has also resulted in the introduction of new pathogens of quarantine importance in India especially lily symptomless *carlavirus* (LSV) which usually cause latent infection and produces no visible symptoms in the imported planting material (bulbs). LSV is of utmost importance in commercial *Lilium* industry as the virus infection renders the flowers completely misshapen and unmarketable. Surveys recorded the incidence of virus disease of LSV ranged from 15-95 per cent in different *Lilium* cultivars in different districts of Himachal Pradesh. The symptoms were mild mottle, curling, interveinal chlorosis, and necrosis of leaves, deformation and virescence of floral buds and flower and reduction in bulb size. The virus isolate was found to be transmissible through sap, bulb and aphids. The virus has a narrow host range restricted to a few members of the family chenopodiaceae, liliaceae and solanaceae. Indirect DAC –ELISA related it to CVB and CLV indicating that the *Lilium* virus isolate under investigation is a member of genus *carlavirus*. Results obtained in DAS-ELISA with LSV antisera revealed that the virus isolate was LSV. Serological detection of tested reactants at different dilutions in DAS-ELISA revealed that LSV could be detected up to a dilution of 10^{-1} for antigen 10^{-4} and 10^{-3} dilution for antibody and enzyme conjugate, respectively. In case of DAC-ELISA it was observed that LSV could be reliably detected at a dilution of 10^{-1} for antigen with 10^{-4} and 2×10^{-3} dilutions of antibody and conjugate, respectively. Further serological studies by DAC-ELISA revealed *Lilium* to be the best maintenance host followed by *Alstroemaria*. DAS-ELISA of different plant parts to study the relative distribution of LSV within *Lilium* plant showed that inner scales in bulbils were virus free. Electron microscopic studies of virus revealed presence of flexuous filamentous virus particle measuring 640 x 17-18 nm. ISEM using LSV resulted in a good decoration and thereby indicated strong positive relationship.

INTRODUCTION

Lily is the common name applied to herbaceous flowering plants belonging to the genus *Lilium* of family liliaceae and has around 100 species and 1,20,000 cultivars. Botanically the true lilies are erect perennial plants with leafy stems, scaly bulbs, usually narrow leaves with solitary or clustered flowers. The flowers of some species are quite fragrant, and they occur in a wide variety of colours. *Lilium* is one of the leading cut flower in the world ranking second next only to tulips in leading bulbous plants and ranks fourth among the top ten cut flowers.

In India, *Lilium* is grown in Nilgiri hills in southern peninsula and Himachal Pradesh in the north western Himalayas. Though, the area under *Lilium* cultivation is meagre, the crop is fast emerging as the most promising cut flower in H.P. *Lilium* is grown on commercial basis in the districts of Solan, Shimla, Kullu, Mandi and Kangra. Unfortunately, the crop falls prey to a number of insect pests and pathogens especially viruses which take a heavy toll of the crop resulting in poor quality of flowers with a significant reduction in flower and bulb yield.

MATERIALS AND METHODS

Survey

Extensive survey was conducted in different *Lilium* growing districts of Himachal Pradesh during the year 2002. Per cent disease incidence was calculated by using the following equation:

$$\text{Per cent disease incidence} = \frac{\text{Number of diseased plants}}{\text{Total number of plants observed}} \times 100$$

Sources of virus and antiserum

Virus isolate from *Lilium* plants was obtained from the germplasm collection of Department of Floriculture and Landscaping, Dr. Y.S. Parmar University of Horticulture and Forestry, Nauni, Solan (H.P.) and the antiserum against LSV was obtained from AGDIA Inc, Elkhart, Indiana, USA.

Virus propagation and maintenance

The *Lilium* virus isolate was propagated and maintained on *Nicotiana tabacum* var. White Burley. The inoculum was prepared by grinding infected leaves in 0.01 M phosphate buffer pH-7.0. The inoculated plants were grown in a glasshouse at 18-25°C.

Transmission

To check the transmission the standard extract was inoculated by the cotton swab method on to the healthy plant. Bulbs were collected from the infected plants of *Lilium* cv. Marco Polo from the experimental field under glasshouse to study transmission of the causal virus through bulbs. Aphid species (*Myzus persicae* Sulz., *Aphis craccivora* Koch. and *Brevicoryne brassicae* L.) maintained on *Lilium* cultivar Marco Polo were used to test their transmission efficacy. The aphids were given one hour pre–acquisition fasting in dark and acquisition access of 30 min on young systemically infected leaves. Ten aphids per plant were given inoculation access for 24 hours.

Biological assay on test plants

Leaves of systemically infected plants on which virus was maintained were used for the preparation of the inoculum. Four plants of *Lilium, Nicotiana tabacum* var. White Burley, *Chenopodium amaranticolor* and *Datura starmonium* were tested. Plants were inspected for virus symptoms up to 4 weeks after inoculation and indexed serologically using DAS-ELISA.

Host range

The plants of each species were inoculated with standard extract of the virus and were kept under observations for 5-6 weeks for development of symptoms.

Serological detection

DAS-ELISA was performed according to Clark and Adams (1977). Indirect DAC-ELISA was performed according to Handa and Bhardwaj (1994). ALP based enzyme system was used in DAS form of ELISA and para-nitrophenyl phosphate (pNPP) was used as a substrate. Rabbit Fc- specific antibodies raised in goats were conjugated to ALP by single step glutaraldehyde method. Incubations for antibodies was done for 4 h at 37°C, overnight for 4 h for antigen (sap extract from infected plants) and 2 h for enzyme-labeled antibodies (enzyme conjugate) at room temperature. Antiserum against LSV was obtained from Agdia Inc., USA.

Five dilutions of extracted sap and antisera were used 10^{-1}, 10^{-2}, 10^{-3}, 10^{-4} and 10^{-5}. The enzyme labeled rabbit Fc- specific antibodies were used at 2 x 10^{-2} , 5 x 10^{-2}, 10^{-3} and 2 x 10^{-3} for direct DAS- ELISA and indirect DAC - ELISA The substrate was prepared fresh every time by dissolving 5 mg tablet of pNPP in 5 ml of substrate buffer. Extracts of healthy *Nicotiana tabacum* var. White Burley were used as control. The experiment was repeated thrice to ascertain the reproducibility of the results. ELISA plates were assessed through PC based microscan plate reader (Microscan MS5605A, Electronics Corporation of India Limited, Hyderabad) at 405 nm after 1 h and the wells showing O.D. values at least 3 times more than that of negative control were considered to be positive.

Electron microscopy

Detection of the virus under electron microscope (JEOL100CXII) was carried out at IARI, New Delhi. About 10 l of sap homogenate was placed on a Para film and then carbon grid was touched to the sap and dried. The grid was negatively stained with 2 per cent aqueous uranyl acetate. After air drying, grid was screened in the electron microscope (EM) to record the shape and size of the virus particles.

ISEM studies included immunosorption and decoration of the virions on the antiserum coated grid i.e. virus–antiserum interaction in liquid phase followed by examination of the coated grid with EM decoration was studied by employing the procedure described by Khurana and Garg (1989).

RESULTS AND DISCUSSION

Survey

Incidence of virus disease(s) in different localities surveyed ranged between 15-95 per cent during the investigations. The symptomatology and laboratory examination of the isolate ruled out the possible association of any fungal or bacterial pathogen indicating thereby the viral etiology of the disease. Surveys conducted in major *Lilium* growing areas of the state resulted in the collection of seven isolates (L_1 - L_7) representing different localities and these were maintained under insect proof glasshouse conditions. After careful comparison of the reaction of these isolates on certain indicator plants, these were considered to be the same isolate as all the isolates infected similar indicator plants and also produced similar type of symptoms in the indicator plants used for studies. These studies indicate that LSV is the only virus infecting *Lilium* in Himachal Pradesh.

Symptomatology

The symptoms evoked by LSV under field as well as under insect proof glasshouse conditions included mild mottling of the leaves (Plate 1), chlorotic streaks between veins, development of small brown spots on the underside of the leaves turning necrotic. The flowers produced in axils of the lower stem leaves dry up prematurely. Virescence and bud/ floral deformation were also commonly observed (Plate 2). Several infected plants exhibited early yellowing of the lower stem leaves resulting in early senescence of the entire plant. The virus was also observed to be reducing the bulb size significantly. Allen (1975) and Boontjes (1978) also observed similar symptoms in *Lilium* infected with LSV.

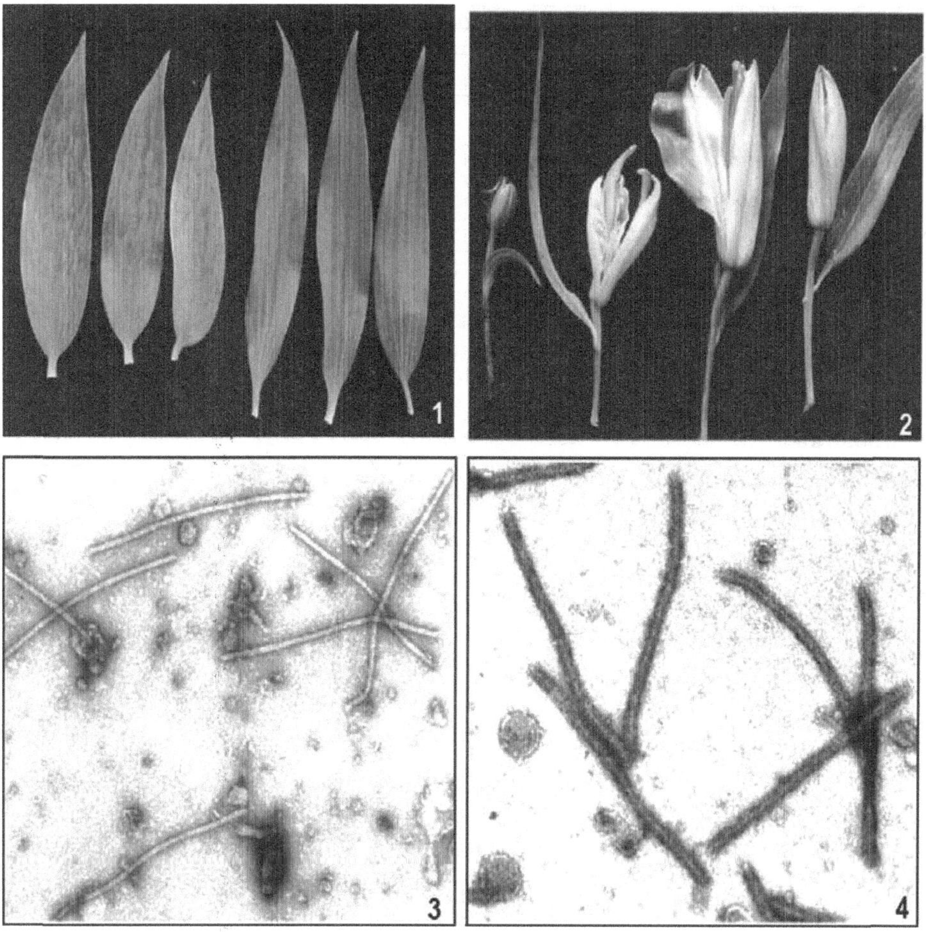

Plate 1. Mottled leaves (test) from infected plants in comparison to the healthy leaves (right); **Plate 2.** Deformation of floral buds in comparison to a healthy bud **Plate 3.** Electron micrograph of *Lilium* virus isolate showing filamentous flexuous virus particles; **Plate 4.** Electron micrograph of *Lilium* virus isolate decorated with lily symptomless *carlavirus* antiserum

Transmission and host range

A detailed study of the transmission characteristics of the virus isolate revealed that the virus was transmissible through sap, bulb and aphid vectors (*Myzus persicae* Sulz., *Aphis craccivora* Koch. and *Brevicoryne brassicae* L.). Results of the experiment conducted to study the host range of virus isolate revealed that the virus had a narrow host range (Table 1) restricted largely to the members of family liliaceae, chenopodiaceae, and solanaceae. These results are in line with the findings of Brierley and Smith (1944) and Allen (1972) who reported LSV to be transmissible through sap. These findings are in accordance with the report suggesting that LSV has a narrow host range often restricted to the family liliaceae only (McWhorter and Allen 1964; Allen, 1972).

Table 1. Host range of *Lilium* virus isolate

Family/host	Symptom
Amaranthaceae	
Amaranthus caudatus L.	—
Gomphrena globosa L.	—
Chenopodiaceae	
Chenopodium album L.	—
C. amaranticolor Coaste and Reyn	LL,CR
C. quinoa Wild	—
C. murale L.	—
Compositae	
Zinnia elegans Jacq.	—
Callistephus chinensis (L.) Nees	—
Calendula officinalis L.	—
Cruciferae	
Brassica oleracea L. var. *capitata* L.	—
B. oleracea L. var. *botrytis* L.	—
Cucurbitacae	
Cucumis sativus L.	—
Luffa actangula L.	—
Leguminosae	
Vicia faba L.	—
Vigna sinensis Savi.	—
Pisum sativum L.	—
Lathyrus odoratus L.	—
Phaseolus vulgaris L.	—
Medicago sativa L.	—
Liliaceae	
Lilium sp.	Mo,FD
Tulipa sp.	PD
Alstroemeria sp.	M
Polemoniaceae	
Phlox drummondii Hook.	—
Ranunculaceae	
Delphinium ajacis L.	—
Solanaceae	
Capsicum annuum L.	—
Datura stramonium L.	—
D. metel L.	—

D. metel L. var *fastuosa* L.	—
Nicotiana glutinosa L.	—
N. tabacum L. var. White Burley	NS
N. xanthi	LNS
N. rustica L.	—
N. occidentalis	NL
N.clevelandii A. Gray	LNS
Lycopersicon esculentum Mill.	—
Physalis floridana Rydb.	—
Solanum nigrum L.	—
S. khasianum Clarke	—
Nicandra physaloides Gaerth	—

LL = Local lesions; PD = Petal deformation; CR = Chlorotic rings; NL = Necrotic lesion; FD = Flower deformation; NS = Necrotic streak; Mo = Mottling; M = Mosaic; LNS = Local necrotic spots

Serology

Indirect DAC-ELISA tests were performed to establish serological relationship of the causal virus with some other viruses belonging to different genera namely, *Carlavirus, Potexvirus, Tobamovirus, Nepovirus, Cucomovirus* and *Tospovirus* reported to be associated with *Lilium* (Table 2).

Table 2. Reaction of *Lilium* virus isolate to different antisera in direct antigen coating indirect ELISA system

Antibody	Genus	Mean O.D. value ($A_{405\ nm}$)
CVB	*Carlavirus*	0.575
PVX	*Potexvirus*	0.144
ToMV	*Tobamovirus*	0.172
CLV	*Carlavirus*	0.610
PVY	*Potyvirus*	0.237
TRSV	*Nepovirus*	0.213
LSV	*Carlavirus*	1.838
CMV	*Cucumovirus*	0.148
TSWV	*Tospovirus*	0.143
Healthy control	—	0.133

Results obtained in DAS-ELISA with LSV antiserum revealed that the *Lilium* virus isolate under investigation reacted positively with LSV (Table 3). The optical density (O.D.) values obtained through PC-based ELISA plate reader confirmed these findings as the infected samples had very high O.D. value in comparison to healthy control. The studies revealed that the virus isolate under investigation reacted positively with antiserum against *carlavirus* and failed to react with antiserum against other viruses representing different genera thereby clearly indicating that *Lilium* virus isolate under investigation is a member of the genus *carlavirus*.

Table 3. Serological detection of LSV using DAS-ELISA in *Lilium* virus isolate

Antibody	Mean O.D. value ($A_{405\ nm}$)	
	Infected	Healthy
LSV	2.218	0.197
CVB	0.201	0.189
CLV	0.224	0.201

A good indicator host yielding high virus concentrations round the year is a pre-requisite for maintaining virus cultures. With this objective in mind, certain indicator plants were inoculated with the virus isolate and indexed serologically used DAC-ELISA. The O.D. values were recorded to establish the most suitable indicator host for maintaining the virus isolate. The study revealed that *Lilium*, tulip and alstroemaria were the best indicator hosts for maintaining the virus isolate under study as they recorded high O.D. values in comparison to *N. tabacum* var. White Burley, *N. glutinosa* and *C. amaranticolor*.

In addition to *Lilium*, DAS-ELISA tests of different plant parts of tulip and alstroemaria were done in order to study the distribution of LSV in these secondary hosts of the virus. The result revealed that tulip flower had a high virus concentration as compared to the other plant parts used for detection.

Optimization of the concentration of the immunoreagents

Experiment was conducted to standardize the concentration of antigen antibody and conjugate for the detection of LSV in DAS-ELISA. Results of the experiment showed that Ag and Ab diluted up to the level of 10^{-1} and 10^{-4}, respectively and the test conjugate could be diluted maximum to the level of 10^{-3} for obtaining positive serological reaction.

For the optimization of concentration of immunoreagents in DAS-ELISA similar dilutions were made and results showed that antigen and antibody diluted to 10^{-1} and 10^{-4}, respectively and the dilution of conjugate at 2×10^{-3} could be used for obtaining positive results.

Distribution of LSV in different plant parts

It is evident from the data in Table 4 that the highest O.D. value of 2.135 was obtained in case of leaves of the virus infected *Lilium* followed by the stem (2.017) and least O.D. of 0.917 was recorded in case of roots.

Table 4. Distribution of LSV in different plant parts of *Lilium*

Plant part	Mean O.D. value ($A_{405\ nm}$)	
	Infected	Healthy
Leaves	2.135	0.213
Flower	1.662	0.209
Stem	2.017	0.245
Bulb	1.915	0.226
Roots	0.917	0.201

The data clearly indicates that LSV can be reliably detected from all plant parts of *Lilium*. Under the present investigations, *Lilium* leaves recorded the maximum concentration of LSV as indicated by high O.D. values in comparison to flowers, stem, bulbs and roots. These results are substantiated by the findings of Kim *et al.* (1995) and Allen (1996) who reported that the leaves of LSV infected plants have higher virus concentration than other plant parts. The study also revealed that LSV was not present in detectable limits in the pollens of *Lilium* flowers which are further substantiated by the fact that LSV is not transmissible through pollen (EPPO, 2002).

Distribution of LSV in propagative organs

In an experiment, different vegetative propagative organs (bulbs, bulbils, bulblets) and seeds were compared to observe the relative distribution of LSV in these propagative organs. The results (Table 5) indicate that bulbs had a higher concentration of virus as indicated by O.D. value of 1.187 in comparison to relatively low O.D. values of 0.820 in case of bulblets and 0.522 in bulbils. The virus was however present in detectable limits in all the propagative organs except for the seeds.

Table 5. Distribution of LSV in propagative organs and seeds

Propagative organ	Mean O.D. value ($A_{405\ nm}$)	
	Infected	Healthy
Bulb	1.187	0.196
Bulbils	0.522	0.189
Bulblets	0.820	0.195
Seeds	0.238	0.206

Distribution of LSV in bulbils of different ages

An experiment was conducted to detect LSV in the bulbils of different age groups (3, 7 and 15 days old) and the scales (outer, middle and inner). The results set out in the Table 6 indicate that 15 days old bulbils recorded highest O.D. value of 1.399.

Table 6. Distribution of LSV in bulbils of different ages

Age of the bulbils	Mean O.D. value ($A_{405\ nm}$)	
	Infected	Healthy
Bulbils(3 days old)	0.279	0.254
Bulbils (7 days old)	0.740	0.214
Bulbils (15 days old)	1.399	0.232
Bulbils (7 days old)		
Outer scales	0.802	0.211
Middle scales	0.353	0.236
Inner scales	0.268	0.225
Bulbils (15 days old)		
Outer scales	0.961	0.241
Middle scales	0.553	0.239
Inner scales	0.368	0.241

However, LSV was not present in detectable limit in 3 days old bulbils. Among different scales, LSV was present in detectable limits in the outer and middle scales of 15 day old bulbils. The virus was not present in detectable limits in the inner scales in bulbils of all age groups. The outer scales had the maximum and the inner scales had the minimum virus concentration. However, the distribution of virus was erratic in different scales. Similar observations have been made by Lawson and Hsu (1996) and Beijersbergen and Van der Hulst (1980).

Electron microscopy

Electron microscopic studies of the virus isolate under investigation revealed the presence of elongated flexuous virus particles (Plate 3) measuring 640 nm in length and 17-18 nm in diameter in negatively stained preparation from epidermal peel of the infected *Lilium* leaves exhibiting mild mottling. These results are substantiated by the findings of Derks and Vink Van den Abeele (1980) who also observed flexuous virus particles in leaf dip preparation from LSV infected *Lilium*. The virus particles contrasted well when stained negatively with 1 per cent uranyl acetate under present investigations. Similar observations have been made by Civerolo *et al.* (1968) who observed that dip preparations negatively stained in 2 per cent sodium phosphate or 1 per cent uranyl acetate yielded well contrasted virus particles when observed under electron microscope.

Immunosorbent electron microscopy (ISEM)

In ISEM studies, the virus particles were well decorated with antiserum against LSV thereby indicating the serological identity of the virus isolate under investigation as LSV (Plate 4). Derks and Vink Van den Abeele (1980) also reported that LSV is most reliably detected by negatively stained preparations and ISEM using LSV antiserum to trap and/ or decorate the virus. Lyons and Allen (1969) observed electron microscopy to a useful aid in diagnosing LSV infections in *Lilium*. Under the present investigations as well, electron microscopic studies hinted at the possible involvement of lily symptomless *carlavirus* in *Lilium* infections in Himachal Pradesh on the basis of particle morphology, size and immunosorbent electron microscopy (decoration).

REFERENCES

Allen, T.C. (1972). Lily symptomless virus. *CMI/AAB Descriptions of Plant Viruses*, Wellesbourne, UK: Association of Applied Biologists. No 96, 4 p.

Allen, T.C. (1975). Viruses on lilies and their control. *Acta Hortic.* 47: 69-75.

Allen, T.C. (1996). Lily symptomless virus. AAB description of plant viruses. http://www.res.bbsrc.ac.uk/webdpr/web/notes.asp? genus=carlavirus.

Beijersbergen, J.C.M. and Van der Hulst, C.T.C. (1980). Detection of lily symptomless virus (LSV) in bulb tissue of lily by means of ELISA. *Acta Hortic.* 109: 487-493.

Boontjes, J. (1978). Vrgelij King van bolprotuctie van virusvrije en lelielatentvir-uszieke lelies. *Bloembollencultuur* 88: 1219.

Brierley, P. and. Smith, F.F. (1944). Studies on lily virus diseases: the necrotic fleck complex of *Lilium longiflorum*. *Phytopathology* 34: 529-555.

Civerolo, E.L., Semancik, J.S. and Weathers, L.G. (1968). Partial purification of virus -like particles associated with the necrotic fleck disease of Easter lily. *Phytopathology* 58: 1481-1486.

Clark, M.F. and Adams, A.N. (1977). Characteristics of microplate method of enzyme linked immunosorbent assay for the detection of plant viruses. *J. Gen. Virol.* 34: 475-483.

Derks, A.F.L.M. and Vink-Van den Abeele, J.L. (1980). Purification of lily symptom-less virus. Use and value of antisera against intact and pyrrolidine- degraded virus for testing lilies and tulips. *Neth. J. Plant Pathol.* 86: 239-250.

EPPO. (2002). PQR Database (Version 4.1). European and Mediterranean Plant Protection Organization, Paris, France.

Handa, Anil and Bhardwaj, S.V. (1994). Comparitive study of the use of alkaline phosphatase and penicillinase based direct antigen coating ELISA for the detection of a *potyvirus* from faba bean. *FABIS* 34/35: 36- 38.

Khurana, S.M.P. and Garg, I.D. (1989). Plant virus, viroid and mycoplasma like organisms detection for quarantine. *Rev. Trop. Plant Pathol.* 6: 251-277.

Kim, J.Y., Shin, H.K., Lee, S.Y., Lim, J.H., Ko, J.Y., Chung, S.K. and Choi, J.K. (1995). Detection of lily symptomless virus, cucumber mosaic virus and tulip breaking virus according to the tissue positions and organs in lily plants by tissue blotting immunoassay. *J. Korean Soc. Hortic. Sci.* 36: 843-850.

Lawson, R.H. and Hsu, H.T. (1996). Lily diseases and their control. *Acta Hortic.* 414: 175-187.

Lyons, A.R. and Allen, T.C. Jr. (1969). Electron microscopy of virus like particles associated with necrotic fleck of *Lilium longiflorum*. *J. Ultrastruc. Res.* 27:198-204.

McWhorter, F.P. and Allen, T.C. (1964). Transfer of lily curl stripe by a leaf union method applicable to monocotyledonous plants. *Nature* 204: 604-605.

Integrated Plant Disease Management (2005)
Eds. Sharma, R.C. and Sharma, J.N.
Scientific Publishers (India), Jodhpur, pp. 265 - 269.

42

Incidence of lily symptomless *carlavirus* in poly-house grown *Lilium* in Himachal Pradesh

BUNTY SHYLLA[1], ASHUTOSH BHAIK, ANIL HANDA and P. D. THAKUR

[1]Department of Soil Science and Water Management
Department of Mycology and Plant Pathology, Dr. Y.S. Parmar University of Horticulture and Forestry, Nauni, Solan – 173 230 (H.P.)

ABSTRACT

Lilium is one of the highly prized cut flowers of commercial market due to its fragrance and long vase life. In India, particularly in Himachal Pradesh, it is fast emerging as the most promising cut flower. However, commercial *Lilium* production is often hampered by lily symptomless *carlavirus* (LSV) infections which result in unmarketable flowers and severe reduction in the bulb size leading to a drastic reduction in economic returns. Extensive surveys were conducted in seven districts of Himachal Pradesh for recording the incidence of LSV in polyhouse grown *Lilium*. The incidence ranged between 15-95 per cent in different localities. Maximum disease incidence of 95 per cent was registered in cultivar Tiber at Mahog in Solan district and a minimum of 15 per cent was registered in cultivar Pisaro at Shamshi in Kullu district. In general, none of the localities surveyed were free from the symptoms of lily symptomless *carlavirus*.

INTRODUCTION

Lilies (*Lilium* sp.) are amongst the most preferred fragrant ornamentals around the world. Lilies are largely grown under polyhouses for producing good quality flowers. In India, particularly in Himachal Pradesh, *Lilium* is fast emerging as the most promising cut flower and is the leading cut flower grown under polyhouse conditions. However, commercial *Lilium* production under protected cultivation is often hampered by lily symptomless *carlavirus* (LSV) infections which result in unmarketable flowers and severe reduction in the bulb size leading to a drastic reduction in economic returns. Extensive surveys were conducted in seven districts of Himachal Pradesh for recording the incidence of LSV in polyhouse grown *Lilium*. The status of lily symptomless *carlavirus* incidence in different parts of the state under polyhouse conditions is presented in this communication.

MATERIALS AND METHODS

Extensive surveys were conducted in different *Lilium* growing areas of different districts of Himachal Pradesh during the year 2002 to ascertain the distribution and incidence of virus diseases of *Lilium*. Incidence counts were made by choosing 4-5 locations at random and observations recorded regularly by counting the number of healthy and diseased plants.

$$\text{Per cent disease incidence} = \frac{\text{Number of diseased plants}}{\text{Total number of plants observed}} \times 100$$

RESULTS AND DISCUSSION

The data in Table 1 indicate that there was variation in the disease incidence in different locations surveyed. The disease incidence ranged from 15-95 per cent during the investigations. Maximum disease incidence of 95 per cent was registered in cultivar Tiber at Mahog and a minimum of 15 per cent in cultivar Pisaro at Shamshi. A high disease incidence ranging from 35-80 per cent in different *Lilium* cultivars was recorded at Nauni. However, none of the localities surveyed were free from the symptoms of virus disease(s).

Various pathological agents namely fungi, bacteria and viruses infects *Lilium* in nature (Lawson and Hsu, 1996). Amongst them, viruses are one of the most important pathogens reported to infect *Lilium* and twenty different viruses are known to be associated with this crop under natural and experimental conditions in different regions of the world (Lee, 1992). Out of these viruses, LSV is the most widely distributed virus infecting *Lilium* around the globe and is also reported to cause huge economic losses (Kim *et al.*, 1995).

Incidence of virus disease(s) in different localities surveyed ranged between 15-95 per cent during the present investigations. Viruses in *Lilium* have been reported to cause moderate to high level of incidence in *Lilium* in many parts of the world. Kim *et al.* (1995) from Korea reported incidence of LSV ranging from 11-98 per cent in different *Lilium* cultivars and observed that almost all the commercial cultivars are susceptible. The varying level of LSV incidence in different *Lilium* cultivars is probably a resultant of several factors such as diversity of *Lilium* cultivars grown, cultural practices followed, sources of propagating material, planting density and most of all presence or absence of efficient aphid vectors.

Table 1. Incidence of virus disease(s) in different *Lilium* growing areas of Himachal Pradesh

District	Locality	Cultivar	Disease incidence (%)
Kullu	Seobagh	Cordiella, Prato	
		London	40
		Casablanca, Star Gazer	35
		Starfighter	45
		Sensor	55
		Snowqueen	65
		Pollyana	30
	Shamshi	Cordiella, Simplan	35
		Bronello, Elite	30
		Star Gazer, Acapulaco	50
		Starfighter	20
		Sensor, Siberia	25
		Pisaro	15
		Novacento	40
		Nerone, Ameron	45
Solan	Mahogbagh	Star Gazer (pink), (white) Le reve, Rosetto	60
		Tiber	40
		Expression, Siberia	45
		Woodriffs Memory, Mediterranean	80
		Acapulaco	90
		Star Fighter, Marco Polo	75
		Policovita	35
		Pesaro, Pisaro Double	70
		Cassandra,Casablanca,Cherry Blossom	65
		Cascade	55
	Mahog	Tiger	40
		Expression, Siberia	45
		Woodrifs Memory, Jolanda, Make up	80
		Tiber	95
		Aucapulaco	90
		Milano	65
		Spring Pink	75
		Apeldoorn	85

Continue.....

District	Locality	Cultivar	Disease incidence (%)
	Nauni	Alaska, Elite	45
		Prato, Mona	50
		Star Gazer	70
		Jollanda	80
		Appeldoorn, Harmony, London	65
		Marco Polo	55
		Alliance	40
		Roseto	35
Mandi	Keran	Alaska	35
		Prato, Siberia	55
		Mona	45
		Sensor	75
		Pisaro	60
		Elite	70
	Chakkar	Novacento, Ameron	35
		Acapulaco	40
		Simplan, Star Gazer	45
		Neron	65
		Jollanda	50
Kangra	Holta	Adelina	60
		Alaska, Cavi	40
		America	85
		Brunello, London, Pollyana	50
		Nerone	20
		Novecento, Romano	45
		Prato, Venix	65
	Bhattu samulan	Galieli	80
		Mediterrannee	55
		Star Gazer Max	45
		White Merostar	60
		Apeldoorn, Cantor	50
		Elite	65
		Kees Nelis	70
		Golden Apeldoorn	65
Shimla	Shoghi	Apeldoorn, Prato	60
		Alasks, Elite	65
		Nove Cento	75

Continue....

District	Locality	Cultivar	Disease incidence (%)
	Mashobara	Mona, Jollanda, Marco Polo	45
		Sensor, Cascade	55
		Star Gazer	60
		Acapulaco	65

REFERENCES

Kim, J.Y., Shin, H.K., Lee, S.Y., Lim, J.H., Ko, J.Y., Chung, S.K. and Choi, J.K. (1995). Detection of lily symptomless virus, cucumber mosaic virus and tulip breaking virus according to the tissue positions and organs in lily plants by tissue blotting immunoassay. *J. Korean Soc. Hortic. Sci.* 36:843-850.

Lawson, R.H. and Hsu, H.T. (1996). Lily diseases and their control. *Acta Hortic.* 414: 175-187.

Lee, K.H. (1992). *Disease Incidence, Symptomatology and Identification of Two New Isometric Viruses from Imported Lilies.* M.Sc. Thesis, Chunnam National University.

Integrated Plant Disease Management (2005)
Eds. Sharma, R.C. and Sharma, J.N.
Scientific Publishers (India), Jodhpur, pp. 271 - 275.

43

Production of *Lilium* hybrid cv. Marco Polo plantlets resistant to fusaric acid through somaclonal variations

MADHURIMA BHATNAGAR, S.V. BHARDWAJ, P.D. THAKUR, POOJA BHATNAGAR and AMIT KUMAR

Department of Biotechnology, Dr. Y.S. Parmar University of Horticulture and Forestry, Nauni, Solan - 173 230 (H.P.)

ABSTRACT

Lilium is one of the leading cut flower all over the world. Fusarium wilt caused by *Fusarium oxysporum* f. sp. *lilii* is a big menace to the cultivation of this crop. While control of this pathogen by fungicides has been reported, this approach is costly and often ineffective. The ideal method of control therefore, is to obtain resistant cultivars. Under present investigations, attempts were made to produce Fusarium wilt resistant plants of *Lilium* hybrids cv. Marco Polo through *in vitro* selection against fusaric acid. For this, callusing protocol of *Lilium* was standardized as a prerequisite. Friable embryogenic and creamish green calli were obtained from bulb scale explants on MS medium supplemented with 2 mg/l 2,4-D and 1 mg/l BA. MS medium supplemented with 1.0 mg/l NAA and 0.5 mg/l BA was used for callus maintenance. Organogenesis took place after 30 days of callus multiplication. Calli were subjected to different concentrations of fusaric acid. Selective dose of fusaric acid was found to be 0.4 mM. Calli survival at selective dose of fusaric acid was 22 per cent. The selected calli were cultured on MS medium containing 0.3 per cent activated charcoal before transferring to shoot regeneration medium. Shoot regeneration occurred in all media tried. Quick organogenesis took place on MS medium supplemented with 2.5 mg/l BA.

INTRODUCTION

Lilium is one of the leading cut flowers all over the world because of its large flowers, a diverse array of colours, wide range of form and long vase life. It ranks fourth among the ten top cut flowers of the world (Anonymous, 1996). The area under flower cultivation in Himachal Pradesh is approximately 154 hectares. Out of this, presently nearly 2 hectares is under *Lilium* cultivation which has drastically reduced during 2001 from 8 hectares in 1995 due to pre and post-harvest diseases and poor bulb management practices. Fusarium wilt caused by *Fusarium oxysporum* Schlechtend ex Fr. f. sp. *lilii* Imle is one such disease inflicting heavy damage to the crop. General symptoms of the disease vary from yellowing and wilting of the foliage to rotting of bulbs.

Plantlets of *Lilium* hybrid cv. Marco Polo resistant to fusaric acid developed through somaclonal variations is reported in the present communications

MATERIALS AND METHODS

Various types of explants such as bulb scales and leaves were subjected to sterilization treatments viz., $HgCl_2$ (0.1%) for 3-4 minutes, a pre-treatment of Bavistin (0.2%) for 8 minutes and rinsing three to four times with sterile distilled water inside a laminar air flow cabinet. The explants were cultured on basic nutrient medium Murashige and Skoog (1962) supplemented with different concentrations of cytokinins viz., BA and auxins (2, 4 - D and NAA). Out of various explants used, the bulb scales proliferated and were found the best. All the cultures were kept in a culture room maintained at 25^0C illuminated by white fluorescent tubes. The surface sterilized explants were inoculated on MS medium supplemented with various combinations of growth regulators i.e. 2, 4 - D, NAA and BA for the establishment of callus cultures. Proliferated calli from the established cultures were transferred to fresh multiplication medium containing the growth supplements i.e. NAA and BA after 4-6 weeks of establishment. The best combination of growth regulators was selected after the first subculture and calli were further maintained on this medium. Repeated subculturings were done to produce somaclones.

Preparation and testing the toxicity of fusaric acid

Sigma branded fusaric acid solution in mM was prepared using autoclaved distilled water. The toxicity of the chemical was tested by normal toxicity assay method (Thakur *et al.*, 2002) before using the same for developing somaclones.

The selective medium was prepared by mixing the filter sterilized fusaric acid with pre-sterilized molten MS medium under aseptic conditions so as to achieve v/v concentrations of fusaric acid ranging from 0.05-1 mM. The calli were chopped into small pieces of approximately 1 mm diameter and inoculated on the selective medium under laminar air flow cabinet. Highest concentration of fusaric acid at which calli survived, was recorded. The surviving calli were further sub-cultured at an interval of 20 days on the same concentration of fresh selective medium for second selection cycle. The selected calli were multiplied on callus maintenance medium and transferred on shoot regeneration medium containing BA and NAA. The regenerated shoots were then multiplied on already standardized multiplication medium BA 1mg/l and NAA-0.5 mg/l (Dilta, 1997) and the plantlets were developed and maintained.

RESULTS AND DISCUSSION

Thakur (2001) worked on *Lilium* micropropagation and found 0.2 per cent Bavistin for 8-9 minutes and 0.1 per cent mercuric chloride for 3-4 minutes to be the best surface sterilants of explants. Under present investigations also same sterilants have been found to be the best. Bulb scale segments of 0.5-0.7 cm size were used for establishment of callus cultures and were found to be the best for callus establishment (Simmonds and Cumming, 1976; Chen, 1983). The basal medium of Murashige and Skoog (1962) was found to give satisfactory results for all stages during *in vitro* culture of *Lilium* callus. The best quality callus which was friable, embryogenic and creamish green in colour, was induced on MS medium containing 2 mg/l 2,4 - D and 1mg/l BA. On this medium, callus initiation took place with in 18.33 days (Table 1).

Table 1. Callus initiation in *Lilium* hybrid Marco Polo in MS medium supplemented with growth regulators

Medium	Days taken for callus initiation
MS medium + 0.25 mg l^{-1} NAA + 0.1 mg l^{-1} BA	42.67*
MS medium + 0.5 mg l^{-1} NAA + 0.25 mg l^{-1} BA	32.34
MS medium + 1.0 mg l^{-1} NAA + 0.25 mg l^{-1} BA	32.67
MS medium + 1.0 mg l^{-1} NAA + 0.5 mg l^{-1} BA	28.67
MS medium + 1.0 mg l^{-1} NAA + 1.0 mg l^{-1} BA	27.00
MS medium + 1.5 mg l^{-1} NAA + 1.0 mg l^{-1} BA	24.33
MS medium + 2.0 mg l^{-1} NAA + 1.0 mg l^{-1} BA	21.67
MS medium + 1.0 mg l^{-1} 2,4-D + 1.0 mg l^{-1} BA	22.33
MS medium + 1.5 mg l^{-1} 2,4-D + 1.0 mg l^{-1} BA	21.34
MS medium + 2.0 mg l^{-1} 2,4-D + 1.0 mg l^{-1} BA	18.33**
$CD_{0.05}$	0.93

* = Least effective; ** = Statistically more effective

This study confirms that 2,4-D is an important factor for callus initiation in *Lilium* but if it is used at high concentration, it has inhibitory effect on subsequent callus growth. These findings are supported by Kato and Yasutako (1977) who reported callus growth inhibition by 2,4-D. After establishment of callus, its multiplication proceeds most luxuriantly on MS medium supplemented with 1.0 mg/l NAA and 0.5 mg/l BA. A change in growth regulators in the multiplication medium i.e. NAA 1 mg/l and BA 0.5 mg/l instead of 2,4-D 2 mg/l and BA 1 mg/l proved to be advantageous showing good callus growth and took only 20 days for multiplication. It was observed that callus maintenance for long period was difficult, as quick organogenesis took place after multiplication. So callus was maintained for long period by excising the bulblets which were developed during the early stages, such results have already been reported in literature (Kato and Yasutako, 1977; Simmonds and Cumming, 1976). Toxicity of fusaric acid was assessed 100 per cent of the cells and these were stained red with erythrocin B, when in fusaric acid.

In vitro selection

The calli were cut into pieces of 1 mm and cultured on MS medium containing different concentrations of fusaric acid. It was observed that per cent survival of cell lines decreased with the increase in concentration of fusaric acid (Table 2). Cell selection was done at 0.4 mM level of fusaric acid where 22.00 per cent survival was observed. The surviving calli were again cultured at 0.4 mM fusaric acid level and it was observed that the cells which survived during first cycle of cell selection remained alive during the second cycle as well. The surviving calli were then multiplied and regenerated into plantlets. In all, 50 plantlets derived of 70 cultures were produced.

Table 2. Selection of resistant cell lines of *Lilium* hybrid Macro Polo against fusaric acid

Fusaric acid concentration (mM)	Survival of cell lines (%)	Callus colour	Growth of callus
0.00	100.00 (90.00)	Light green	++++
0.05	88.83 (70.73)	Yellowish green	+++
0.1	77.77 (61.96)	Yellowish green	++
0.25	69.40 (56.47)	Dark green	++
0.4	22.00 (28.02)	Dark blackish green	—
0.75	0.00 (0.00)	Dark blackish green	—
1.0	0.00 (0.00)	Dark blackish green	—
CD $_{0.05}$	6.38 (4.69)		

++++ = Normal growth of cell colonies; +++ = Satisfactory growth; ++ = Inhibition of cell lines; — = Complete inhibition

Fusaric acid is one of the most studied pathogen produced wilt toxins classified as a non-specific vivotoxin (Singh, 1984). Though, the literature available indicates involvement of a combination of several toxins and metabolites to produce the wilt syndrome, however, fusaric acid has been designated as the principal toxin. It is therefore, assumed and expected that plants which are resistant to fusaric acid, may also possess resistance against wilt. As far as selective dose of fusaric acid is concerned there is an isolated report of selection of resistant clones of gladiolus at 0. 12 mM (Remotti *et al.*, 1997).

The lower concentration may have been successful because of different plant materials or procedural variations. The selected callus was unable to regenerate shoots in the normal regeneration medium. So after desirable callus multiplication, the calli were transferred to MS medium containing 3 g/l activated charcoal to leach out inhibitors from the callus (Zaghmout and Torello, 1988). Shoot regeneration was observed in all the media tried but fastest growth took place in MS medium + BA 2.5 mg/l with in 10 days. The results are almost similar to those reported by (Kato and Yasutako, 1977; Mullin, 1970; Hussey, 1975). However, Okazaki and Koizumi (1995) could not get effective regeneration with BA. Difference in genotypes may have been responsible for such variations in *in vitro* studies.

REFERENCES

Anonymous. (1996). International flower trade show. Flower Auction Market Aalsmeer, Holland, 8-12 Nov., 1996.

Chen, W.M. (1983). Plant regeneration from bulb-scales and seedlings of *Lilium brownii* var. *vividulum in vitro. Plant Physiol. Commun. Zhiwu Shenglixue Tongxun.* 3: 46.

Dilta, B.S. (1997). *In vitro Propagation Studies in Lilium Hybrids.* M.Sc. Thesis, Dr. Y.S. Parmar University of Horticulture and Forestry, Nauni, Solan (H.P.).

Hussey, G. 1975. Totipotency in tissue explants and callus of some members of the liliaceae, iridacae and amaryllidaceae. *J. Exp. Biol.* 26: 253-262.

Kato, Y. and Yasutako, Y. (1977). Plantlet formation and differentiation of epidermal tissues in green callus cultures from excised leaves of lilium. *Phytomorphology.* 27: 390-396.

Mullin, M. (1970). Tissue culture of some monocotyledons plants. *Aust. J. Biol. Sci.* 23: 473-477.

Murashige, T. and Skoog, F. (1962). A revised medium for rapid growth and bioassays with tobacco tissue cultures. *Physiol. Plant.* 15: 473-479.

Okazaki, K. and Koizumi, M. (1995). Callus formation and regeneration of some species of lilium. *Acta Hortic.* 392: 97-105.

Remotti, P.C., Van Harmelen, M.J. and Loffler, H.J.M. (1997). *In vitro* selection for *Fusarium* resistance in gladiolus. *Acta Hortic.* 430: 605-610.

Simmonds, J.A. and Cumming, B.G. (1976). Propagation of lilium hybrids. II. Production of plantlets from bulb-scale callus cultures for increased propagation rates. *Sci. Hortic.* 5: 161-170.

Singh, R.S. 1984. *Introduction to Principles of Plant Pathology.* Oxford & IBH Publishing Co. Pvt. Ltd., New Delhi, 125 p.

Thakur, M., Sharma D.R. and Sharma, S.K. (2002). *In vitro* selection and regeneration of carnation (*Dianthus caryophyllus* L.) plants resistant to culture filtrate of *Fusarium oxypsorum* f.sp. *dianthi. Plant Cell Rep.* 20: 825-828.

Thakur, N. (2001). *Studies on Cost Effective Micropropagation of Lilium.* M.Sc. Thesis, Dr. Y.S. Parmar University of Horticulture and Forestry, Solan, H.P.

Zaghmout, O.M.F. and Torello, W.A. (1988). Enhanced regeneration in long term callus cultures of red fescue by pretreatment with activated charcoal. *Hortic. Sci.* 23: 615-616.

CEREAL, PULSE AND OILSEED CROPS

Integrated Plant Disease Management (2005)
Eds. Sharma, R.C. and Sharma, J.N.
Scientific Publishers (India), Jodhpur, pp. 279 - 287.

44

Status of resistance in popular wheat cultivars against major diseases in India

D.P. SINGH, A.K. SHARMA, M.S. SAHARAN, J. KUMAR and JAG SHORAN

Crop Protection Programme, Directorate of Wheat Research, Post Box No. 158, Karnal - 132 001 (Haryana)

ABSTRACT

Host resistance has been utilized to the maximum in wheat to manage the important diseases like stem, leaf and stripe rust, leaf blight, Karnal bunt, powdery mildew and loose smut. Breeding for rust resistance is the top priority and any advanced line proposed for All India Coordinated Yield Trials should possess the durable level of rust resistance. Further, it should also show tolerance to other diseases. The testing of promising entries is done against diseases under artificially inoculated conditions at hot spot locations simultaneously along with their testing in yield trials. The data thus generated is used effectively in promotion of particular entry in yield trials as well as during identification and release of varieties. The popular released varieties of wheat are always kept as check for comparison. Most of the old varieties like, HS 240, Sonalika, HD 2329, HD 2285, C 306 and LOK 1 have become susceptible to rusts and other diseases. Diseases like Karnal bunt and leaf blight need attention of breeders to utilize the resistant parents identified against these diseases in the crossing programme, so that resistance is incorporated in future varieties against these diseases. Resistant cultivars thus developed can be deployed in the field, wherever a particular disease is endemic and weather is favourable, to contain losses in yield and grain quality. During past three crop seasons (2000-01, 2001-02 and

2002-03), popular wheat cultivars were tested and their status of resistance against different diseases is discussed in the paper.

INTRODUCTION

Wheat (*Triticum aestivum* and *T. durum*) along with rice serves as life-sustaining crops for large Indian population. With increasing trend towards urbanisation and increase in the consumption of wheat, well being of Indian society is interlinked with the productivity of wheat crop. Organized wheat research in India is a century old, primarily initiated at the Imperial Agricultural Research Institute, Pusa, Bihar (now Indian Agricultural Research Institute, New Delhi). Against the current wheat production level of about 73 million tonnes, the projected demand for the 2020 AD is over 109 million tonnes. With increased wheat production, India attained status of the second largest wheat producing country in world and is next only to China. Wheat is the most important cereal in India, which ushered in an era of Green Revolution in the country. This happened mainly due to introduction of high yielding dwarf cultivars from Mexico, along with adoption of improved agronomic practices and curtailing the yield losses due to diseases by effectively managing them through host resistance. Major diseases of wheat in India are leaf rust, stripe rust, stem rust, foliar blights, Karnal bunt, powdery mildew and loose smut. Diseases of importance to a specific region and environment are hill bunt, foot rot and flag smut. Leaf or brown rust caused by *Puccinia recondita* f. sp. *tritici* Rob. ex Desm. (Fig. 1) is the most important rust, widely distributed and most frequently occurring in the Indian sub-continent (Joshi, 1976). Stripe or yellow rust caused by *Puccinia striiformis* West. (Fig. 2) is of importance to the hills as well as in the north-western plains. Stem or black rust caused by *P. graminis* ex Pers. f. sp. *tritici* Eriks. & Henn. (Fig. 3) is quite important and known to cause epidemics in the peninsular and central India. Karnal bunt (*Tilletia indica* Mitra) is a major seed-borne disease which is endemic in the north-western India, though it may be found sometimes in traces in eastern parts of India and the northern parts of central India. Loose smut caused by *Ustilago segetum* (Pres.) Russel var. *tritici* Jensen (Fig. 4) is important in the hills as well as the northern states of India. Foliar blight caused by *Bipolaris sorokiniana* (Sacc.) Shoemaker (Fig. 5), *Pyrenophora tritici repentis* (Died.) Drechs. and *Alternaria triticina* Prasada and Prabhu, is a complex disease and has been a major biotic stress in eastern parts of India due to high humidity and warmer temperatures, although, present in all the six agroclimatic zones in India. Powdery mildew caused by a biotrophic parasitic fungus *Erysiphe graminis* DC. f. sp. *tritici* Em. Marchal (Fig. 6) is a destructive foliar disease of wheat in the hilly and sub-mountainous regions of northern India (Goel *et al.*, 1986; Singh and Sood, 1977; Singh *et al.*, 1986). In favourable weather, it also spreads into Indo-Gangetic plains (Joshi *et al.*, 1978). Losses from this disease have increased in recent years due to the reason that most of the present day varieties including the most dominant variety PBW 343 do not possess sufficient level of resistance against this important disease. Although, the diseases of wheat are capable of causing serious yield losses, but no serious disease epidemic has occurred in India during last four decades due to the efforts taken by wheat scientists working under the umbrella of All India Co-ordinated Wheat Improvement Project. The resistant varieties thus developed can be best utilized and deployed stategically in problem areas of diseases for combating the wheat diseases. The present paper deals with the methodology of screening of entries of yield trials against diseases and about the status of resistance to diseases in present day popular wheat cultivars.

Fig. 1. Leaf rust

Fig. 2. Stripe rust

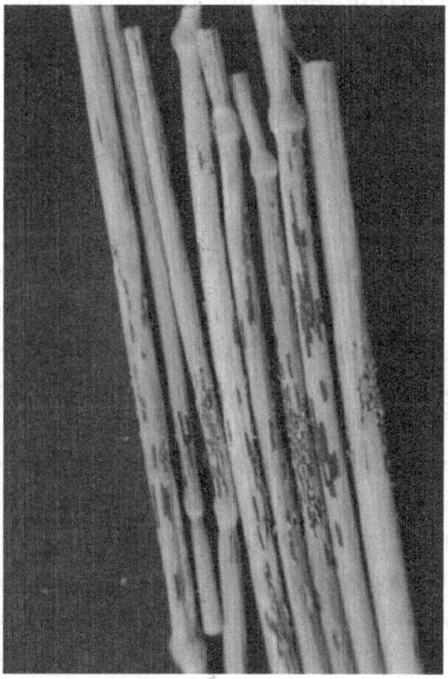

Fig. 3. Stem rust

Fig. 4. Loose smut

Fig. 5. Leaf blight - inital symptoms

Fig. 6. Powdery mildew

MATERIALS AND METHODS

As a well set procedure, the advanced lines of wheat along with standard check varieties of each zone are tested against diseases at hot spot multilocations under artificially inoculated conditions. The information presented here is based on the evaluation done during 2000-01, 2001-02 and 2002-03 crop seasons. Wheat lines were first tested against three rusts and leaf blight in Initial Plant Pathological Screening Nursery (IPPSN) before these are entered in the co-ordinated varietal trials, i.e. National Initial Varietal Trials (NIVT). The materials are promoted from NIVT to Advance Varietal Trials (AVTs) based on the performance of yield and resistance to major diseases. The entries of yield trials are simultaneously tested against diseases in PPSN (Plant Pathological Screening Nursery) at 10-12 hot spot locations all over the country. The emphasis on resistance against each rust varies from one agro-climatic zone to another.

For creation of artificial epiphytotics for rusts, inocula of leaf and stripe rusts, maintained at DWR Regional Research Station, Flowerdale, Shimla, were provided to the cooperators during the crop season for inoculation in field. The inoculum of different pathotypes of rusts was further multiplied at respective centres in polythene houses by sowing the crop in advance (Sep.-Oct.). The test entries of AVTs and NIVTs were planted in a definite design so that the nursery blocks have infector rows all around as well as after every 20 test entries to create proper infection pressure. The pots with infected plants raised in the polythene tunnels were also kept in the field close to the infector rows at seedling stage. The plants were inoculated using sprays of inoculum suspension as well as through syringe inoculation in the evening so that enough moisture is maintained due to dew. Under such conditions, only genetically resistant varieties remain free from diseases. For evaluation against Karnal bunt, five ear heads were syringe inoculated at crop growth stage 49 (Zadoks *et al.*, 1974) during late evening. The susceptible checks (HD 2009, WL 711) were planted after every 20 test entries in the plot. The perfo sprays were used in evening for creating high humidity for one month after inoculation. The disease severity was recorded at crop maturity and the per cent coefficient of infection was calculated. For blights, inoculum raised on sorghum seeds was sprayed on the plants and warm and humid environment conducive for disease development was created. For loose smut evaluation, lines were inoculated with loose smut teliospores at growth stage 59 of Zadok's scale (Zadoks *et al.*, 1974) using modified 'Go-go' method (Joshi *et al.*, 1988). Inoculation for powdery mildew was done by dusting the inoculum and scoring was done on 0-9 scale. Entries having average coefficient of infection up to 10.0 for rusts and 5 per cent infected grains due to Karnal bunt and 5 per cent loose smut and hill bunt infected tillers are considered resistant. Likewise for powdery mildew, the resistant entries have powdery mildew score up to 3, whereas for leaf blight, the average score is up to 35 on top two leaves, to qualify for resistance.

RESULTS AND DISCUSSION

The data received from all the hot spot locations were compiled at DWR, Karnal and entries as well as check varieties were categorized for resistance against one or more diseases. Based on three years data, following released varieties popularly grown in different agroclimatic zones have been identified as resistant against diseases mentioned against each of it:

(i) Resistant to leaf and stripe rust

Northern hill zone (NHZ): HPW 42, HS 375

Moderately resistant to powdery mildew: HS 420

Resistant to hill bunt: VL 804

North western plains zone(NWPZ): PBW 343, PBW 373, PBW 498, PBW 502, Raj 3765,WH 896 durum (d)

Resistant to Karnal bunt and loose smut: PBW 34 (d)

Karnal bunt, loose smut and flag smut: PDW 233 (d)

(ii) Resistant to leaf rust

North eastern plains zone (NEPZ): HUW 468, K 9107

Moderately resistant to leaf blight: HD 2733, HUW 533

(iii) Resistant to stem and leaf rust

Central zone (CZ): GW 322, DL 788-2

Peninsular zone (PZ): NIDW 295 (d), GW 322, HD 2189, HUW 510, LOK 45, NIAW 34, MACS 2846

(iv) Resistant to all three rusts

Southern hills zone (SHZ): HW 2044

All zones:

+ Soil alkalinity and salinity: KRL 19

d - *T. durum*

Most of the old varieties like, HS 240, Sonalika, HD 2329, HD 2285, C 306 and LOK-1 have become susceptible to rusts and other diseases. The diseases like Karnal bunt and leaf blight need attention of breeders to utilize the resistant parents identified against these diseases in the crossing programme, so that resistance is incorporated in future varieties against these diseases. The resistant cultivars thus developed can be deployed in the field wherever a particular disease is endemic and weather is favourable to contain losses in yield and grain quality. Based on the rigorous screening at hot spot multilocations over the years, the released varieties have stood the best of the times (Sharma *et al.,* 2001). The varieties recommended for various zones are enlisted below in Table 1.

Table 1. Some recommended wheat varieties for different agro-climatic zones of India

Agroclimatic zone	Recommended varieties
I. NORTHERN HILL ZONE	HS 240, HS 277, HS 295, HS 365, HS 375, HS 420, VL 616, VL 738, VL 804, HPW 42
II. NORTH WESTERN PLAIN ZONE	HD 2687, PBW 34, PBW 175, PBW 343, PBW 373, PBW 396, PBW 502, PDW 233, UP 2338, UP 2425, WH 542, WH 896, RAJ 3765
III. NORTH EASTERN PLAIN ZONE	DBW 14, HD 2733, HP 1731, HP 1761, HUW 468, HUW 533, HW 2045, K 8027, K 9107, NW 1012, NW 1014, PBW 343
IV. CENTRAL ZONE	DL 788-2, GW 173, GW 273, GW 322, GW 337, GW 1139, HD 4672, HI 8498, HW 2004, MP 4010, RAJ 1555, RAJ 4037
V. PENINSULAR ZONE	DWR 1006, GW 322, GW 1189, HD 2189, HD 2781, HI 977, HUW 510, K-9644, LOK 45, MACS 1967, MACS 2496, MACS 2846, NIAW 34, NIDW 15, NI 5439
VI. SOUTHERN HILL ZONE	HW 1085, HW 2044

The emphasis on breeding for disease resistance, especially for rusts is the top priority of the Indian wheat programme. The varieties released during the past, have different resistance gene combinations and the targeted gene deployment (Table 2) has helped in avoiding the outbreak or epidemic of the diseases especially the rusts during the last few decades. The regular surveillance and surveys should be continued in strategic zones to monitor the occurrence of new pathotypes of rusts especially of stripe rust along with Indo-Pak borders in NWPZ. The monoculture of one variety like PBW 343 and its sister lines should be discouraged over large areas to avoid major epidemic of rusts and other diseases. The cultivation of varieties with diverse genetic make up should be encouraged in each zone to create mosaic of resistant genes at village level.

Table 2. Recommended wheat varieties with different combinations of rust resistant genes

Varieties	Postulated genes		
	Lr	Sr	Yr
I. NORTHERN HILL ZONE			
HS 240	26+1+34+	31+2+	9+18+
HS 277	26+34+	31+	9+18+
HS 295	23+34+	8b+2+	3+18+
HS 365	26+1+	31+2+	9+18+
HS 375	26+34+	31+5+2+	9+18+
HS 420	13+10+1+34+	—	18+
VL 616	3+	11+	2+
VL 738	26+23+34+	31+	9+18+
VL 804	26+34+	31+5+2+	9+18+
HPW 42	26+1+34+	31+	9+18+
II. NORTH WESTERN PLAIN ZONE			
HD 2687	26+23+	31+2+	9+
PBW 343	26+34+	31+5+2+	9+18+
PBW 396	26+	31+2+	9+
WH 896	—	9e+2+	—
PBW 373	26+34+	31+5+	9+18+
PBW 34	—	9e+2+	—
PBW 175	23+24+	2+	2+18+
UP 2425	26+23+	2+31+	9+
UP 2338	26+34+	2+5+31+	9+18+
WH 542	26+34+	31+	9+18+
RAJ 3765	13+10+	2+	2ks+
PBW 502	—	31+5+2+	9+
PDW 233	—	9e+2+	9+18+
III. NORTH EASTERN PLAIN ZONE			
NW 1014	13+	11+	—
K-9107	34+	5+2+	2+18+
HUW 468	13+	2+	
HD 2733	26+	31+	—
HW 2045	—	2+	—
DBW 14	—	2+	—
HUW 533	—	2+	—
IV. CENTRAL ZONE			
GW 173	23+34+	11+9b+2+	2ks+18+
GW 273	13+10+	—	—
GW 322	13+	11+2+	—
HI 8498	—	2+	
Raj 1555	—	9e+	—
DL 788 -2	24+	24+5+2+	—
HW 2004	24+	24+2+	—

Continue....

Varieties	Postulated genes		
	Lr	Sr	Yr
HD 4672	—	9e+2+	
V. PENINSULAR ZONE			
HD 2189	13+34+	—	2ks+18+
MACS 1967	—	11+	—
MACS 2496	26+23+	31+2+	9+
MACS 2846	23+	9e+	—
HI 977	23+34+	11+9b+	18+
HD 2781	—	2+	2ks+18+
NIAW 34	13+34+	—	18+
NI 5439	—	11+	2+
DWR 1006	—	9e+	—
K-9644	13+1+	2+	—
VI. SOUTHERN HILL ZONE			
HW 1085	26+	31+2+	9+
HW 2044	—	2+	—

Source: Directorate of Wheat Research, Karnal

Therefore, The present status of resistance against diseases in popular wheat cultivars in different agroclimatic zones is satisfactory to avoid any out-break of epidemic. It is achieved due to rigorous testing of varieties even before their entry into the yield trials and thereafter for at least three crop seasons at hot spot multilocations. The strict compliance of norms for minimum level resistance at the time of identification and release of wheat varieties during past few decades is in real sense paying the dividends to the farmers and greatly helping in increasing and sustaining the wheat production in India. The released/identified varieties are evaluated as checks along with the new entries at hot spot locations and under artificially inoculated conditions. This type of multilocation testing acts as a 'wake up alarm' well in advance on breakdown of resistance in existing varieties and also on the appearance of new race(s) or pathotype(s). This helps in replacing the vulnerable varieties with new resistant ones, before the new pathotype or race becomes prevalent in the region.

REFERENCES

Goel, L.B., Singh, S. and Sinha, V.C. (1986). The mildew of wheat. In: *Problems and Progress of Wheat Pathology in South Asia* (Eds. Joshi, L.M., Singh, D.V. and Srivastava, K.D.), Malhotra Publishing House, New Delhi, pp. 176-190.

Joshi L.M., Singh, D.V. and Srivastava, K.D. (1988). *Manual of Wheat Diseases.* Malhotra Publishing House, New Delhi, 75 p.

Joshi, L.M. (1976). Recent contributions towards epidemiology of wheat rusts in India. *Indian Phytopathol.* 29: 1-16.

Joshi, L.M., Srivastava, K.D., Singh, D.V., Goel, L.B. and Nagarajan, S. (1978). *Annotated Compendium of Wheat Diseases in India.* Indian Council of Agricultural Research, New Delhi, 331 p.

Sharma, A.K., Singh, D.P. and Kumar, J. (2001). Concept and use of multilocation hot spot testing for identifying potential donor lines/varieties in wheat. In: *Role of*

Resistance in Intensive Agriculture (Eds. Nagarajan, S. and Singh, D.P.), Kalyani Publishers, Ludhiana, pp. 223-232.

Singh, B.M. and Sood, A.K. (1977). Seedling reaction of spring and winter wheat cultivars to powdery mildew. *Indian Phytopathol.* 30: 277-279.

Singh, S., Goel, L.B. and Sinha, V.C. (1986). Fungal diseases and their management. In: *Twenty Five Years of Coordinated Wheat Research 1961-86* (Eds. Tandon, J.P. and Sethi, A.P.), Wheat Project Directorate, AICWIP, New Delhi, pp. 127-157.

Zadoks, J.C., Chang, T.T. and Konzak, C.E. (1974). A decimal code for the growth stages of cereals. *Weed Res.* 14: 415-421.

Integrated Plant Disease Management (2005)
Eds. Sharma, R.C. and Sharma, J.N.
Scientific Publishers (India), Jodhpur, pp. 289 - 294.

45

Method for scoring of leaf blight of wheat caused by *Bipolaris sorokiniana* (Sacc.) Shoem. at adult stage on top two leaves and its validation

D.P. SINGH and PANKAJ KUMAR

Crop Protection Programme, Directorate of Wheat Research, Post Box No. 158, Karnal - 132 001 (Haryana)

ABSTRACT

Leaf blight of wheat caused by *Bipolaris sorokiniana* is a destructive disease of wheat. The disease is present in all the six agro-climatic zones in India and capable of causing losses in grain yield up to 50 per cent besides adversely affecting the seed quality in case of susceptible wheat cultivars. Most of the cultivars presently grown fall under moderately resistant to moderately susceptible category against leaf blight. The segregating breeding populations are later evaluated against leaf blight and other diseases at shuttle breeding centres which are also hot spots for diseases. A new double digit scoring method (0-9 scale) has been developed based on per cent area blighted on flag and a leaf below it at dough stage and also validated under controlled environmental conditions of polyhouse. The disease progress was checked at different levels of growth by protecting it with fungicide. Higher losses in yield due to blight on top two leaves has been estimated indicating the major share of flag (F) and F-1 leaf in grain setting and grain filling in wheat. The losses were 61.0, 33.7, 26.1 and 21.7 per cent in case of numbers of grains/spike, whereas those in respect of grain yield/spike were 80.2, 48.0, 37.6 and 24.4 per cent in the plants having leaf blight on the leaves except flag (F) leaf, except F and F-1 leaves and except F, F-1 and F-2 leaves respectively, as compared to completely healthy tillers. The new double digit method of recording of leaf blight of wheat is therefore, quite perfect and easy for rapid screening of breeding population of wheat.

INTRODUCTION

Leaf blight (spot blotch) of wheat caused by *Bipolaris sorokiniana* (Sacc.) Shoem. is an important disease of wheat India (Fig. 1) and analysis of blighted samples of wheat confirmed its presence in all the six agroclimatic zones. It is known to cause losses up to 50 per cent in grain yield as well as seed quality in susceptible cultivars (Singh *et al.*, 2002). Most of the cultivars presently grown are moderately resistant to moderately susceptible to leaf blight (Singh *et al.*, 2003). The resistant parents are now being used in breeding programme and material is screened under artificially created conditions at hot spot locations under shuttle breeding programme (Fig. 2). A standarized disease scoring system can be used for proper measurement of losses, epidemiological studies and to identify the suitable varieties. The degree of accuracy desired in assessment of disease however, varies according to the particular objectives of the research programme. In the past, a number of scales for the measurement of foliar diseases have been proposed in case of wheat and other cereals (Chester, 1950; James, 1971; Large, 1966). These are of detailed nature and time consuming for the preliminary evaluation of large populations for resistance. James (1971) proposed 23 keys based on the percentage of severity for measurement of several diseases of cereals. He used logarithmic scales and indicated that most of pathogens grow up in logarithmic rates. Horsfall and Barratt (1945) reported that the grades detected by the human eye are approximately equal divisions on a log scale and generally follow the Webber-Fechner law which states that visual acuity depends on the logarithm of the intensity of the stimulus. The disease rating as per scale (1-9) of Saari and Prescott (1975) is based on the progress of blight vertically and extent of leaf area blighted in double digit scale. The 0-2 score comes under resistant category, whereas 3 is moderately resistant. The scale is good for epidemiological studies and inadequate to evaluate a large number of breeding populations. The main shortcoming is that diseased area and type of lesion are not rated. Further, the lower and oldest leaves are first and most severely infected and become dry at dough stage. By using this scale, it was difficult to get a genotype with a score up to 3 at dough stage and most of these were of intermediate type with score of blight up to 5. The old scoring scale based on spread of disease vertically on plant height and scoring of average per cent leaf area has not proved useful in screening breeding population. Therefore, a new scoring scale based on per cent area blighted of flag and a leaf below it at dough stage has been proposed and validated.

MATERIALS AND METHODS

The present study was planned to validate the new double digit method of scoring the leaf blight based on per cent leaf area infected on flag and a leaf below it at dough stage. In this method, the data are recorded on flag (F) and one leaf below flag (F-1) at 71-77 stages on Zadoks' scale (1974), on double digit scale (0-9) as follows:

0 = No blight , 1 = Up to 10 per cent area of leaf blighted, 2 = 11-20 per cent area of leaf blighted, 3 = 21-30 per cent area of leaf blighted, 4 = 31-40 per cent area of leaf blighted, 5 = 41-50 per cent area of leaf blighted, 6 = 51-60 per cent area of leaf blighted, 7 = 61-70 per cent area of leaf blighted, 8 = 71-80 per cent area of leaf blighted, 9 = >81 per cent area of leaf blighted. The right hand digit indicates score of blight on F-1 and left hand digit represents the score on flag leaf. The average score of multilocation data was calculated by taking the average of both digits separately. Based on average score, the entries were classified in to following categories:

00-13 (resistant), 14-35 (moderately resistant), >35 (susceptible)

Fig. 1. Leaf blight caused by *B. sorokiniana*

Fig. 2. Leaf blight resistant (R) and susceptible (S) genotypes of wheat

The highest score of these entries amongst multilocation hot spots never exceeded 57 over years (Singh *et al.*, 2003). The estimates were made on the per cent share of top most two leaves in grain development in wheat. For this purpose, the plants of two highly blight susceptible wheat varieties Sonalika and Raj 4015 were grown inside the polyhouse. The congenial conditions of humidity (>90%) and temperature (25-30°C) for the development of leaf blight were created inside the polyhouse. The plants were inoculated with *B. sorokiniana* starting from 3-4 leaf stage till flag leaf. The control plants were kept healthy by spraying of Tilt 25 EC (propiconazole) @ 0.1 per cent as well as different levels of leaf blight were created by avoiding development of disease on flag (F), a leaf below flag leaf (F-1) and two leaves below flag leaf (F-2). Each tiller was tagged separately and ten replications were taken. The data on per cent leaf area infected, grains/ear head as well as grain weight/spike were measured. The data were analysed statistically.

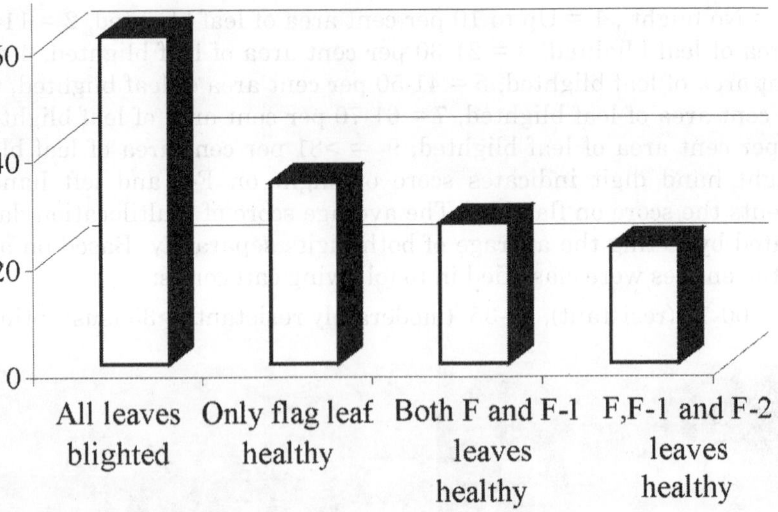

Fig. 3. Per cent losses in numbers of grains/spike due to different levels of leaf blight

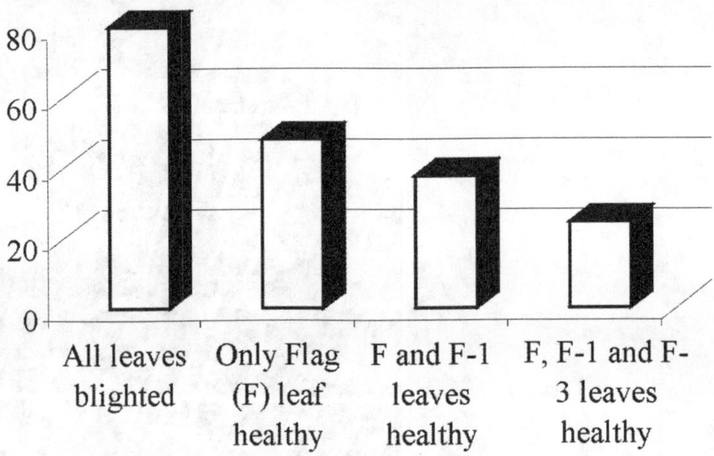

Fig. 4. Per cent losses in grain wt./spike under different levels of leaf blight

RESULTS AND DISCUSSION

The average per cent loss of grains/spike over healthy tiller, in case of plants having all the leaves blighted, only F leaf healthy, F and F-1 leaves healthy and F, F-1 and F-2 leaves healthy was 61.0, 33.7, 26.1 and 21.7 per cent (Fig. 3), whereas the losses in grain weight/spike were 80.2, 48.0, 37.6 and 24.4 per cent (Fig. 4), respectively. The losses in grain weight/spike as well as number of grains/spike were significantly reduced in plant where flag leaf was healthy as compared to total leaves blighted and gain was 32.2 and 26.3 per cent, respectively. Per cent gain in yield parameters like grain weight/spike and numbers of grains/spike was 42.6 and 35.0, respectively, in case of plants having F and F-1 leaves healthy in comparison to plants with all leaves blighted. Higher yield losses due to blight on top leaves indicates the major share of flag leaf in grain setting and grain filling in wheat. The number of discoloured seeds/spike (>80% seeds) were significantly increased in case of tillers showing leaf blight up to flag leaf (Fig. 5). Results of this study indicate a major role of top two leaves of plant at adult stage in photosynthesis and their greater share in contributing in grain filling. The early blight appearance on flag leaf had most adverse effect on grain yield. Perello *et al.* (1998) also proposed a 1-9 scale of recording of leaf blight of wheat caused by *Alternaria triticimaculans*, based on magnitude of disease on top leaf at flag leaf stage. The score 1 denotes no disease, 2 = Up to 5 per cent, 3 = 6 – 12 per cent, 4 = 13 – 20 per cent, 5 =21-35 per cent, 6 =36 – 45 per cent, 7 = 46 - 60 per cent, 8 = 61 – 80 per cent and 9 = >80 per cent area of leaf affected due to blight. The diseased leaf area in relation to the total leaf area was determined in laboratory by bringing the leaf samples after flag leaf stage and measuring the leaf using millimeter paper. This is however, difficult to use for screening the large numbers of breeding population.

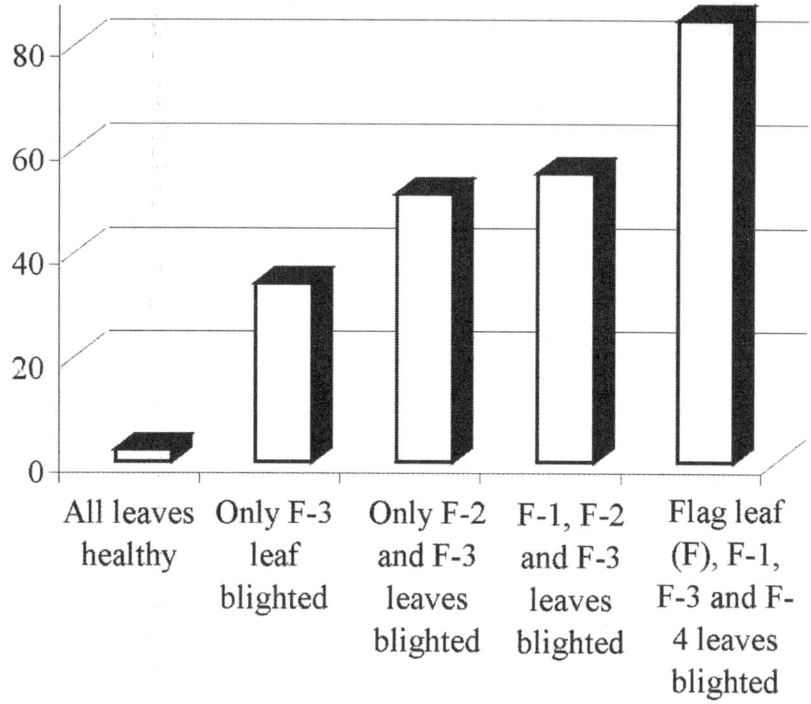

Fig. 5. Percentage of discoloured grains/ spike under different levels of leaf blight infection

REFERENCES

Chester, K.S. (1950). Plant disease losses : their appraisal and interpretation. *Plant Dis. Rep.* 55: 319-322.

James, W.C. (1971). An illustrated series of assessment keys for plant disease, their preparation and usage. *Can Plant Dis. Surv.* 51: 39-65.

Large, E.C. (1966). Measuring plant diseases. *Annu. Rev. Phytopathol.* 4: 9-28.

Horsfall, J.G. and Barratt, R.W. (1945). An improved grading system for measuring plant diseases. *Phytopathology* 35: 655.

Perello, A., Sisterna, M.N. and Cortese, P. (1998). A scale for appraising the leaf blight of wheat caused by *Alternaria triticimaculans*. *Cereal Res. Commun.* 26: 189-194.

Saari, E.E. and Prescott, J.M. (1975). A scale for appraising the foliar intensity of wheat diseases. *Plant Dis. Rep.* 59: 377-380.

Singh, D.P., Chand, R., Dodan, D.S., Amerika Singh, Singh, K.P., Tewari, A.N., Singh, K.M.P., Satvinder Kaur, Singh, R.N., Singh, A.K., Singh, S.P., Singh, V.K., Brahma, R.N., Kalappanawar, I.K., Solanki, V.A., Pathak, R.K., Pant, S.K., Das, S.Y. and Chaudhary, A.K. (2003). Evaluation of wheat genotypes for resistance to leaf blight caused by *Bipolaris sorokiniana* and *Alternaria triticina*. *Indian Phytopathol.* 56: 473-475.

Singh, D.P., Sharma, A.K., Nagarajan, S., Goel, L.B., Basant Ram, Amerika Singh, Singh, R.V., Tewari, A.N., Singh, A.K., Singh, R.N., Singh, S.P., Khanna, B.M., Dharam Singh Dodan, Bagga, P.S. and Kalappanavar, I. (2002). Losses caused due to leaf blight in wheat in different agro-climatic zones in India. *Plant Dis. Res.* 17: 313-317.

Zadoks, J.C., Chang, T.T. and Konzak, C.E. (1974). A decimal code for the growth stages of cereals. *Weed Res.* 14: 415-421.

Integrated Plant Disease Management (2005)
Eds. Sharma, R.C. and Sharma, J.N.
Scientific Publishers (India), Jodhpur, pp. 295 - 303.

46

Characterization of *Alternaria brassicae* isolates from Haryana on the basis of nutritional behaviour

NARESH MEHTA, ANIL K. KHURANA and M.S. SANGWAN

Department of Plant Pathology, CCS Haryana Agricultural University, Hisar -125 004 (Haryana)

ABSTRACT

Fourteen isolates of *Alternaria brassicae* (Berk.) Sacc. responsible for Alternaria blight in rapeseed–mustard were collected from different parts of Haryana. The isolates were characterized by their responses to various carbon and nitrogen sources as well as on the basis of pH. The results revealed that all the isolates behaved differentially in growth and sporulation in relation to different carbon and nitrogen sources. Isolates KTL-I and BWL showed significantly higher growth on all the carbon sources, whereas isolate CHR-I had minimum growth and responded differentially to different carbon sources. Similarly, isolate REW showed more variation in sporulation whereas, isolates SPT and HSR had least variation in sporulation. Amongst the nitrogen sources evaluated, none of the isolates showed growth on potassium nitrite, however, growth was maximum on sodium nitrate followed by potassium nitrate, ammonium nitrate and glycine. Isolates KTL-I and CHR-II had maximum average radial growth whereas SRS and JD-II had minimum radial growth irrespective of different nitrogen sources. Isolates BWL, CHR-I, CHR-II, RTK, KTL-I and SPT responded maximum to KNO_3 amended medium whereas, REW responded poorly to glycine. Isolates BWL, CHR-II, JD-I, SRS and SPT showed maximum sporulation on KNO_3 whereas,

HSR had minimum sporulation with glycine. All the isolates showed better growth at pH 7.5 and sporulation at 5.0 after 21 days of incubation. On the basis of nutritional behaviour, all the isolates could be categorized into two major groups i.e. isolates BHI, JD-I, JHR, REW, RTK, SRS, CHR-I, CHR-II, JD-II and HSR could be placed in first group whereas, isolates KTL-I, KTL-II and BWL constituted the second group.

INTRODUCTION

The oilseed crops especially brassicas play a vital role in agricultural economy of the world and rapeseed-mustard is one of them. The average productivity of rapeseed-mustard in India is very low as compared to the developed countries. There are a number of factors responsible for low productivity and diseases are one of the major contributing factors. Among the various diseases, Alternaria blight incited by *Alternaria brassicae* (Berk.) Sacc. is one of the most important and devastating resulting up to 70 per cent yield loss (Chahal, 1986; Kolte *et al.*, 1987; Verma and Saharan, 1994) coupled with the deterioration of quality of the produce (Kadian and Saharan, 1983; Ansri *et al.*, 1988).

The effect of different nutritional sources on the growth and sporulation of *Alternaria* spp. has been studied by various workers (Gupta *et al.*, 1969; Ansari *et al.*, 1989; Khatun *et al.*, 1996), however, the variation in *Alternaria brassicae* isolates on the basis of their nutritional requirements and pH has not been studied so far. Therefore, present investigations were undertaken with a view to distinguish various isolates of *A. brassicae* on the basis of their cultural behaviour.

MATERIALS AND METHODS

Fourteen isolates of *A. brassicae* responsible for Alternaria blight in rape seed- mustard were collected from different locations of Haryana on the basis of symptomatology (Table 1).

All the isolates were purified by single spore technique as per the method of Toussoun and Nelson (1976) and maintained on PDA slants in refrigerator at 5±2°C for further studies. The various isolates collected form different locations were designated with a code (Table 1).

Effect of carbon and nitrogen sources

Five carbon sources viz., dextrose, fructose, lactose, mannitol and sucrose were used in an amount equivalent to carbon present in Richard's medium (sucrose @ 50 g/l). A carbon free medium served as control. Similarly, various nitrogen sources viz., ammonium nitrate, glycine, potassium nitrate, potassium nitrite and sodium nitrate were evaluated in an amount equivalent to nitrogen present in Richard's medium (potassium nitrate @ 10 g/l). Medium without nitrogen served as control.

Table 1. *Alternaria brassicae* infecting rapeseed – mustard collected isolates from various locations of Haryana

Location	Variety	Isolate code	Symptoms on leaf/pod
Bhiwani	Local	BHI	Light brown spot with prominent concentric rings.
Hisar	YSPb-24	HSR	Dark brown concentric rings with yellowish margin.
Jind	Local	JD-I	Brown spot.
Jind	Varuna	JD-II	Light brown spot with some greyish appearance.
Jhajjar	Local	JHR	Light brown concentric rings with papery growth.
Kaithal	Local	KTL-I	Dark brown eye shaped rings on the pod/blackish spot.
Kaithal	BSH-I	KTL-II	Light brown spot with yellowish margin around the spot.
Rewari	RH-30	REW	Dark brown concentric rings.
Rewari (Bawal)	RH-8113	BWL	Dark brown spot with greenish margin.
Rohtak	Local	RTK	Light brown spot with greyish margin.
Rohtak (Chiri)	RH-30	CHR-I	Light brown spot.
Rohtak (Chiri)	RH-30	CHR-II	Dark brown spot.
Sirsa	Laxmi	SRS	Dark brown prominent concentric rings with yellow halo.
Sonipat	Local	SPT	Light brown concentric rings with irregular shape.

Inoculum discs of 5 mm diameter from 10 days old culture were placed in the Petri plates containing the desired medium and each treatment was replicated four times. Such plates were incubated at 22±1°C and radial growth (cm) was recorded on 20 days after inoculation. Number of spores produced on various media were also recorded 20 days after inoculation. To count the number of spores, three discs each of 5 mm diameter were selected at random from the outer periphery of the fungal colony and transferred to 5 ml of distilled water. A suspension of 0.1 ml was placed on a acid clean slide and examined under the compound microscope. The number of spores per microscopic field (45 x 10 X) was counted and average spore production was calculated.

Effect of pH

To identify the variation among different isolates on the basis of their growth in relation to hydrogen ion concentration (pH) study was also undertaken. The pH in PDA medium was adjusted at 4.0, 5.0, 7.5, 9.0 and 10.0 with 1N sodium hydroxide (NaOH) or 1N hydrochloric acid (HCl) before autoclaving with a pH meter (Century CP- 901). Discs of 5 mm diameter from 10 days old culture were used for inoculation of Petri plates containing medium of desired pH. Each treatment was replicated four times and these Petri plates were incubated in BOD at 22±1°C and the radial growth (cm) was recorded after 20 days of inoculation. The number of spores produced by each isolate at each pH was recorded after 20 days as per the procedure described in nutritional variation study.

RESULTS AND DISCUSSION

Carbon source

The study revealed that all isolates differed in their growth with different carbon sources (Table 2). The analysis of data revealed significant differences among the isolates with a particular carbon source after 20 days of incubation. Isolate KTL-I had maximum (8.65 cm) radial growth followed by isolates BWL, SPT, JHR and KTL-II (8.61, 8.55, 8.45 and 7.31 cm, respectively) whereas, isolate CHR-II showed minimum (2.93 cm) radial growth with dextrose amended medium. Isolate BWL recorded maximum (8.89 cm) radial growth followed by isolates KTL-I, SPT and KTL-II (8.88, 8.53 and 7.15 cm, respectively) and isolate CHR-I recorded the minimum (1.90 cm) radial growth in the medium amended with fructose. Other isolates such as HSR, JHR, RTK and SRS had less growth. Isolate KTL-I had maximum (8.96 cm) radial growth followed by isolates BWL, SPT and KTL-II (8.61, 7.51 and 7.28 cm, respectively) on lactose amended medium and on this medium also isolates HSR, RTK and CHR-I had less growth and the minimum being in isolate CHR-I (2.56 cm). Isolate KTL-I exhibited maximum (8.98 cm) radial growth followed by isolates BWL and KTL-II (8.83 and 8.58 cm, respectively) in mannitol amended medium whereas isolate CHR-I showed minimum (2.36 cm) radial growth. Similarly, isolates BWL, JD-II and KTL-I exhibited maximum (8.63, 8.61 and 8.58 cm, respectively) radial growth in sucrose amended medium and minimum (2.16 cm) growth was recorded in isolate CHR-I. Isolates KTL-I and BWL showed average maximum radial growth (8.47 and 8.37 cm, respectively) and isolate CHR-I, the minimum (2.31 cm) among all the isolates irrespective of the carbon sources in the medium.

Table 2. Effect of various carbon sources on the radial growth (cm) of different isolates of *Alternaria brassicae*

Carbon source	Radial growth of different isolates (cm)															
	BHI	BWL	CHR-I	CHR-II	HSR	JD-I	JD-II	JHR	KTL-I	KTL-II	REW	RTK	SRS	SPT	MEAN	SEM ±
Dextrose	5.26	8.61	3.30	2.93	4.68	3.21	6.30	8.45	8.65	7.31	4.18	4.21	5.93	8.55	5.82	0.17
Fructose	5.16	8.89	1.90	5.41	2.73	4.58	5.40	3.55	8.88	7.15	4.10	3.35	3.18	8.53	5.19	0.19
Lactose	5.18	8.61	2.56	5.28	3.53	5.26	5.26	4.46	8.96	7.28	5.25	3.98	6.43	7.51	5.70	0.18
Mannitol	5.08	8.83	2.36	5.50	4.30	5.63	5.40	6.13	8.98	8.58	5.33	5.76	5.93	7.25	6.07	0.16
Sucrose	2.90	8.63	2.16	5.20	2.73	4.76	8.61	5.86	8.58	6.23	5.10	3.00	3.45	7.30	5.32	0.16
Control	3.06	6.45	1.63	2.13	2.63	5.85	6.28	3.61	6.78	6.58	2.88	4.63	1.93	6.35	4.34	0.17
Mean	4.44	8.37	2.31	4.40	3.43	4.88	6.20	5.34	8.47	7.18	4.47	4.15	4.47	7.58		

$CD_{0.05}$

Isolate = 0.14

Carbon source = 0.09

Isolate x Carbon source = 0.34

Spore suspension made after 20 days of incubation depicted that maximum spores per microscopic field were recorded in mannitol amended medium (6.91) followed by fructose (6.51) and least (1.03) in control irrespective of the isolates. Isolate REW showed maximum average spores (9.58) per microscopic field and isolate SPT (4.16), the minimum. All the isolates could produce moderate sporulation in all the media amended with different carbon sources. Isolate REW showed extremely good sporulation on mannitol, sucrose and fructose, however, it responded poor to lactose. Similarly, isolate CHR-I and RTK showed good sporulation at lactose and sucrose, respectively as compared to other carbon sources (Table 3). The higher growth in mannitol amended medium is contrary to the earlier reports (Gupta *et al.*, 1979; Ansari *et al.*, 1989). It may be due to that isolates vary in their behaviour and characteristics etc. A fairly good growth of isolates KTL-I and SPT on all the carbon sources also indicated the existence of different groups of the isolates. Similarly, isolate CHR-I showed minimum growth on all the media amended with different carbon sources except on dextrose and lactose which further gave an indication of the existence of variation in isolates. The possible reason may be that different isolates behave differentially with various carbon sources.

Table 3. Effect of various carbon sources on the sporulation of different isolates of *Alternaria brassicae*

Carbon source	Isolates															
	BHI	BWL	CHR-I	CHR-II	HSR	JD-I	JD-II	JHR	KTL-I	KTL-II	REW	RTK	SRS	SPT	MEAN	SEM ±
Dextrose	6.50	3.75	4.25	6.25	6.00	5.25	6.25	4.50	3.25	5.75	7.25	7.00	5.25	1.50	5.41	0.89
Fructose	7.00	6.75	5.75	4.50	5.00	6.75	6.25	7.25	4.75	5.00	14.00	6.25	7.00	5.00	6.51	1.01
Lactose	2.75	7.25	10.50	2.75	4.75	7.25	5.00	2.50	4.75	3.00	3.50	4.00	6.00	4.75	4.91	0.88
Mannitol	6.50	6.50	6.75	6.75	6.25	5.00	7.25	4.75	6.25	6.50	16.00	6.75	7.00	4.50	6.91	1.14
Sucrose	5.00	5.75	5.25	4.00	4.75	4.00	2.75	6.50	6.25	4.75	14.25	9.75	4.75	5.50	5.94	0.89
Control	2.50	0.25	1.00	1.25	0.75	0.50	1.00	0.75	0.50	0.25	2.50	2.00	1.50	0.75	1.03	0.52
Mean	4.87	5.04	5.58	4.25	4.58	4.79	4.75	4.37	4.29	4.20	9.58	5.95	5.25	4.16		

$CD_{0.05}$

Isolate = 0.73

Carbon source = 0.47

Isolate x Carbon source = 1.78

Nitrogen nutrition

The results in Table 4 indicated a significant difference among the isolates with a particular nitrogen source. In glycine amended medium, isolate HSR showed maximum (4.05 cm) radial growth followed by isolates KTL-I and SPT (3.38 and 3.30 cm, respectively) whereas, the minimum growth was recorded in case of isolate REW (1.95 cm). Isolate BWL showed maximum (7.23 cm) radial growth closely followed by isolates KTL-I, CHR-II and SPT (7.21, 7.15 and 7.13 cm, respectively) in potassium nitrate (KNO_3) amended medium as compared to 4.41 cm, the minimum in case of SRS.

Similarly, isolate HSR afforded maximum (7.53 cm) radial growth followed by isolates BWL, CHR-I and CHR-II (7.50, 7.43 and 7.43 cm, respectively) with sodium nitrate ($NaNO_3$) amended medium. The minimum radial growth was recorded in case of isolate SRS (4.01 cm). The average growth of *Altenaria brassicae* on ammonium nitrate (NH_4NO_3) amended medium was in general poor. However, isolates HSR and CHR-I showed maximum (4.03 cm and 4.01 cm, respectively) radial growth followed by BWL and CHR-II (3.85 cm each). The isolates CHR-II and KTL-I responded well to all of the nitrogen sources under study, though statistically at par with isolates CHR-I, HSR and BWL.

Table 4. Effect of various nitrogen sources on the radial growth (cm) of different isolates of *Alternaria brassicae*

Nitrogen source	Growth of different isolates (cm)															
	BHI	BWL	CHR-I	CHR-II	HSR	JD-I	JD-II	JHR	KTL-I	KTL-II	REW	RTK	SRS	SPT	MEAN	SEM ±
Glycine	2.51	3.06	2.48	2.91	4.05	3.10	2.78	2.56	3.38	3.20	1.95	2.46	2.95	3.30	2.90	0.07
KNO_2	0.00	0.00	0.00	0.00	0.00	0.00	0.00	0.00	0.00	0.00	0.00	0.00	0.00	0.00	0.00	0.00
KNO_3	5.13	7.23	7.11	7.15	5.55	5.63	4.60	6.41	7.21	6.65	4.51	7.00	4.41	7.13	6.12	0.10
$NaNO_3$	6.10	7.50	7.43	7.43	7.53	6.20	6.00	6.45	7.35	6.61	5.01	6.48	4.01	6.55	6.47	0.09
NH_4NO_3	3.65	3.85	4.01	3.85	4.03	3.68	2.51	3.75	3.45	3.58	3.15	2.48	3.05	3.56	3.47	0.11
Control	3.50	3.85	4.46	4.48	4.31	4.13	3.31	4.31	4.31	3.40	5.51	3.93	3.68	4.13	4.09	0.08
Mean	3.57	4.34	4.34	4.39	4.33	3.88	3.29	4.00	4.37	3.99	3.44	3.81	3.10	4.20		

$CD_{0.05}$

Isolate	= 0.07
Nitrogen source	= 0.04
Isolate x Nitrogen source	= 0.17

The results in Table 5 indicated significant differences among the isolates on sporulation with a particular nitrogen source. The data also revealed that irrespective of the isolates, average maximum spores (9.55) were produced on medium containing potassium nitrate and minimum (1.10) on control medium. Isolates SRS and JD-I showed maximum average spores (4.97 each) per microscopic field followed by isolate KTL-I (4.50), irrespective of the nitrogen sources. Isolate JD-II showed minimum average spores (3.66) per microscopic field. However, not much variation among all the isolates in terms of sporulation could be derived.

The good amount of growth in case of *A. brassicae* has already been reported in nitrate form of nitrogen by Ansari *et al.* (1989). Similar behaviour was also recorded in most of the isolates in potassium nitrate amended medium except isolates JD-II, REW and SRS. Similar opinion has already been given by other workers that nitrate form of nitrogen behaves well towards growth of *Alternaria* (Khatun *et al.*, 1996). Good amount of sporulation was also observed in sodium nitrate and potassium nitrate amended media whereas very poor sporulation was recorded in glycine amended medium and nil with potassium nitrite. It has already been reported by other workers that nitrate form of nitrogen enhanced the sporulation of *A. brassicae* (Ansari *et al.*, 1989) and *A. brassicicola* (Khatun *et al.*, 1996).

Table 5. Effect of various nitrogen sources on the sporulation of different isolates of *Alternaria brassicae*

Nitrogen source	Isolate															
	BHI	BWL	CHR -I	CHR -II	HSR	JD -I	JD -II	JHR	KTL -I	KTL -II	REW	RTK	SRS	SPT	MEAN	SEM ±
Glycine	1.50	2.25	2.50	2.75	1.00	2.75	2.00	3.50	2.25	1.75	2.25	2.25	3.00	1.75	2.25	0.64
KNO_2	0.00	0.00	0.00	0.00	0.00	0.00	0.00	0.00	0.00	0.00	0.00	0.00	0.00	0.00	0.00	0.00
KNO_3	8.75	10.00	8.50	10.00	9.75	11.50	8.25	9.25	9.75	9.50	7.50	9.25	11.25	10.50	9.55	1.19
$NaNO_3$	7.75	6.75	7.25	8.25	7.50	8.50	7.00	7.50	8.25	8.25	7.50	7.00	8.25	7.25	7.64	0.90
NH_4NO_3	5.25	5.50	6.25	4.00	4.25	4.25	4.75	3.75	5.50	4.50	3.75	4.00	5.25	4.25	4.66	0.75
Control	1.25	2.25	1.75	1.25	1.50	1.75	0.00	1.50	1.25	0.00	2.50	0.00	0.00	1.00	1.10	0.58
Mean	4.08	4.45	4.37	4.37	4.25	4.79	3.66	4.25	4.50	4.00	3.91	3.75	4.79	4.12		

$CD_{0.05}$

Isolate = 0.62

Nitrogen source = 0.40

Isolate x Nitrogen source = 1.53

Table 6. Effect of pH on the radial growth (cm) of different isolates of *Alternaria brassicae*

pH	Growth of different isolates (cm)															
	BHI	BWL	CHR-I	CHR-II	HSR	JD-I	JD-II	JHR	KTL -I	KTL -II	REW	RTK	SRS	SPT	MEAN	SEM ±
4.0	2.41	2.71	5.53	5.25	2.71	2.18	4.41	3.80	6.81	3.76	2.16	2.98	2.75	2.73	3.58	0.06
5.0	2.15	3.10	5.51	6.63	4.05	6.10	2.36	3.98	6.63	5.41	2.70	3.70	2.28	2.10	4.05	0.08
7.5	5.15	5.88	7.66	7.56	6.46	5.43	5.08	6.71	8.83	7.56	6.05	5.48	5.61	4.41	6.27	0.10
9.0	6.16	4.83	7.53	7.98	6.18	4.50	4.08	5.21	8.90	7.18	5.18	5.46	3.98	4.33	5.82	0.09
10.0	5.01	3.15	6.05	6.70	3.65	2.36	2.48	4.80	7.66	5.96	2.96	3.93	2.95	3.13	4.34	0.07
Mean	4.17	3.93	6.45	6.82	4.61	4.11	3.68	4.90	7.76	5.97	3.81	4.31	3.51	3.34		

$CD_{0.05}$

Isolate = 0.30

pH = 0.18

Isolate × pH = 0.67

Effect of pH

Effect of different levels of pH on the growth and sporulation of *A. brassicae* was studied *in vitro* to know the variations among isolates, if any. The perusal of data in Table 6 revealed that there were significant variations in growth of various isolates at a particular pH. The critical analysis of the data revealed that isolates KTL-I, CHR-I and CHR-II exhibited fairly good growth (>5.0cm) at all pH levels. However, growth of these isolates was further better at higher pH levels (9.0 and 10.0) as compared to lower ones (pH 4.0 and 5.0). All the isolates on an average had better growth at pH 7.5 whereas less at extreme pH (4.0 and 10.0).

Sporulation of the pathogen was maximum at pH 5.0 (9.44 per microscopic field) followed by pH 4.0 and 7.5. At higher pH i.e. 9.0 and 10.0, very little or no sporulation was recorded (Table 7).

Table 7. Effect of pH on the sporulation of different isolates of *Alternaria brassicae*

pH	Sporulation of different isolates (spores/microscopic field)															
	BHI	BWL	CHR-I	CHR-II	HSR	JD-I	JD-II	JHR	KTL-I	KTL-II	REW	RTK	SRS	SPT	MEAN	SEM ±
4.0	5.60	6.25	9.75	7.25	6.50	7.75	9.50	6.50	8.75	7.50	6.00	6.75	7.25	6.25	7.25	0.92
5.0	8.50	10.50	8.50	7.50	8.25	10.00	8.25	10.25	8.25	11.50	11.75	9.25	12.25	7.50	9.44	1.03
7.5	4.00	4.75	7.75	5.00	5.25	5.50	7.00	5.00	7.00	6.00	4.50	5.00	5.00	4.75	5.46	0.96
9.0	1.75	2.50	1.75	1.75	3.25	3.00	3.75	2.50	2.50	1.25	2.25	3.00	4.75	5.50	3.03	0.62
10.0	0.00	0.00	0.00	0.00	0.00	0.00	0.00	0.00	0.00	0.00	0.00	0.00	0.00	0.00	0.00	0.00
Mean	3.85	5.70	5.55	4.30	4.65	5.00	5.70	4.90	5.30	5.25	4.90	4.80	5.28	4.80		

$CD_{0.05}$

Isolate = 0.73

pH = 0.44

Isolate x pH = 1.65

Significant variations in growth and sporulation of each isolate at a particular pH have been recorded. However, there are no earlier reports on differentiation of isolates on the basis of their response to different levels of pH. It has been reported that pathogen can grow and sporulate best between pH 5.5 to 7.2 (Gupta *et al.*, 1969; Ansari *et al.*, 1989; Kumari *et al.*, 1998). The data on sporulation also suggested that maximum sporulation was at pH 5.0 which further confirms the earlier findings (Li *et al.*, 1991; Kumari *et al.*, 1998). Variation in growth and sporulation of isolates at a particular pH may be due to fact that a particular isolate grows better at a specific pH. The other possible reason may be that isolates collected from a particular location may have adjusted themselves according to the conditions prevailing in that area and adopted a particular pH for growth and sporulation.

On the basis of nutritional behaviour of isolates, all the isolates could be categorized into two major groups. Isolates BHI, JD-I, JHR, REW, RTK, SRS, CHR-I, CHR-II, JD-II and HSR showed average growth could be placed into first group whereas isolates KTL-I, KTL-II, SPT and BWL showing good growth could be placed in second group. Similarly, on the basis of growth of isolates in relation to different pH levels could also be categorized into two groups. Isolates BHI, BWL, HSR, JD-I, JD-II, JHR, KTL-II, REW, RTK, SRS and SPT showing average growth (up to 6.0 cm) were placed in the first group whereas, isolates CHR-I, CHR-II and KTL-I showing average good growth (more than 6.0 cm) irrespective of the pH levels were placed in the second group. Some of the isolates behaved similarly to carbon and nitrogen sources as well to different pH level fall in the same group. Similar variation in the isolates have already been reported by Kumar *et al.* (2003) on the basis of pathogenicity on seventeen host differentials and all the isolates were categorized into eight groups. In the present studies also, similar groups can be proposed but all the isolates have been grouped broadly into two groups on the basis of nutritional behaviour.

REFERENCES

Ansari, N.A., Khan, M.W. and Muheet, A. (1988). Effect of Alternaria blight on oil content of rapeseed and mustard. *Curr. Sci.* 57: 1023-1024.

Ansari, N.A., Khan, M.W. and Muneet, A. (1989). Nutritional requirements of *Alternaria brassicae* for growth and sporulation. *Indian J. Plant Pathol.* 7: 127-135.

Chahal, A. S. (1986). Losses and chemical control of *Alternaria* in rapeseed- mustard in Punjab. *Plant Dis. Res.* 1: 46-50.

Gupta, R.B.L., Desai, B.G. and Pathak, V.N. (1969). Nutritional studies on *Alternaria brassicae* (Berk.) Sacc. I. Influence of different nutrient media, pH, nitrogen and carbon/nitrogen ratio. *Phyton* 26: 201-205.

Gupta, Y.K., Gupta, M.N. and Seth, A.K. (1979). Effect of different carbon compounds on the growth and sporulation of *Alternaria alternata* (Fr.) Keiseler, causing leaf blight of sunflower. *Proc. Indian Natl. Sci. Acad.Part B Biol. Sci.* 45: 190-192.

Kadian, A.K. and Saharan, G.S. (1983). Symptomatology, host range and assessment of yield losses due to *Alternaria brassicae* infection in rapeseed and mustard. *Indian J. Mycol. Plant Pathol.* 13: 319-323.

Khatun, F., Hossain, M.A., Hossain, M.M. and Dey, D.K. (1996). Effect of culture media, temperature, pH and nitrogen sources on growth and sporulation of *Alternaria brassicicola*. *Bangladesh J. Plant Pathol.* 12: 29-32.

Kolte, S.J., Awasthi, R.P. and Vishwanath (1987). Assessment of yield losses due to Alternaria blight in rapeseed and mustard. *Indian Phytopathol.* 40: 209-211.

Kumar Satish, Sangwan, M.S., Mehta, Naresh and Kumar Rakesh. (2003). Pathogenic diversity in isolates of *Alternaria brassicae* infecting rapeseed and mustard. *J. Mycol. Plant Pathol.* 33: 59-64.

Kumari, Annapurna, Jha, A.K., Sinha, S.K. and Ojha, K.L. (1998). Cultural and physiological studies on cauliflower leaf blight fungus *Alternaria brassicae* (Berk.) Sacc. *J. Appl. Biol.* 8: 65-68.

Li, M.Y., Ke, C.Q. and Zeng, L. (1991). Effects of environmental factors on the development of *Alternaria brassicae. Acta Phytophylactica Sinica* 18: 317-322.

Toussoun, T.A. and Nelson, P.E. (1976). *A Pictorial Guide to the Identification of Fusarium Species According to the Taxonomic System of Synder and Hanson.* University Park, Pennsylvnia, USA, Pennsylvnia State University Press, 43 p.

Verma, P.R. and Saharan, G.S. (1994). *Monograph on Alternaria Diseases of Crucifers.* Research Branch, Agriculture and Agri-Food Canada, Saskatoon Research Centre Canada. *Tech. Bull.* 1994-6E: 162 p.

Integrated Plant Disease Management (2005)
Eds. Sharma, R.C. and Sharma, J.N.
Scientific Publishers (India), Jodhpur, pp. 305 - 309.

47

Linear regression model for yield-loss determination of foliar diseases in black gram and common bean

C.L. BHARDWAJ, B.S.THAKUR*, J.C. GUPTA and KISHORE KHOSLA*

CSK HPKV, Hill Agriculture Research and Extension Centre, Bajaura-175 125 (H.P.)
*Dr. Y.S. Parmar University of Horticulture and Forestry, RHRS Bajaura-175 125 (H.P.)

ABSTRACT

Application of different fungicides stabilized dry-grain yield against foliar diseases of black gram and common bean dominated by *Cercospora cruenta* and angular leaf spot, respectively. This was corresponding to the reduction of terminating disease severity visualized using 0 to 9 rating score at physiological maturity of the crops. The yield was used as independent variable and disease severity as dependent variable to workout relationships and draw the conclusion. There was a highly significant negative correlation between yield and disease severity afforded by the efficacy of fungicides against the two foliar diseases r =-0.846 and -0.946, respectively. Further, the goodness of fit of regression equation also confirmed the relationship (Y=10.1-2 (X), R^2=0.99% and Y=14.04–4 (X), R^2=0.89, respectively). Fungicide Bayleton was the most effective fungicide to manage leaf spot of black gram stabilizing yield 8.11q/ha, whereas Baynate and Bavistin were the most effective to manage angular leaf spot of common bean yielding 8.51 and 8.05 q/ha, respectively.

INTRODUCTION

The complexity of disease-yield loss relationship is increased with the occurrence of multiple diseases on a single host or when other pests are examined as explanatory variables of yield loss. Further, it is also important that the underlying biology of relationship between variables of disease parameters and corresponding yield loss are not ignored for statistical reasons and the relationship must be synthesized into a common model (James and Teng, 1979). Foliar diseases on black gram dominated by *Cercospora cruenta* Sacc. and angular leaf spot caused by *Phaeoisariopsis griseola* (Sacc.) Ferraris on common bean occur frequently under sub-temperate and sub-humid agro-climate of Bajaura in Himachal Pradesh (Bhardwaj, 1998). Disease severity variables corresponding to yield loss variables as afforded by comparative efficacy of the test fungicides inclusive of control without fungicidal sprays, were examined for establishing a common model for the assessment of yield loss relationship of the diseases under study.

MATERIALS AND METHODS

A field plot technique was used to generate disease yield-loss database afforded by efficacy of fungicide formulations, viz., Bavistin (0.1%), Bayleton (0.1%), Baynate (0.1%), Score (0.1%), Ovis (0.1%) including the control plots without spray of any fungicide. Susceptible varieties of crops, black gram (*Phaseolus mungo* L. cv. PDV1) and common bean (*Phaseolus vulgaris* L. cv. Kanchan) were raised in randomized block design (RBD) keeping three replications with a plot size of 5.40m^2 under each treatment. The crops were raised separately adjacent to each other in the same field. Each crop was sown by mid-June keeping row to row distance of 30 cm and plant to plant distance of 10 cm. Fertilizers @15 kg N and 45 kg P_2O_5/ha to black gram and @20 kg N and 60 kg P_2 O_5/ha to common bean were applied as basal dose through urea and single super phosphate. The fungicides were applied at 15 day intervals starting from flowering in each case.

Data were recorded on disease severity of leaf spot dominated by *Cercospora cruenta* on black gram and angular leaf spot caused by *Phaeoisariopsis griseola* on common bean as terminating disease severity at physiological maturity. Disease severity was assessed on each crop based on visual 0-9 rating scale. Similarly, yield data were recorded for each plot on dry grain yield after harvesting of the crop. Data on healthy and diseased pods due to angular leaf spot were also recorded. The mean per cent of infected pods was converted into logarithmic transformations for purpose of its relationship with yield variables. Similarly, mean disease severity of the foliar diseases was transformed into square root values. Adequate disease pressure was recorded during year, 1999 and was considered for working out yield-loss relationship by running computer-loaded programme using standard statistical methods.

RESULTS AND DISCUSSION

Yield loss relationship of *Cercospora* leaf spot on black gram

A perusal of Table 1 clearly depicts that the disease severity decides the yielding capacity of the crop, which increases or decreases with the severity of the disease afforded by the comparative efficacy of the fungicides/ formulations and the control without sprays. The relationship of two variables yield and diseases severity was nearly perfect and negative (r = -0.997), clearly indicating that with the increase in disease severity there will be a proportionate decrease in the yield (Table 2). Further analysis of regression equation {Y= 14.03-4 (X)} and its significance as judged through coefficient of determination (R^2 = 0.99) also confirms nearly 100 per cent accountability of the disease severity as assessed on the basis of visual 0-9 rating scale in controlling the yield of crop with minimum chance of error (0.21%). This precision was less with inclusion of efficacy of all the five fungicide formulations and the severity afforded by them including that of treatment without spray. Such variation could be attributed to added or decreased effect of fungicide/formulation application on the yield or expression of the disease as evident from comparatively low expression of disease with decreased yield with application of Bavistin compared to Bayleton application (Table 1). Such results are in conformation with the study conducted by different researchers (Jenkins *et al.*, 1972; Melville and Jemmett, 1971). However, except for the slight variation in the observed value of severity over the expected values for the two fungicides Bavistin and Bayleton, the regression on other fungicidal treatments inclusive of no fungicidal spray gave results on expected lines indicating that regression equation {Y= 14.03-4 (X)}, is the true indicator of increased severity with decreased yield for assessing loss due to Cercospora leaf spot of black gram. Yield loss relationship can be worked out with greater precision with the help of this regression equation. Further, the use of Bavistin and Baynate for creating different disease severity levels would be appropriate since they did not amount to any effect on yield irrespective of any appreciable variation of the disease severity.

Yield loss relationship of angular leaf spot on common bean

The data presented in Table 1 on the yield and disease severity recorded on the basis of 0-9 scale, including per cent pods infected by angular leaf spot afforded by use of different fungicides/formulations indicate that Bavistin and Bayleton stabilized the yield of common bean.

A negative and significant correlation between yield and two parameters of disease severity on foliage (-0.946) and pods (-0.819) was obtained (Table 2). This indicates that disease severity significantly influences the yield afforded by differential efficacy of test fungicide formulations. An analysis of variance for regression with significant mean sum of squares further confirms the negative relationship between yield and severity of angular leaf spot on foliage as well as its incidence on pods. The coefficient of determination was highly significant (R^2 = 0.89) only for disease severity on foliage. This reveals that severity on foliage accounts for 89 per cent accuracy in influencing the disease to be assessed through regression equation {Y= 14.03-4 (X)} given in Table 2. The measure of yield loss relationship using extents of per cent pods infected due to angular leaf spot did not prove of much value.

Table 1. Efficacy of fungicide formulations to manage *Cercospora* leaf spot of black gram and angular leaf spot of common bean

Fungicide	Grain-yield of black gram (q/ha)	[1]Disease severity (0-9 score)	Grain-yield of common bean (q/ha)	Disease severity (0-9 score)	[2]Disease incidence of angular leaf spot on pods (%)
Bavistin 50 WP (0.1%)	6.01	1.00 (1.22)	8.05	1.66 (1.47)	3.12 (0.49)
Bayleton 25 WP (0.1%)	8.11	1.83 (1.53)	6.10	5.33 (2.41)	85.66 (1.93)
Baynate 70 WP (0.1%)	6.85	1.33 (1.35)	8.51	1.66 (1.47)	5.78 (0.76)
Ovis (0.1%)	3.61	6.50 (2.64)	6.10	5.16 (2.38)	88.42 (1.94)
Score 25 EC (0.1%)	5.62	4.33 (2.19)	3.08	7.16 (2.76)	97.75 (1.99)
Control	3.70	6.90 (2.72)	2.08	7.66 (2.85)	100.0 (2.00)

[1]Figures in parentheses are square root transformation values of disease severity of respective crops

[2]Figures in parentheses are logarithmic transformation values of disease incidence on pods of common bean

Table 2. Correlation co-efficient and regression equation of yield and disease severity/incidence

Disease	Correlation coefficient	R^2	Regression equation
Cercospora leaf spot of black gram	-0.846	0.72	Y = 10.01-2 (X)
	-0.997	0.99	Y* = 14.03-4 (X)
Angular leaf spot of common bean	-0.946	0.89	Y = 14.04-4 (X)
	-0.815	0.66	Y* = 10.04-3 (X)

Y* = Regression equation based on the efficacy of Bayleton, Ovis, Score and the control;

X = Disease severity; Y = Yield

It is therefore, concluded from the present study that yield loss relationship can be assessed with great precision from the regression equation derived each for Cercospora leaf spot on black gram and angular leaf spot on common bean as indicated by highly significant relationship between yield and disease severity. Fungicide Bavistin is effective against both the foliar diseases of respective crops and can find a common use for management of the disease. However, other fungicides/formulations attribute towards differential efficacy to manage two different diseases on these two crops.

REFERENCES

Backman, P.A. and Crawford, M.A. (1984). Relationship between yield-loss and severity of early and late leaf spot diseases of peanut. *Phytopathology* 74: 1101-1103.

Bhardwaj, C.L. (1998). Agro ecological yield stability of wet season pulses for quantum production. In: *Ecological Agriculture and Sustainable Development,* Vol. I (Eds. Dhaliwal, G.S., Randhawa, N.S., Arora, R. and Dhawan, A.K.), Digigrafics, D-69, Gulmohar Park, New Delhi , pp 394-396.

James, W.C. and Teng, P.S. (1979). The quantification of production constraints associated with plant diseases. In: *Applied Biology,* Vol. IV (Ed. Coaker, T.H.), Academic Press, London. 285 p.

Jankins, J.E.E., Melville, S.C. and Jemmett, J.L. (1972). The effect of fungicides on leaf diseases and on yield in spring barley in South West England. *Plant Pathol.* 21: 49-58

Melville, S.C. and Jemmett, J.L. (1971). The effect of glume blotch on the yield of winter wheat. *Plant Pathol.* 20: 14-17.

Roming, R.W. and Calpouzous, L. (1970). The relationship between stem rust and loss in yield of spring wheat. *Phytopathology* 60: 1801-1805.

Schneider, R.W., Williams, R.J. and Sinclair., J.B. (1976). *Cercospora* leaf spot of cowpea: Models for estimating yield loss. *Phytopathology* 66:384-388.

Integrated Plant Disease Management (2005)
Eds. Sharma, R.C. and Sharma, J.N.
Scientific Publishers (India), Jodhpur, pp. 311 - 317.

48

Performance of promising common bean cultivars under natural fertility and biotic disease stress condition in dry temperate agroclimate

C.L. BHARDWAJ, A. SINGH, B.S. THAKUR and J.C. GUPTA

CSK HPKV, Hill Agriculture Research and Extension Centre, Bajaura- 175 125 (H.P.)

ABSTRACT

Common bean genotypes from different sources comprising indeterminate, semi-determinate and determinate types were field evaluated under natural fertility, sprinkler irrigation and disease stress conditions of web blight and common bacterial blight in dry temperate agroclimate of Kukumseri (Lahaul and Spiti) in Himachal Pradesh. The results showed that among promising common bean genotypes semi-determinate genotypes viz., HPR 407 (22.90 q/ha) and HPR 423 (22.65 q/ha) out yielded the indeterminate and determinate genotypes. Significantly, more pods per plant were recorded for semi-indeterminate genotype, AFR 234 (27.63) and minimum number in determinate genotype LRK 435 (10.97), whereas the highest 50 dry-grain seed weight was recorded for indeterminate genotype HPR 75 (33.78 g) and the minimum for determinate variety Him-1 (16.48 g). A negative but nonsignificant correlation of yield with pods per plant and 50 dry grain seed weight was observed among tested genotypes with relatively higher magnitude for 50 dry grain weight indicating that an increase in seed weight would attribute to decrease in yield of common bean cultivars or genotypes. However, the relationship between pods per plant with 50 dry grain weight was positively correlated which indicated that normally an increase either in grain weight or number of pods would attribute to increased yield. No apparent incidence of web blight was recorded on AFR 234, which was highest on determinate genotype, HPR 418 (24.60, post-emergence plant mortality) followed by determinates Kanchan (22.66 %) and HPR 28 (22.66%). Similarly, no apparent incidence of common bacterial blight was recorded on determinate genotypes HPR 421 and LRK 425, which was recorded highest on determinate variety Kanchan (5.33, score) followed by Him-1 (4.33) and HPR 28 (3.33). It is inferred that both semi-indeterminate and determinate genotypes or cultivars possessed higher yield potential and medium dry-grain weight, increased number of pods per plant, increased pod length coupled with resistance/tolerance to abiotic and biotic stress.

INTRODUCTION

Dry beans also referred as common beans (*Phaseolus vulgaris* L.) are grain legume or pulse crop in India popularly known as Rajmash. Abiotic factors such as location, habitat, plant densities, soil types, soil fertility conditions as well as biotic stress factors such as weeds, diseases and insects are important factors limiting bean production (Bhardwaj, 1998; Bhardwaj and Thakur, 2000; Schwartz and Galvez, 1980; Skorch *et al.*, 1998; Nienhuis and Singh, 1985). Two biotic stresses among diseases and pests namely, web blight caused by a fungal pathogen, *Thanatephorus cucumeris* (Frank.) Donk. (imperfect state, *Rhizoctonia solani* Kuhn.) and bacteriosis by the variants/ pathotypes of bacterium, *Xanthomonas axonopodis* pv. *phaseoli* (Smith) Vauterin *et al.* are of versatile occurrence. The former is a potential soil-borne (Anderson, 1982; Ogoshi, 1976, 1987; Dillard, 1987; Galindo, 1982) pathogen, whereas later was known to be potential seed-borne (Valladares-Sanchez *et al.*, 1983; Malin *et al.*, 1983). Both stress factors pose a threat of heavy yield losses (Singh *et al.*, 1998; Wallen and Jackson, 1975) in different parts of the world. Yield is a complex and polygenically inherited trait for every crop. Dry bean is also known to possess large variation in growth habit, ecological adaptation and grain characters such as size, shape and colour (Singh, 1982) classifiable into distinct market classes relative to consumer preferences (Nienhuis and Singh, 1985) as well as quality standards (Tobin and Carpenter, 1978; Ma and Bliss, 1978; Lindgren, 1975). Medium and large seeded beans are generally grown in cooler highland agro-climates of the world, whereas small seeded predominate in relatively warmer lowland areas. Seed yield in dry bean can be expressed as product of three components viz., pods per plant, seeds per pod and seed weight (Nienhuis and Singh, 1986) out of which number of pods per plant has often been recommended as an indirect selection criteria for increasing yield primarily because of higher and more consistent correlation with yield (Bennet *et al.*, 1977; Coyne *et al.*, 1965; Sarafi, 1978). The challenge facing stability and sustainability of dry bean production is to identify, develop and deploy disease resistant/tolerant high yielding bean varieties (Singh *et al.*, 1998; Jung *et al.*, 1996; Muyolo *et al.*, 1993; Webster *et al.*, 1980, 1983) with precision and rapid growth rates by employing conventional and non-conventional tools. Therefore, taking into consideration the factors of attaining and stabilizing higher production of dry beans, the present study was conducted to find relationship of 50 grain weight and number of pods per plant with dry-grain yield among dry bean cultivars of different growth habits under natural fertility and incidence of important biotic disease stress(es) in temperate highland agroclimate of Himachal Pradesh.

MATERIALS AND METHODS

One hundred forty five common bean genotypes from different sources were field evaluated during summer seasons of 2000 and 2001 in dry temperate agroclimate of Kukumseri (Lahaul and Spiti) in Himachal Pradesh. The genotypes comprised of indeterminate, semi-determinate and determinate types. Sixteen promising genotypes were evaluated under natural fertility and disease stress conditions of web blight and common bacterial blight in a completely randomized block design (CRBD) replicated thrice. A plot size of 2.70 m^2 was maintained to accommodate indeterminate, semi-indeterminate and determinate genotypes at a (row x plant) spacing of 60x20 cm, 40x15 cm and 30x10 cm, respectively. Sprinkler irrigation approximately at week intervals was applied for 30-45 min to raise the crop. Data on dry-grain yield per plot, number of pods from five randomly selected plants and 50 grain weight from dry seed lots was recorded at appropriate timings. Similarly, data on post-emergence disease incidence of web blight caused by fungal pathogen *Rhizoctonia solani* Kuhn was recorded on the basis of plant mortality expressed as per cent data into arcsine transformed values. The severity of common bacterial blight due to *Xanthomonas axonopodis* pv. *phaseoli* (Smith) Vauterin *et al.* was recorded per plot on the basis of 0-9 score (Van Schoonhoven and Pastor-Corrales, 1987) at pod filling stage. The data so obtained were computed to standard units and analysis of variance was performed for CRBD as well as correlation matrix on a computer run programme.

RESULTS AND DISCUSSION

Among promising common beans, semi-determinate genotypes viz., HPR 407 (22.90 q/ha) and HPR 423 (22.65 q/ha) out yielded the indeterminate and determinate genotypes and gave yield at par with only one determinate variety i.e. Triloki (20.49 q/ha). These results (Table 1) on the superior yield performance of semi-determinates and determinates compared to indeterminates or tall ones are in confirmation of observations made by many researchers (Singh, 1982; Nienhuis and Singh, 1985; Brothers and Kelly, 1993; Jung *et al.*, 1996). Strengthening interpretations based on phenotypic correlations rather than significance of genetic correlations has been suggested better in predicting correlated responses to selection (Pandey and Gritton, 1975). A negative but non-significant correlation of yield with pods per plant and 50 dry grain seed weight was observed among tested genotypes with relatively higher magnitude for 50 dry grain weight (-0.426) indicating that an increase in seed weight would attribute to decrease in yield of common bean cultivars or genotypes.

Table 1. Dry-grain yield of promising common bean cultivars under natural biotic disease stresses and common bacterial blight in dry temperate agro-climate

Cultivar	Dry-grain wt (q/ha)	Number of pods/ plant	50 grain wt (g)	Disease incidence or severity	
				Web blight (% post emergence plant mortality)	Common bacterial blight (0-9 score)
Indeterminate					
HPR 148	15.53	19.77	26.01	12.10	3.66
HPR 75	12.96	19.73	33.78	9.70	2.00
HPR 234	10.39	23.30	28.93	9.80	1.00
Semi-indeterminate					
Kalara local	19.94	12.43	31.73	9.00	0.66
Dhara local	21.49	14.70	25.43	4.50	1.00
LRK 429	19.37	11.93	26.29	14.20	0.66
HPR 423	22.65	18.13	25.22	10.30	1.33
HPR 207	22.90	15.91	22.44	14.10	2.00
Determinate					
HPR 421	17.53	13.20	22.37	16.33	0.00
AFR 234	16.94	27.63	22.21	0.00	0.66
LRK 425	16.30	10.97	25.31	19.01	0.00
HPR 418	18.58	14.77	21.77	24.60	3.00
HPR 28	13.33	12.10	24.20	22.66	4.33
Kanchan	17.79	14.23	24.20	22.66	5.33
Triloki	20.49	17.90	21.98	19.33	3.33
Him-1	19.10	12.13	16.48	20.33	4.33
$CD_{0.01}$	2.84	2.73	0.07	—	—

Significantly, highest number of pods per plant were recorded for semi-indeterminate genotype AFR 234 (27.63) and minimum for determinate genotype LRK 435 (10.97), whereas significant highest 50 dry grain seed weight was recorded for indeterminate genotype, HPR 75 (33.78 g) and minimum for determinate variety Him-1 (16.48 g). Seed size and yield potential of common bean was negatively correlated (Sexton *et al.*, 1994; White *et al.*, 1994). However, irrespective of seed weight a few cultivars/ genotypes under study gave higher yield indicating influence of other factors on the performance of individual genotype, especially the environment (Schwartz and Galvez, 1980) interactions suitable for the responsiveness of genotypes to higher yield potential. However, the relationship between pods per plant with 50 dry grain weight was positively correlated (+0.118), which indicated that normally an increase either in grain weight or number of pods would attribute to increased yield (Table 2). However, contribution of pod length or number of seeds per pod inter alia cannot be ignored as an important yield-contributing factor (Nienhuis and Singh, 1986; Bennet *et al.*,

1977). Moreover, genetic relationships among yield components tend to be negative especially in stress environment, which promote within plant by exhibiting competition for metabolites (Adams, 1967; Denis and Adams, 1978).

Table 2. Correlation matrix of number of pods and 50-grain weight with dry-grain yield in common bean cultivars varying in height parameter

Parameter	Yield	Number of pods
Yield	1.00	-
Number of pods	(-) 0.264	1.00
50-grain weight	(-) 0.426	(+) 0.118

Data given in Table 2 revealed that no apparent incidence of web blight was observed on AFR 234, which was highest on determinate genotype HPR 418 (24.60, post-emergence plant mortality) followed by determinate variety Kanchan (22.66%) and genotype HPR 28 (22.66%). Similarly, no apparent incidence of common bacterial blight was recorded on determinate genotypes HPR 421 and LRK 425, which was recorded highest on determinate variety Kanchan (5.33 score) followed by Him-1 (4.33) and HPR 28 (3.33). Such differences for reaction of common bean cultivars/genotypes have been reported in literature for web blight pathogen (Anderson, 1982; Ogoshi, 1976, 1987; Dillard, 1987; Galindo *et al.*, 1982) as well as common bacterial blight pathogen by various workers (Denis and Adams, 1978; Bennet *et al.*, 1977; Coyne *et al.*, 1965).

Irrespective of biotic disease stresses, both semi-indeterminate and determinate genotypes or cultivars possessed higher yield potential than indeterminate ones. Further, moderate or medium dry-grain weight, increased number of pods per plant, and increased pod length coupled with resistance/tolerance to abiotic and biotic stress would serve as an ideal selection criteria for selecting semi-indeterminate and determinate genotypes/cultivars for dry temperate agro-climate.

REFERENCES

Adams, M.W. (1967). Basis of yield component compensation in crops with special reference of field bean, *Phaseolus vulgaris* L. *Crop Sci.* 27: 637-640.

Anderson, N.A. (1982). The genetics and pathology of *Rhizoctonia solani. Annu. Rev. Phytopathol.* 20: 329-347.

Bennet, J.P., Adams, M.W. and Burga, C. (1977). Pod yield component variation and intercorrelation in *Phaseolus vulgaris* L. as affected by planting density. *Crop Sci.* 17: 73-75.

Bhardwaj, C.L. (1998). Agroecological yield stability of wet season pulses for quantum production. In: *Ecological Agriculture and Sustainable Development,* Vol. I (Eds. Dhaliwal, G.S., Randhawa, N.S., Arora, R. and Dhawan, A.K.), Digigrafics, D-69, Gulmohar Park, New Delhi, pp. 394-398.

Bhardwaj, C.L. and Thakur, D.R. (2000). Stability in dry-grain yield of *Phaseolus vulgaris* relative to meteorological factors and incidence of angular leaf spot. In: *Proceedings International Conference on Integrated Plant Disease Management for Sustainable Agriculture,* Vol. II. Indian Phytopathological Society, IARI, New Delhi, pp. 1018-1020.

Brothers, M.E. and Kelly, J.D. (1993). Interrelationship of plant architecture and yield components in the pinto bean ideotype. *Crop Sci.* 32: 1234-1238.

Coyne, D.P., Schuster, M.L. and Harris, L. (1965). Inheritance, heritability and response to selection for common blight (*Xanthomonas phaseoli*) to tolerance in *Phaseolus vulgaris,* field bean crosses. *Proc. Am. Hortic. Sci.* 86: 373-378.

Denis, J.C. and M.W. Adams, L. (1978). A factor analysis of plant variables related to yield in dry beans. I. Morphological traits. *Crop Sci.* 18: 74-78.

Dillard, H.R. (1987). Characterization of isolates of *Rhizoctonia solani* from lima beans grown in New York state. *Phytopathology* 77: 748-751.

Galindo, J.J., Abawi, G.S. and Thurston, H.D. (1982). Variability among isolates of *Rhizoctonia solani* associated with snap bean hypocotyls and soil in New York. *Plant Dis.* 66: 390-394.

Jung, G.W., Coyne, D.P., Skroch, P.W., Nienhuis, J., Arnaud-Santana, E., Bokosi, J., Ariyarathne, H.M., Steaqdman, J.R. Beaver, J.S. and Kaepper, S.M. (1996). Molecular markers associated with plant architecture and resistance to common blight, web blight and rust in common beans. *J. Am. Soc. Hortic. Sci.* 121: 794-803.

Lindgren, E. (1975). The nutritive value of peas and field beans for hen. *Swed. J. Agric. Res.* 5: 159-161.

Ma, Y. and Bliss, F.A. (1978). Tannin content and inheritance in common beans. *Crop Sci.* 18: 201-204.

Malin, E.M. , Roth, D.A. and Belden, E.L. (1983). Indirect immunoflorescent staining for detection and identification of *Xanthomonas campestris* pv. *phaseoli* in naturally infected bean seed. *Plant Dis.* 67: 645-647.

Muyolo, N.G., Lipps, P.E. and Schmitthenner, A.F. (1993). Anastomosis grouping and variation in virulence among isolates of *Rhizoctonia solani* associated with dry bean and soybean in Ohio and Zquire. *Phytopathology* 83: 438-444.

Nienhuis, J. and Singh, S.P. (1986). Combining ability analysis and relationships among yield components and architectural traits in dry beans. *Crop Sci.* 26: 21-27.

Nienhuis, J. and Singh, S.P. (1985). Effects of location and plant density on yield and architectural traits in dry beans. *Crop Sci.* 25: 579-584.

Ogoshi, A. (1976). Studies on the grouping of *Rhizoctonia solani* Kuhn with hyphal anastomosis and on the perfect stages of groups. *Bull. Natl. Inst. Agric. Sci. C.* 30: 1-63.

Ogoshi, A. (1987). Ecology and pathogenicity of anastomosis interaspecific groups of *Rhizoctonia solani* Kuhn. *Annu. Rev. Phytopathol.* 25: 125-143.

Pandey, S. and Gritton, E.T. (1975). Genotypic and phenotypic variances and correlations in peas. *Crop Sci.* 15: 353-356.

Sarafi, A. (1978). A yield component selection experiment involving American and Iranian cultivars of common bean. *Crop Sci.* 18:5-7.

Schwartz, H.F. and Galvez, G.E. (1980). *Bean Production Problems : Diseases, Insects, Soil and Climate Constraints of Phaseolus vulgaris.* CIAT, Cali, Colombia, 424 p.

Schwartz, H.F., Pastor-Corrales, M.A. and Singh, S.P. (1982). New sources of resistance to anthracnose and angular leaf spot of bean (*Phaseolus vulgaris*). *Euphytica* 31: 741-754.

Sexton, J.P., White, J.W. and Boote, K.J. (1994). Yield -determining progress in relation to cultivar seed size of common bean. *Crop Sci.* 34: 84-91.

Singh, S.P. (1982). A key for identification of different growth habits of *Phaseolus vulgaris* L. *Bean Improv. Coop.* (New York) 25: 92-94.

Singh, S.P., Cardona, C., Morales, F.J., Pastor-Corrales, M.A. and Voysest, O. (1998). Gamete selection for upright Carioca bean with resistance to five diseases and a leafhopper. *Crop Sci.* 38: 666-672.

Skorch, P.W., Nienhuis, J., Beebe, S.E., Tohme J. and Pedraza, F. (1998). Comparison of Mexican common bean (*Phaseolus vulgaris* L.) core and reserve germplasm collections. *Crop Sci.* 38: 488-496.

Tobin, G. and Carpenter, K.J. (1978). The nutritional value of the dry beans (*Phaseolus vulgaris*): A literature review. *Nutr. Abstr-Rev. Series - A: Human and Experiment* 48: 919-936.

Valladares-Sanchez, N.E., Coyne, D.P. and Mumm, R.P. (1983). Inheritance and association of leaf external and internal pod reactions to common blight bacterium in *Phaseolus vulgaris* L. *J. Am. Soc. Hortic. Sci.* 108: 272-278.

Van Schoonhoven, A. and Pastor-Corrales, M.A. (1987). *Standard System for the Evaluation of Germplasm.* CIAT, Cali, Colombia, 53 p.

Wallen, V.R. and Jackson, H.R. (1975). Model for yield loss determination of bacterial blight of field beans utilizing infrared photography combined with field plot studies. *Phytopathology* 65: 942-948.

Webster, D.M., Temple, S.R. and Galvez, G. (1983). Expression of resistance to *Xanthomonas campestris* pv. *phaseoli* in *Phaseolus vulgaris* under tropical conditions. *Plant Dis.* 67: 394-396.

Webster, D.M., Temple, S.R. and Schwartcz, H.F. (1980). Selection for resistance to *Xanthomonas phaseoli* in dry beans. *Crop Sci.* 20: 519-522.

White, J.W., Ochoa, M.R., Ibarra-P. F.J. and Singh, S.P. (1994). Inheritance of seed yield, maturity and seed weight of common bean (*Phaseolus vulgaris*) under semi-arid rainfed conditions. *J. Agric. Sci.* 122: 265-273.

FOREST AND MEDICINAL PLANTS

Integrated Plant Disease Management (2005)
Eds. Sharma, R.C. and Sharma, J.N.
Scientific Publishers (India), Jodhpur, pp. 321 - 324.

49

Biological control of damping-off disease caused by *Fusarium* species in *Dalbergia sissoo* Roxb.

J.C. KAUSHIK, B.S. MANDAL and SUSHIL SHARMA[1]

Department of Forestry, CCS Haryana Agricultural University, Hisar-125 004 (Har.)
[1]HAU Regional Research Station, Bawal - 123 501 (Haryana)

ABSTRACT

Survey of forest nurseries for disease prevalence was conducted during the month of March-April 2001. Isolations from the infected seedlings of sissoo frequently yielded three species viz., *Fusarium solani, F. moniliforme* and *F. oxysporum* f.sp. *dalbergiae*. All the three species were found pathogenic. Six biological control agents viz., *Trichoderma harzianum, T. viride, Gliocladium virens, Bacillus subtilis, B. ceres* and *Pseudomonas fluroscens* were evaluated against all the three pathogens under *in vitro* and *in vivo* conditions. Results indicated that *G. virens* and *B. subtilis* were the most effective antagonists that significantly controlled the disease.

INTRODUCTION

Dalbergia sissoo Roxb. is an important timber tree species of India. It suffers from many fungus diseases, especially Fusarium wilt in plantations and damping off and root rots in nurseries. Though, the actual loss due to damping-off has not been assessed but generally it causes 15-20 per cent loss of planting stock and in severe conditions the losses may be as high as 80-90 per cent (Kaushik and Singh, 1996; Harsh *et al.*, 1992). During a survey in

March-April 2001, three species of *Fusarium* viz., *F. solani* (Mart.) Sacc., *F. moniliforme* J. Sheld. and *F. oxysporum* Schlechtend ex Fr. f.sp. *dalbergiae* were isolated from the infected seedlings of sissoo. Earlier, *Fusarium solani* was reported to be associated with the dying/infected seedlings of sissoo (Bakshi, 1954; Bakshi and Singh, 1959; Browne, 1968). Chemical control of soil-borne pathogens is costly and not environment friendly thus biological control of these pathogens has been suggested with antagonist fungi and bacteria (Baker, 1987; Weller, 1988; Sen, 2000; Mukhopadhyay *et al.*, 1998). The present investigation embodies biocontrol potential of some antagonistic fungi and bacteria against *Fusarium* species causing damping-off in sissoo.

MATERIALS AND METHODS

The damping-off pathogens *Fusarium oxysporum* f.sp. *dalbergiae*, *F. solani* and *F. moniliforme* were isolated from infected seedlings of sissoo, purified and maintained on potato dextrose agar medium . In order to prove the pathogenicity of all the three species of *Fusarium*, cultures were multiplied separately on oatmeal: sand medium (2:5) for 20 days. The cultures thus obtained were mixed thoroughly with sterilized loam soil @ 20 g/kg of soil. Four replications for each treatment were maintained. Seedling mortality was recorded 30 days after sowing. Three species of antagonistic fungi viz., *Trichoderma viride*, *T. harzianum* and *Gliocladium virens* and three species of antagonistic bacteria viz., *Bacillus subtilis*, *B. ceres* and *Pseudomonas fluroscens*, were obtained from Department of Plant Pathology/Microbiology, CCS Haryana Agricultural University, Hisar. Antagonistic activity of all the six isolates was tested by dual culture technique (Hazarica and Das, 1998). Growth inhibition of *Fusarium* spp. in dual culture by antagonistic fungi and bacteria over check was calculated after 5 days of incubation at 25±2°C.

To study biocontrol potential of all antagonistic fungi, an experiment was conducted under nursery conditions in polythene bags of (9 x 4 cm^2 size). Antagonistic agents were used both for seed and soil applications. Seed treatment was done by coating the seeds with spore suspension (3.4 x 10^8/ml) and soil application was carried out by adding mass culture of antagonistic fungi developed on wheat bran (5.2x10^8 cfu) @ 50 g/kg of soil one week before seed sowing. The pathogen was multiplied on sand oatmeal for 15 days at room temperature. The soil was made sick by inoculating *Fusarium* spp. individually @ 20 g/kg of soil. Seeds of sissoo with 95 per cent germination were sown in polythene bags @ 1 seed/bag coating with antagonistic fungi or bacteria separately. Surface sterilized seeds sown in *Fusarium* spp. infested soil in polythene bags served as control. The observations on seedling survival, disease control, seedling stand over control were taken after 30 days of sowing as suggested by Kumar and Mukhopadhyay (1994).

RESULTS AND DISCUSSION

Pathogenicity of the three *Fusarium* spp. isolated from infected seedlings of *Dalbergia sissoo* was tested under nursery conditions. *Fusarium moniliforme* was found to be the most pathogenic causing 70.0 per cent seedling mortality. *F. solani* and *F. oxysporum* caused 65.0 and 40.09 per cent seedling mortality, respectively. *Fusarium* spp. have also been recorded causing rotting of seed and seedlings in sissoo (Bakshi and Singh, 1959; Kaushik and Singh, 1996; Harsh *et al.*, 1992). The growth of all the three species viz., *Fusarium oxysporum* f. sp. *dalbergiae*, *F. moniliforme* and *F. solani* was significantly inhibited by all the antagonistic fungi over control in dual culture (Table 1) and was completely over-grown by *Trichoderma* spp. Among fungi, *Gliocladium virens* was most effective in inhibiting mycelial growth (>75%) followed by *Trichoderma viride*. Amongst the bacterial antagonists *Bacillus subtilis* was the most effective whereas *Bacillus ceres* was the least effective.

Table 1. Growth inhibition of *Fusarium* species by different antagonists

Antagonist	Per cent inhibition*		
	Fusarium solani	*Fusarium moniliforme*	*Fusarium oxysporum* f. sp. *dalbergiae*
Trichoderma viride	75.7	70.9	65.70
T. harzianum	70.0	70.0	65.0
Gliocladium virens	76.0	79.8	72.1
Bacillus subtilis	73.0	70.5	70.0
B. ceres	42.6	38.4	41.4
Pseudomonas fluroscens	71.1	70.4	75.6
$CD_{0.05}$	2.82	6.48	4.42

*Average of four replications

Antagonism of *Trichoderma harzianum*, *T. viride* and *G. virens* against *Fusarium* spp. has also been reported (Abo Ellil *et al.*, 1998; Okhavvat and Karampour, 1996). Bacterial antagonism against *Fusarium* spp. has also been recorded (Knaflewski and Weber, 1998).

Seed coating with conidial/bacterial suspension of all the antagonist recorded reduction in disease incidence. Among antagonists, *Gliocladium virens* and *B. subtilis* effectively controlled damping-off disease followed by *Pseudomonas fluroscens* which was at par with *B. subtilis* (Table 2).

Table 2. Biological control of damping-off disease in sissoo seedlings by different antagonists

Antagonist	Per cent disease control		
	Fusarium solani	*Fusarium moniliforme*	*Fusarium oxysporum* f.sp. *dalbergiae*
Trichoderma viride	52.2	40.8	49.5
T. harzianum	54.3	31.1	50.5
Gliocladium virens	57.2	42.8	49.1
Bacillus subtilis	53.6	60.4	46.9
B. ceres	36.2	24.5	46.1
Pseudomonas fluroscens	49.7	55.2	44.1
$CD_{0.05}$	4.6	11.2	NS

NS = Non-significant

It has been well documented that seed and soil application of *Trichoderma* spp., *G. virens, B. subtilis* and *Pseudomonas fluroscens* successfully managed damping-off diseases in a number of host plants (Kumar and Mukhopadhyay, 1994; Mukhopadhyay *et al.*, 1992; Pedersen *et al.*, 1998). Per cent disease control in *F. oxysporum* f. sp. *dalbergiae* infected plants ranged between 44.1 to 49.5 per cent but statistically non-significant.

REFERENCES

Abo Ellil, A.H.A., Elhaleem, S.T.A. and Awad, N.G.H. (1998). Biocontrol of vegetable root rot disease by *T. harzianum* and *T. viride. Afr. J. Mycol. Biotechnol.* 6: 41-55.

Baker, K.F. (1987). Evolving concepts of biological control of plant pathogens. *Annu. Rev. Phytopathol.* 25: 67-85.

Bakshi, B.K. (1954). Wilt disease of shisham (*Dalbergia sissoo* Roxb.) due to *Fusarium solani. Nature* 174: 278-279.

Bakshi, B. K. and Singh, Sujan. (1959). Root diseases of shisham. Inoculation studies on wilt. *Indian For.* 85: 415-421.

Browne, F. G. (1968). *Pests and Diseases of Forest Plantation Trees.* Oxford, Clarendon Press. 1330 p.

Harsh, N. S. K., Tewari, C. K. and Nath, V. (1992). *Fusarium* wilt of *Dalbergia sissoo* Roxb. seedlings. *Indian J. For.* 15: 64-67.

Hazarica, A. K. and Das, K. K. (1998). Biological management of root rot of French bean caused by *Rhizoctonia solani. Plant Dis.Res.* 13: 101-105.

Kaushik, J.C. and Singh, A. (1996). Effect of soil texture and moisture on seedling root rot of *Dalbergia sissoo* and its control. In: *Proceedings of the IUFRO Symposium on Impact of Diseases and Insects Pests in Tropical Forests.* 1996, pp. 209-216.

Knaflewski, M. and Weber, Z. (1998). Biological protection of asparagus against *Fusarium* fungi. *Peznaniu Ogrodnictwo.* 27: 151-157.

Kumar, D.E.R. and Mukhopadhyay, A.N. (1994). Biological control of tomato damping off by *Gliocladium virens. J. Biol. Control* 18: 34-40.

Mukhopadhyay, A.N., Shrestha, S.M. and Mukherjee, P.K. (1992). Biological seed treatment for control of soil borne plant pathogens. *FAO Plant Prot. Bull.* 40: 21-30.

Okhavvat, M. and Karampour, F. (1996). Effect of some isolates of antagonistic fungi on chickpea black root rot caused by *F. solani* under greenhouse conditions. *Iran. J. Agric. Sci.* 27: 37-43.

Pedersen, E.A., Reddy, M.S. and Chakravarty, P. (1998). Effect of three species of bacteria on damping off, root rot development and ectomycorrhizal colonization of lodgepole pine and white spruce seedlings. *Eur. J. For. Pathol.* 29: 135-142.

Sen, B. (2000). Biological control: A success story. *Indian Phytopathol.* 53: 243-249.

Weller, D.M. (1988). Biological control of soil borne plant pathogens in the rhizosphere with bacteria. *Annu. Rev. Phytopathol.* 26: 379-407.

Integrated Plant Disease Management (2005)
Eds. Sharma, R.C. and Sharma, J.N.
Scientific Publishers (India), Jodhpur, pp. 325 - 329.

50

Evaluation of plant extracts against *Fusarium solani* causing root rot of khair (*Acacia catechu*)

MANICA TOMAR and R.C. SHARMA

Department of Mycology and Plant Pathology, Dr. Y.S. Parmar University of Horticulture and Forestry, Nauni, Solan -173 230 (H.P.)

ABSTRACT

Different plant extracts were tested to see their efficacy against *Fusarium solani* causing root rot of khair. Out of twenty different aqueous plant extracts, garlic (*Allium sativum*) aqueous extract showed a maximum inhibition (64.79%) followed by *Aloe barbedensis* (62.89%). Out of the ten different acetone based plant extracts tested at different concentrations against *F. solani* under *in vitro* conditions, *Ocimum sanctum* gave the overall maximum (81.97%) mycelial inhibition followed by *A. sativum* with a mean mycelial inhibition of 78.43 per cent. Higher concentration of both aqueous and acetone based plant extracts were effective in inhibiting the radial growth of *Fusarium solani.*

INTRODUCTION

The use of plant extracts for curing various human ailments has been in practice since ages. Of late, the revival of use of organic products in place of chemicals has led to explore the plant genera having anti-microbial properties. Many workers have successfully inhibited the growth of different pathogens by using plant extracts. Root rot of *Acacia catechu* (khair), caused by *Fusarium solani* (Mart.) Sacc. causes huge losses in the nurseries of Himachal Pradesh. (Kaushal *et al.*, 2003). In *Acacia* seedlings, the root rot causes yellowing of leaflets followed by severe necrosis. The roots become shriveled, dark and often resemble a thread. In severe conditions, white mycelial growth can be seen outside the roots. Ultimately plant droops down and dies. In the present study, extracts of some of the plants have been evaluated against *Fusarium solani*, the cause of root rot of khair.

MATERIALS AND METHODS

Extracts of twenty different plant species viz., aloe (*Aloe barbedensis*), aonla (*Phyllanthus emblica*), artemesia (*Artemesia* sp.), basuti (*Adathoda vesica*), banah (*Vitex negundo*), bhang (*Cannabis sativus*), safeda (*Eucalyptus globulus*), garlic (*Allium sativum*), ginger (*Gingiber officinale*), lantana (*Lantana camara*), mentha (*Mentha viridis*), murraya (*Murraya koengii*), tulsi (*Ocimum sanctum*), chirpine (*Pinus roxburghii*), marigold (*Tagetus erecta*), neem (*Azadirachta indica*), harad (*Terminalia chebula*), turmeric (*Curcuma longa*), bahera (*T. belerica*) and valeriana (*Valeriana jatamansi*) were tested against *F.solani*. In most of the cases the extracts of leaves and flowers were used. In all the cases the plant material of required quantity (200 g) was washed in tap water and then crushed in pestle and mortar by adding small quantity of distilled water. In cold water extract, distilled water of required quantity (200 ml) was put in the ground material and then homogenized to give 100 per cent botanicals. After heating for 2 minutes, all the extracts were filtered through double layered muslin cloth and refrigerated till further use. These botanicals extracted in cold water were tested at 100, 75, 50, 25 per cent concentration.

In acetone extract, the dried and crushed plant material was passed through Soxlett apparatus. The acetone was used as medium for homogenizing. The distilled acetone vapours passing through crushed plant material were collected and kept for further use. The concentrate available was mixed with equal quantity of distilled water to make 100 per cent concentration. The acetone based extracts were tested at 1, 2, 3 and 4 per cent concentrations.

The extracts were tested *in vitro* by poisoned food technique to study the inhibitory effect of botanicals on the mycelial growth of *F. solani*. The Petri plates were incubated at 25°C. The data on the radial growth was noted after seven days and per cent mycelial inhibition was calculated as:

$$\text{Per cent inhibition} = \frac{C - T}{C} \times 100$$

Where,

C = Growth of mycelium in control (mm)

T = Growth of mycelium in treatment (mm)

RESULTS AND DISCUSSION

Aqueous extracts

Aqueous extracts of twenty different plant species were tested against *F. solani* under *in vitro* conditions. The plant extracts were tested at 100, 75, 50 and 25 per cent concentration (Table 1).

Table 1. *In vitro* effect of aqueous plant extracts on the radial growth of *Fusarium solani*

Treatment	Per cent mycelial inhibition at concentration (%)				Mean
	100	75	50	25	
Valeriana jatamansi	100 (90.00)	13.41(21.47)	0.00 (0.00)	0.00 (0.00)	28.35 (27.87)^{a}
Tagetus erecta	100 (90.00)	84.23 (66.75)	26.02 (30.51)	0.00 (0.00)	52.56 (46.82)^{b}
Lantana camara	100 (90.00)	93.96 (75.85)	19.75 (26.34)	0.00 (0.00)	53.43 (48.05)^{b}
Adathoda vesica	100 (90.00)	94.96 (79.28)	49.75 (41.12)	0.00 (0.00)	54.21 (33.83)^{ab}
Vitex negundo	86.32 (71.88)	59.12 (58.49)	11.64 (4.93)	0.00 (0.00)	44.29 (41.12)^{b}
Cannabis sativus	100 (90.00)	96.53 (79.28)	43.27 (41.12)	0.00 (0.00)	59.95 (52.60)^{bc}
Aloe barbedensis	100 (90.00)	100 (90.00)	26.29 (23.51)	0.00 (0.00)	62.89 (50.87)^{bc}
Artemesia sp.	100 (90.00)	72.73 (58.53)	11.10 (4.93)	0.00 (0.00)	50.19 (37.36)^{ab}
Eucalyptus globulus	83.23 (65.90)	46.58 (43.00)	11.66 (19.96)	0.00 (0.00)	35.17 (36.54)^{ab}
Pinus roxburthii	54.60 (47.67)	9.68 (17.48)	0.55 (2.46)	0.00 (0.00)	16.35 (17.52)^{d}
Allium sativum	100 (90.00)	100 (90.00)	36.92 (35.45)	9.98 (11.20	64.79 (54.48)^{bc}
Mentha viridis	76.42 (61.06)	33.92 (35.46)	10.42 (12.36)	0.00 (0.00)	28.82 (28.67)^{a}
Phyllanthus emblica	92.97 (74.86)	31.81 (34.22)	8.12 (16.05)	0.00 (0.00)	32.16 (31.28)^{ab}
Terminalia belerica	100 (90.00)	96.53 (79.28)	43.27 (41.12)	0.00 (0.00)	59.95 (52.60)^{bc}
Ocimum sanctum	100 (90.00)	100 (90.00)	25.29 (18.61)	3.58 (10.58)	60.16 (52.29)^{bc}
Azadirachta indica	100 (90.00)	97.13 (80.25)	42.01 (40.39)	0.00 (0.00)	59.78 (52.66)^{bc}
Murraya koengii	48.67 (44.24)	6.57 (14.63)	0.00 (0.00)	0.00 (0.00)	13.80 (14.71)^{d}
Curcuma longa	100 (90.00)	93.75 (75.53)	15.39 (22.97)	0.00 (0.00)	53.38 (47.13)^{b}
Gingiber officinale	100 (90.00)	93.88 (75.69)	25.19 (30.12)	3.30 (10.47)	55.59 (51.57)^{bc}
Terminalia chebula	100 (90.00)	88.89 (70.70)	26.84 (31.20)	0.00 (0.00)	53.39 (41.98)^{b}
Mean	92.06 (81.28)^{a}	70.68 (61.79)^{b}	21.67 (20.75)^{c}	0.82 (1.61)^{d}	-

The values in parentheses are arcsine transformed values

In each column or row the values represented by same alphabet do not differ significantly

$CD_{0.05}$

Treatment	= 6.41
Concentration	= 4.26
Treatment x Concentration	= 2.41

The perusal of data presented in Table 1 indicated that the higher concentrations of plant extracts were able to inhibit the mycelial growth effectively. Garlic (*Allium sativum*) aqueous extract showed a maximum inhibition (64.79%) followed by *Aloe barbedensis* (62.89%). Tulsi (*Ocimum sanctum*) showed a mean mycelial inhibition of 60.16 per cent. Aqueous extract of *Murraya koengii* was least effective (13.80%) in inhibiting the mycelial growth of *F. solani*.

Out of the different concentrations tested against *F. solani*, maximum mycelial inhibition (92.06 %) occurred at 100 per cent concentration followed by 75 per cent concentration (70.68 %). Twenty five per cent concentration showed a least mycelial growth inhibition of 0.82 per cent. The interaction between concentration and treatments showed that only *A sativum, Ocimum sanctum* and *Gingiber officinale* were able to inhibit the mycelium of *F. solani* at the lowest concentration of 25 per cent. All other plant extracts were effective at higher concentration.

Acetone based extracts

Ten different acetone based plant extracts were tested at different concentrations against *F. solani* under *in vitro* conditions. The perusal of data presented in Table 2 reveal that all the plant extracts were able to inhibit the fungal growth at 4 per cent concentration. However, *Ocimum sanctum* gave the overall maximum (81.97%) mycelial inhibition followed by *A. sativum* with a mean mycelial inhibition of 78.43 per cent.

Table 2. Effect of acetone based plant extracts on the radial growth of *Fusarium solani*

Treatment	Per cent mycelial inhibition at concentration (%)				Mean
	1	2	3	4	
Aloe barbedensis	19.22(17.30)	50.22(45.01)	100.00(90.00)	100.00(90.00)	67.80(60.57)[a]
Azadirachta indica	0.00(0.00)	20.00(29.24)	90.70(72.25)	100.00(90.00)	58.90(47.87)[b]
Adathoda vesica	0.00(0.00)	11.64(19.95)	63.00(52.62)	92.40(74.28)	42.80(36.71)[c]
Vitex negundo	0.00(0.00)	4.51(12.25)	77.09(61.48)	91.58(73.26)	42.71(36.75)[c]
Cannabis sativus	0.00(0.00)	40.01(39.23)	93.62(75.43)	100.00(90.00)	58.41(51.17)[d]
Eucalyptus globulus	0.00(0.00)	38.19(38.09)	82.62(65.44)	100.00(90.00)	55.20(48.38)[b]
Allium sativum	10.68(18.92)	73.71(59.17)	100.00(90.00)	100.00(90.00)	78.43(59.79)[e]
Gingiber officinale	0.00(0.00)	28.87(32.48)	94.74(77.47)	100.00(90.00)	55..90(49.99)[d]
Curcuma longa	0.00(0.00)	42.19(40.50)	95.11(77.31)	100.00(90.00)	59.32(51.95)[b]
Ocimum sanctum	28.21(18.58)	77.67(61.82)	100.00(90.00)	100.00(90.00)	81.97(65.10)[e]
Mean	5.81(5.48)[a]	38.70(37.77)[b]	89.68(75.19)[c]	98.40(86.75)[d]	-

The values in parentheses are arcsine transformed values

In each column or row the values represented by same alphabet do not differ significantly

$CD_{0.05}$

Treatment	=	3.19
Concentration	=	2.30
Treatment x Concentration	=	4.76

The results clearly showed that higher concentration of acetone based plant extracts were more efficient in inhibiting the fungal growth. Only *Adathoda vesica* and *Vitex negundo* gave the mycelial inhibition of 92.48 per cent and 91.58 per cent, respectively, while rest of the plant extracts showed 100 per cent mycelial inhibition at 4 per cent concentration.

Contrarily, only *Aloe barbedensis* and *Ocimum sanctum* were able to inhibit the mycelial growth at 1 per cent concentration with 19.22 and 28.21 per cent mycelial inhibition, respectively. The interaction between concentration and plant extract showed only aloe, *A. sativum* and *O. sanctum* were able to inhibit the mycelial growth at 1 per cent concentration with 19.22, 10.68 and 28.21 per cent mycelial inhibition, respectively. Rest all the extracts inhibited the mycelial growth at higher concentrations.

The present investigations gave an overview that garlic, tulsi and aloe extracts were quite effective against *F. solani* causing seedling root rot of *A. catechu*. Bambarkar (1990) emphasized the importance of neem as a potential plant against many pathogenic fungi. Similarly, garlic extract was found to inhibit many pathogenic fungi and *Xanthomonas campestris* pv. *vesicatoria* (Mangamma and Sreeramulu, 1991). Garlic extract has been found to be effective against many pathogens viz., *Pyricularia oryzae, Corticium sasakii, Fusarium udum, F. oxysporum* f.sp. *lentis* and *F. solani* (Tewari and Dath., 1984; Singh and Tripathi, 1992). The present investigations also presented similar results where *Ocimum sanctum* (81.97%) and garlic plant extracts (78.43 %) were found to be effective against *F. solani*.

Methanol-water extracts of tomato were found effective against *F. solani, Rhizoctonia solani* and *Septoria nodorum* (Jiratko, 1993). Similarly, Bansal and Gupta (2000) found *Azadirachta indica, Ocimum bacillum* and *Lantana camara* plant extracts highly toxic to *Fusarium oxysporum* causing fenugreek wilt. Many plant extracts have been found to be effective against number of other pathogenic fungi (Pans *et al.*, 1985; Singh and Tripathi, 1995).

REFERENCES

Bambarkar, S. (1990). Neem – A vast potential for agrochemicals. *Pesticides* 24: 36-38.

Bansal, R.K. and Gupta, R. (2000). Evaluation of plant extracts against *Fusarium oxysporum* wilt pathogen of fenugreek. *Indian Phytopathol.* 53: 107-108.

Jiratko, J. (1993). Comparison of antifungal activity of tomatine and tomato extract. *Ochrana Rostli* 29: 93-98.

Kaushal, A.N., Sharma, K.R. and Sharma, R.C. (2003). *Khair- A Multipurpose Tree.* Dr. Y.S. Parmar University of Horticulture and Forestry, Solan, 104 pp.

Mangamma, P. and Sreeramulu, K. (1991). Garlic extract inhibitory to growth of *Xanthomonas campestris* pv. *vesicatoria. Indian Phytopathol.* 44: 372-374.

Pans, S., Mukherjee, B., Ganguly, A., Mitra, S.R. and Bhattacharya, A. (1985). Use of plant extracts against root rot of soyabean. *Z. Pflanzenkr. Pflanzenschutz* 92: 392-395.

Singh, Jaspal and Tripathi, N.N. (1995). Evaluation of botanicals against soil-borne pathogens. *Indian Phytopathol.* 48: 363-364.

Singh, Jaspal and Tripathi, N.N. (1992). Efficacy of some angiospermic plants against *Fusarium oxysporum* f.sp. *lentis* Schl. *Nat. Acad. Sci. Lett.* 15: 277-278.

Tewari, S.N. and Dath, Premlata A. (1984). Effect of leaf extract media of some plants on the growth of three fungal pathogens of rice. *Indian Phytopathol.* 37: 458-461.

Integrated Plant Disease Management (2005)
Eds. Sharma, R.C. and Sharma, J.N.
Scientific Publishers (India), Jodhpur, pp. 331 - 337

51

Development of leaf rust (*Melampsora ciliata* Barclay) in *Populus ciliata* hybrids

SANJEEV SHARMA and R.C. SHARMA

Department of Mycology and Plant Pathology, Dr. Y. S. Parmar University of Horticulture and Forestry, Nauni, Solan - 173 230 (H.P.)

ABSTRACT

Progress of Melampsora leaf rust was studied on 19 hybrids of *Populus ciliata* (*P. ciliata* x *P. maximowiczii*). The disease incidence varied from 36.19 to 79.08 per cent at the time of first observation with an average of 58.57 per cent. The mean disease incidence increased to 84.90 per cent on the last date of observation. Hybrids 42M1 and CM2-26 had least mean disease incidence of 58.42 and 60.35 per cent, respectively. Urediniopustules per leaf were minimum in CM1-54 (14.03) and maximum in CM2-42 (114.98). Apparent infection rate irrespective of the hybrids was more between July 25 – August 14 (0.0383) as compared to September 6 to 26 (0.0119). Hybrids CM26 and 42M1 had lesser area under the disease progress curve.

INTRODUCTION

Species of the genus *Populus* grow primarily in the temperate and cold regions of the northern hemisphere and are among the most important species for plantation forestry worldwide. Owing to its rapid growth, genetic diversity, colourless wood, ease of vegetative propagation and suitability for wood based industries; the genus is well suited for management and utilization. *Populus* spp. have resulted in highly productive clones that have been deployed in many countries. Out of more than 110 organisms reported pathogenic on *Populus*, the leaf rust caused by *Melampsora* species is the most common, serious and widely distributed disease in the nurseries as well as in plantations and native forests throughout the world (Berbee, 1964). The indigenous species of poplar (*Populus ciliata* Wall. ex Royle) is widely distributed in Indian temperate Himalayas between 1300 to 3500 m above sea level from Kashmir to Arunachal Pradesh and is severely attacked by *Melampsora ciliata* Barclay, the indigenous and microcyclic rust which is prevalent between 700 to 3130 m above sea level (Bor, 1953; Sharma and Sharma, 2000, 2002; Sharma *et al.*, 2001a). *Populus ciliata* clones are highly susceptible to this rust and none was found immune to this disease (Sharma, 2000). Large number of hybrids (*P. ciliata* x *P. maximowiczii*) have been developed but a few of them have shown immunity to rust. Therefore, investigations were planned to study the rust development in hybrids of *P. ciliata* on various epidemiological parameters with the progress of time.

MATERIALS AND METHODS

The studies were carried out in Nauni nursery (1275 m asl) of Dr. Y.S. Parmar University of Horticulture and Forestry, Solan during 1999 and 2000. The cuttings (18-23 cm length) of 19 *Populus ciliata* hybrids (*P. ciliata* x *P. maximowiczii*) were planted in the beds at a spacing of 45 x 45 cm in the month of February in a randomized block design with five replications each. The data on disease incidence, number of urediniopustules per leaf and disease severity were recorded at 21 day intervals between July and September 1999 and 2000. Per cent disease incidence was worked out by counting the infected and total number of leaves on each sapling.

Urediniopustules per leaf were calculated by counting the urediniopustules of ten tagged leaves of each sapling and then average was taken. The apparent infection rate (r) for different hybrids was calculated as per the method of Van der Plank (1963) using the logistic equation:

$$r = \frac{2.3}{t_2 - t_1} \left(\log_{10} \frac{X_2}{1 - X_2} - \log_{10} \frac{X_1}{1 - X_1}\right)$$

where, r denotes the apparent infection rate per unit per day, t_2-t_1 is the time interval between first and last observations, X_1 and X_2 represent the proportion of diseased plant parts, and 1-X_1 and 1-X_2 represent the proportion of healthy plant parts at t_1 and t_2 time intervals, respectively.

Disease severity was rated on the basis of leaf area covered by urediniopustules using the scale of Sharma *et al.* (2001b) and from the severity data, area under the disease progress curve (AUDPC) was calculated in accordance to Shaner and Finney (1977) as follows:

$$\text{AUDPC} = \sum_{i=1}^{n} [(y_{i+1} + y_i)/2]\,(X_{i+1} - X_i)$$

where y_i denotes disease severity (%) at the ith observation, x_i denotes time (days) at the ith observation and n denotes total number of observations.

The data pertaining to disease incidence, urediniopustules per leaf, apparent infection rate and area under the disease progress curve were first analysed in usual way for each year separately and then homogeneity of error variances was tested using F-test. The error variances were found to be homogeneous so the pooled analysis for 1999 and 2000 was carried out accordingly. The data on per cent disease incidence was subjected to arcsine transformation before the analysis of variance. The least significance difference was used to compare the values (Gomez and Gomez, 1984).

RESULTS AND DISCUSSION

Disease incidence, irrespective of hybrids, ranged from 36.19 per cent in 221M2 to 79.37 per cent in 11M1 on first date of observation (July 25) with mean disease incidence of 58.57 per cent (Table 1). After three weeks (August 14), mean disease incidence was 71.00 per cent which increased to 78.94 per cent at the time of third observation (September 5) and 84.90 per cent on the date of fourth observation (September 26). Disease incidence in the month of August ranged from 54.25 per cent in 42M1 to 87.06 per cent in CM2-62, while in September it ranged from 62.80 per cent in 42M1 to 97.25 per cent in CM2-62. Irrespective of the time intervals, disease incidence ranged from 58.42 to 88.22 per cent being minimum in 42M1 and statistically at par with CM2-26 and 221M2 while maximum was recorded in CM1-8 which did not differ significantly from 11M1.

Table 1. Melampsora leaf rust progress (disease incidence) in hybrids of *Populus ciliata* during 1999 and 2000 (pooled)

Hybrid	Disease incidence (%) on				Mean
	July 25	August 14	September 5	September 26	
11M1	79.37 (63.23)	85.38 (67.61)	89.95 (71.53)	96.25 (78.90)	87.73 (70.32)
CM2-83	57.66 (49.41)	71.44 (57.79)	80.51 (64.07)	88.31 (70.20)	74.48 (60.37)
CM2-49	48.85 (44.33)	72.19 (58.20)	84.19 (66.72)	90.31 (72.17)	73.89 (60.35)
CM2-62	62.30 (51.97)	87.06 (68.97)	95.09 (78.98)	97.25 (81.86)	85.44 (70.45)
CM1-54	51.33 (48.07)	65.19 (53.85)	68.29 (55.74)	75.35 (60.29)	66.04 (54.49)
CM2-46	57.85 (49.52)	64.37 (53.36)	68.54 (55.91)	75.89 (60.63)	66.66 (54.86)
CM1-8	79.08 (62.83)	85.25 (67.43)	91.09 (72.89)	97.44 (82.63)	88.22 (71.44)
42M1	46.68 (43.10)	54.25 (47.45)	62.80 (52.45)	69.94 (56.78)	58.42 (49.95)
CM2-53	56.96 (49.00)	65.38 (53.98)	72.25 (58.26)	77.11 (61.49)	67.93 (55.68)
324M2	74.43 (59.70)	83.25 (65.83)	88.12 (70.20)	93.50 (77.31)	84.81 (68.27)
113M1	60.01 (50.79)	73.19 (58.82)	83.61 (66.26)	91.11 (72.66)	77.01 (62.14)
221M2	36.19 (36.87)	61.86 (51.93)	73.04 (58.84)	80.38 (63.95)	62.87 (52.90)
CM2-60	57.00 (49.03)	72.49 (58.40)	83.31 (65.91)	88.75 (70.42)	75.39 (60.94)
390M1	65.62 (54.14)	73.74 (59.20)	81.30 (64.45)	84.50 (66.88)	76.29 (61.16)
6M1	54.85 (47.79)	69.81 (56.10)	81.13 (64.29)	85.39 (67.57)	72.80 (59.09)
CM2-26	50.00 (45.00)	57.06 (49.07)	64.56 (54.50)	69.75 (56.68)	60.35 (51.06)
CM2-39	57.81 (49.50)	71.74 (57.90)	78.13 (62.12)	84.81 (67.12)	73.12 (59.16)
CM2-95	55.15 (47.96)	69.69 (56.79)	80.25 (63.83)	85.74 (68.16)	72.71 (59.19)
84M1	57.70 (49.40)	67.56 (55.29)	73.56 (59.08)	81.25 (64.37)	70.02 (57.04)
Mean	58.57 (50.09)	71.10 (57.82)	78.94 (63.42)	84.90 (68.42)	

Figures in parentheses are arcsine transformed values

$CD_{0.05}$

Hybrid	= 3.55
Interval	= 1.63
Hybrid x Interval	= 7.09

Disease progress with respect to production of urediniopustules per leaf is presented in Table 2. At the time of first observation (July 25), urediniopustules number per leaf ranged from 6.50 in CM1-54 to 53.50 in CM2-46. Irrespective of hybrids, mean number of urediniopustules per leaf on first date of observation was 20.92 which increased to 44.83 on second date of observation (August 14), 63.16 on third date (September 5) and 77.96 on the last date of observation (September 26). Number of urediniopustules per leaf ranged from 10.88 to 16.00 in CM1-54 and 93.25 to 165.95 in CM2-46 in the month of August and September, respectively. Irrespective of the intervals, number of urediniopustules per leaf ranged from 14.03 to 114.98, it being minimum in CM1-54 not differing significantly from CM2-26, 221M2, 84M1 and 390M1. On the other hand, highest number was recorded in CM2-46 followed by CM1-8.

Table 2. Melampsora leaf rust progress (urediniopustules/ leaf) in hybrids of *Populus ciliata* during 1999 and 2000 (pooled)

Hybrid	Urediniopustules/ leaf on				Mean
	July 25	August 14	September 5	September 26	
11M1	22.56	54.31	77.00	86.44	60.08
CM2-83	10.13	26.38	33.25	44.50	28.56
CM2-49	25.87	66.00	110.00	111.10	83.28
CM2-62	47.63	72.63	89.25	100.13	77.41
CM1-54	6.50	10.88	16.00	22.75	14.03
CM2-46	53.50	93.25	147.15	165.95	114.98
CM1-8	32.80	81.69	112.95	129.05	89.14
42M1	16.25	27.00	33.56	40.40	29.33
CM2-53	23.63	47.25	65.00	81.44	44.33
324M2	25.88	79.50	108.60	128.00	85.50
113M1	9.25	78.86	114.81	134.80	84.42
221M2	8.19	22.75	30.19	40.81	25.49
CM2-60	24.67	39.73	51.83	80.87	49.28
390M1	12.93	26.61	33.42	41.25	28.55
6M1	24.31	33.03	38.40	45.71	35.37
CM2-26	6.82	12.25	19.90	30.54	17.38
CM2-39	20.00	27.75	43.84	89.25	45.21
CM2-95	17.54	35.25	40.63	43.52	34.24
84M1	9.00	16.78	34.35	44.50	26.16
Mean	20.92	44.83	63.16	77.96	

$CD_{0.05}$

Hybrid	= 15.17
Interval	= 7.48
Hybrid x Interval	= 30.33

Rate of spread of disease and area under the disease progress curve were also studied as disease development parameters and results are presented in Table 3. Rust developed at the rate of 0.0383 per unit per day between July 25 to August 14 which reduced to 0.0178 per unit per day between August 15 to September 5 and 0.0119 per unit per day between September 6 to 26. Irrespective of the hybrids, the rate of spread of disease between July 25 to August 14 ranged from 0.0100 per unit per day in 6M1 to 0.0566 per unit per day in 324M2. In second interval (August 15 – September 5), the rate of spread of disease ranged from 0.0059 to 0.0288 per unit per day, it being minimum in CM2-95 and maximum in CM2-26. Between September 6 to 26, the rate of spread of disease ranged from 0.0051 (6M1) to 0.0174 per unit per day (84M1). Irrespective of intervals, lowest rate of spread of disease (0.0115 per unit per day) was recorded in 6M1 followed by CM2-62 and 42M1 which did not differ significantly from one another, while highest was recorded in 113M1 (0.0269 per unit per day) followed by 324M2, 84M1 and CM2-26 the latter statistically at par with each other.

Table 3. Apparent infection rate and area under the disease progress curve of Melampsora leaf rust in *Populus ciliata* hybrids during 1999 and 2000 (pooled)

Hybrid	Apparent infection rate between			Mean	AUDPC
	July 25–Aug. 14	Aug.15 –Sep.5	Sep.6– Sep. 26		
11M1	0.0436	0.0181	0.0036	0.0217	19.82
CM2-83	0.0469	0.0132	0.0121	0.0240	17.03
CM2-49	0.0453	0.0222	0.0107	0.0260	16.12
CM2-62	0.0233	0.0090	0.0056	0.0127	20.48
CM1-54	0.0251	0.0175	0.0164	0.0196	15.00
CM2-46	0.0314	0.0242	0.0064	0.0206	15.07
CM1-8	0.0474	0.0174	0.0059	0.0235	19.91
42M1	0.0210	0.0096	0.0091	0.0132	13.25
CM2-53	0.0307	0.0167	0.0057	0.0193	15.47
324M2	0.0566	0.0164	0.0075	0.0268	19.24
113M1	0.0548	0.0191	0.0069	0.0269	17.40
221M2	0.0444	0.0146	0.0157	0.0239	14.75
CM2-60	0.0327	0.0190	0.0086	0.0201	16.01
390M1	0.0257	0.0207	0.0134	0.0199	16.11
6M1	0.0100	0.0195	0.0061	0.0115	15.48
CM2-26	0.0287	0.0288	0.0211	0.0261	12.76
CM2-39	0.0390	0.0139	0.0230	0.0252	15.48
CM2-95	0.0341	0.0059	0.0048	0.0149	15.43
84M1	0.0335	0.0281	0.0174	0.0263	14.74
Mean	0.0383	0.0178	0.0119		

$CD_{0.05}$

Hybrid	= 0.0130	AUDPC	= 0.96
Interval	= 0.0052	Hybrid x Interval	= 0.0224

In the present studies, response of *P. ciliata* hybrids (*P. ciliata* x *P. maximowiczii*) to *Melampsora ciliata* was evaluated by four different criteria to determine which individual criterion might provide a consistent measure to relative susceptibility or resistance to poplar leaf rust. There was rapid development of rust with the progress of time, which is evident from disease incidence and number of urediniopustules per leaf and were direct indicators of rust development as they assessed primary symptoms at the site of infection. Progress of disease from July to September revealed that the disease incidence and number of urediniopustules per leaf increased from 58.57 to 84.90 per cent and 20.92 to 77.96, respectively. The other two criteria measured disease by assessing the rate of spread of disease. In the present studies, apparent infection rate and AUDPC in different hybrids were found to be variable. The apparent infection rate is a measure of the spread at which an epidemic develops. The apparent infection rate decreased with progress of time. The rate of spread of disease was 0.0383 per unit per day between July to August, which decreased to 0.0178 and 0.0119 per unit per day between August 15 to September 5 and in between September 6 to 26, respectively. The decline in the rate of spread of disease with progress of time may be attributed to the reduction in healthy tissue for the new infection. Area under the disease progress curve is a reliable parameter to estimate and rank the performance of various host genotypes depending on their ability to record the rate of disease development. The value of AUDPC in the present investigations varied from 12.76 to 20.48. The variation in the degree of susceptibility of *P. ciliata* hybrids to *M. ciliata* is attributed to genetic composition of the plants. Pinon and Teissier Du Cros (1976) studied the relative susceptibility/resistance of *P. nigra, P. deltoides, P. trichocarpa* and *P. x euramericana* to *M. larici-populina* and *M. allii-populina* and found varying degree of susceptibility to both the rust species. Similarly, Sharma *et al.* (2001a) reported considerable variation in the disease levels in *Populus* species, hybrids and cultivars due to *M. ciliata*. The disease incidence, number of urediniopustules per leaf, infection rate and AUDPC all separated the least susceptible ones (CM2-26, 42M1 and 221M2) from the highly susceptible hybrids (11M1 and CM2-62). Sharma and Sharma (2003) also found CM2-26 as the least susceptible when tested under *in vitro* conditions.

REFERENCES

Berbee, J.G. (1964). Diseases of *Populus*. In: *Diseases of Widely Planted Forest Trees*. FAO/IUFRO Sec 24. Oxford. pp. 168-183.

Bor, N.L. (1953). *Manual of Indian Forest Botany*. Oxford Univ. Press, 441 pp.

Gomez, K.A. and Gomez, A.A. (1984). *Statistical Procedures for Agricultural Research*. John Wiley & Sons, New York, Inc. 680 p.

Pinon, J. and Teisser Du Cros, E. (1976). Susceptibility to rusts in miscellaneous poplar species to *Melampsora larici-populina* Kleb. and *M. allii-populina*. *Ann. Sci. For.* 33: 49-55.

Shaner, G. and Finney, R.E. (1977). Effect of nitrogen fertilization on the expression of slow mildewing resistance in Knox wheat. *Phytopathology* 67:1051-1056.

Sharma, R.C. and Sharma, Sanjeev. (2000). Status and distribution of foliar diseases of poplar in Himachal Pradesh. *Indian Phytopathol.* 53: 57-60.

Sharma, R.C., Khan, Y., Sharma, Sanjeev and Malhotra, R. (2001a). Development of *Melampsora ciliata* rust on nursery-grown poplars in north-western Himalayas. *For. Pathol.* 31: 313-319.

Sharma, Sanjeev. (2000). *Studies on Epidemiology and Management of Poplar Leaf Rust.* Ph.D. Thesis, Dr. Y.S. Parmar University of Horticulture and Forestry, Solan, (H.P.), 130 p.

Sharma, Sanjeev and Sharma, R.C. (2002). Prevalence of poplar leaf rust in Himachal Pradesh. *Indian Phytopathol.* 55: 81-83.

Sharma, Sanjeev and Sharma, R.C. (2003). Influence of host genotype and pathogen isolate in the development of poplar leaf rust. *Z. Pflanzenkr. Pflanzenschutz* 110: 359-365.

Sharma, Sanjeev, Sharma, R.C. and Sharma J.N. (2001b). Morphological variability in *Melampsora ciliata*- the incitant of poplar leaf rust. *Indian For.* 127: 242-248.

Van der Plank, J.E. (1963). *Plant Diseases: Epidemics and Control.* Academic Press, New York, 349 p.

Integrated Plant Disease Management (2005)
Eds. Sharma, R.C. and Sharma, J.N.
Scientific Publishers (India), Jodhpur, pp. 339 - 345.

52

In vitro inhibitory effect of some fungal antagonists against *Sclerotium rolfsii* from *Glaucium flavum* Crantz

S. KANSAL and R.C. SHARMA[1]

Department of Vegetable Crops, [1]Department of Mycology and Plant Pathology, Dr.Y.S. Parmar University of Horticulture and Forestry, Solan – 173 230 (H.P.)

ABSTRACT

The antagonistic properties of different species/isolates of *Gliocladium* and *Trichoderma* were evaluated against *Sclerotium rolfsii* under *in vitro* experiments. *Trichoderma harzianum, T. viride* and *Gliocladium virens* proved strongly antagonistic to *S. rolfsii* in dual culture. The non-volatile toxic substances of *T. harzianum* and *G. virens* revealed more inhibitory effect on the test pathogen both under agar plate and well methods. The toxic effect of fungal antagonists was more deleterious in agar plate than in well method. The volatile toxicants of *T. harzianum* were significantly inhibitory to *S. rolfsii,* followed by the toxicity of *T. viride* isolates and *G. deliquescens,* respectively. Nine day old cultures of antagonists produced toxic gaseous metabolites that affected maximum inhibition of the test pathogen.

INTRODUCTION

Yellow horned poppy (*Glaucium flavum*) is a binnial medicinal herb introduced in Himachal Pradesh from Bulgaria. It yields glaucine a potential substitute for codeine in the formulation for bronchitis and various cough syrups. The cultivation of this medicinal plant is facing a serious root rot problem due to *Sclerotium rolfsii* (Gupta, 1987). Chemical control measures of Sclerotium root rot have often met with limited success (Punja, 1985). The practical use of antagonistic micro-organims in disease management in the form of biocontrol agents has a great potential. *In vitro* evaluation of some fungal antagonists against different host interaction with *S. rolfsii* have been investigated (Dennis and Webster, 1971a,b,c; Mathur and Sarbhoy, 1978; Chamswarng and Sangkaha, 1988; Ikoton and AdeKunle, 1990). The present investigations were undertaken to ascertain the *in vitro* effectivity of different species/isolates of *Gliocladium* and *Trichoderma* against *S. rolfsii* isolate of yellow horned poppy.

MATERIALS AND METHODS

Sclerotium rolfsii was isolated from diseased portions of yellow horned poppy and maintained on potato dextrose agar medium (PDA). Different species/isolates of *Gliocladium* and *Trichoderma* were isolated by dilution plate method (Johnson, 1957) and tested for their antagonistic behaviour against the test pathogen.

The inhibitory effect of these fungal antagonists on the mycelial growth of *S. rolfsii* was evaluated by using dual culture method. Radial growth of both the antagonists and the test pathogen was recorded after 72 h incubation at 25±1°C, and growth inhibition of the test fungus over control was calculated in accordance with the formula suggested by Vincent (1947).

The efficacy of non-volatile toxicants produced by these fungal antagonists on the growth of *S. rolfsii* was assayed by using well method (Singh and Webster, 1973) and agar plate method (Dennis and Webster, 1971a). In the well method, the inhibitory action of non-volatile secretion of the fungal antagonist of different ages ranging from 24 to 144 h was tested, whereas in agar plate method the test ages ranged from 12 to 48 h. Each treatment including the control was replicated four times. Observations on radial growth of the test pathogen were recorded after 72 h of incubation at 25±1°C and per cent growth inhibition was calculated.

The fungal antagonists were also tested for the inhibitory action of their volatile toxicants against *S. rolfsii* by using the method suggested by Dennis and Webster (1971b). Potato dextrose agar plates (75 mm dia) were inoculated separately in the centre with 5 mm disc(s) cut from the margin of vigorously growing culture of each antagonist and kept for incubation at 25±1°C. At interval of 0, 3, 6, 9, 12 and 15 days of incubation, the lid of each plate was replaced by bottom containing PDA (15 ml) inoculated centrally with the growth culture of the test pathogen. The plates were sealed together with adhesive tape. The lids of the control plates which had not been inoculated with the antagonists were also replaced in the same way. The test and control plates were replicated four times. Observations on the radial growth of test pathogen after 72 h of incubation were recorded in each treatment and per cent growth inhibition of test pathogen was calculated.

RESULTS AND DISCUSSION

It is evident from the data given in Table 1 that amongst different species/isolates of *Trichoderma* and *Gliocladium* evaluated by dual culture,

T. harzianum exhibited a significantly higher inhibitory effect against *S. rolfsii,* while *T. viride* –1, *T. viride* –2 and *G. virens* were also potent. Growth inhibition was also recorded with *T. harzianum,* while *G. deliquescens* reflected an intermediate efficacy. *Gliocladium roseum* provided the minimum inhibitory effect.

Table 1. *In vitro* efficacy of fungal antagonists against *Sclerotium rolfsii* in dual culture

Antagonist	Per cent growth inhibition
Trichoderma viride –1	79.50 (63.10)[b]
Trichoderma viride –2	75.60 (60.43)[c]
Trichoderma harzianum	84.00 (66.45)[a]
Trichoderma hamatum	66.50 (54.65)[d]
Gliocladium virens	73.10 (58.74)[c]
Gliocladium deliquescens	42.60 (40.72)[e]
Gliocladium roseum	28.50 (32.25)[f]
$CD_{0.01}$	1.780

The data presented in Table 2 indicated that under well method, non-volatile substances of *T. harzianum* were most inhibitory out of the seven fungal antagonists tried at test ages ranging from 24 to 144 h.

Table 2. Effect of non-volatile toxic substances produced by fungal antagonists on the growth of *Sclerotium rolfsii* (well method)

Antagonist	Per cent growth inhibition by fungal antagonist of age (h)						Mean
	24	48	72	96	120	144	
Trichoderma viride –1	4.80 (12.10)	15.60 (23.24)	40.00 (39.23)	40.30 (39.39)	35.50 (36.57)	30.00 (33.21)	27.70 (30.62)[d]
T. viride –2	4.70 (12.52)	15.90 (23.45)	42.20 (40.51)	42.30 (40.67)	36.60 (37.24)	30.00 (33.21)	28.60 (31.26)[c]
T. harzianum	8.40 (16.79)	25.60 (30.38)	44.40 (41.80)	46.70 (43.11)	48.90 (44.37)	48.90 (44.37)	37.20 (36.80)[a]
T. hamatum	0.00 (2.87)	11.10 (19.45)	25.60 (30.38)	30.00 (33.21)	27.80 (31.82)	25.30 (30.18)	20.00 (24.65)[e]
Gliocladium virens	7.30 (15.61)	22.20 (28.11)	44.40 (41.80)	47.80 (43.74)	42.20 (40.51)	37.80 (37.94)	33.60 (34.62)[b]
G.deliquescens	0.00 (2.87)	8.90 (17.34)	17.80 (24.95)	22.20 (28.11)	22.20 (28.11)	20.00 (26.57)	15.20 (21.33)[f]
G. roseum	0.00 (2.87)	5.00 (12.90)	10.00 (18.42)	12.20 (20.43)	14.40 (22.31)	14.40 (22.31)	9.30 (16.54)[g]
Mean	3.60 (9.38)[u]	14.90 (22.12)[t]	32.10 (33.87)[r]	34.50 (35.52)[p]	32.50 (34.42)[q]	29.50 (32.54)[s]	-

Figures in parentheses are angular transformed values

Means represented with same letter do not differ significantly

$CD_{0.01}$

Antagonist = 0.51

Age = 0.48

Antagonist x Age = 1.26

Next efficacious ones were *G. virens* and the isolates of *T. viride*. *T. hamatum* provided some inhibition. The inhibitory effect of both *G. deliquescens* and *G. roseum* was quite low. The overall inhibitory response of the antagonists ageing 72 h or more was significantly higher as compared to those ageing less than 72 h. The non-volatile secretion of the antagonists ageing 96 h were most inhibitory. It was followed by ageing periods of 120, 72 and 144 h, respectively. The filtrates obtained from 48 h old cultures could not inhibit the growth considerably. The inhibitory influence of antibiotics produced by antagonists ageing 24 h was quite low.

The non-volatile toxicants of the test antagonists ageing 12 to 48 h provided a varied inhibitory response under agar plate method (Table 3). The inhibitory action of *G. virens,* irrespective of different incubation treatments, was most deleterious. It was closely followed by the inhibitory potency of *T. harzianum* and the isolates of *T. viride*. The growth inhibition with *T. hamatum* treatments remained at a moderate level and with *G. deliquescens* and *G. roseum* the response was poor. A significant increase in the inhibitory effect of the toxic substances produced by the antagonists was noticed with their ageing increasing from 12 to 48 h. Complete inhibition was caused by non-volatile secretions of 48 h old cultures of *G. virens, T. harzianum* and isolates of *T. viride*. The antibiotics released by 12 h old antagonists exerted a low level of inhibition.

Table 3. Effect of non-volatile toxic substances produced by fungal antagonists on growth of *Sclerotium rolfsii* (agar plate method)

Antagonist	Per cent growth inhibition by fungal antagonist of age (h)				Mean
	12	24	36	48	
Trichoderma viride-1	13.30(21.38)	40.80(39.71)	75.30(60.22)	100.00(87.13)	57.40(52.11)[c]
T. viride –2	12.20(20.43)	40.30(39.39)	72.80(58.56)	100.00(87.13)	56.30(51.37)[c]
T. harzianum	14.70(22.54)	44.40(41.80)	77.00(61.34)	100.00(87.13)	59.00(53.20)[b]
T. hamatum	5.00(12.62)	18.90(25.76)	27.10(31.64)	47.20(43.41)	24.60(28.36)[d]
Gliocladium virens	16.20(23.69)	48.10(43.90)	80.80(64.04)	100.00(87.13)	61.30(54.69)[a]
G. deliquescens	0.00(2.87)	0.00(2.87)	14.70(22.54)	26.40(30.91)	10.30(14.80)[e]
G. roseum	0.00(2.87)	0.00(2.87)	10.80(19.18)	16.70(24.11)	6.90(12.26)[f]
Mean	8.80(15.20)[s]	27.50(28.04)[r]	51.20(45.36)[q]	70.00(63.85)[p]	

Figures in parentheses are angular transformed values

Means represented with same letter do not differ significantly

$CD_{0.01}$

Antagonist	= 0.84
Age	= 0.63
Antagonist x Age	= 1.69

It is obvious from data (Table 4) that volatile toxicants released by *T. harzianum* was highly inhibitory to *S. rolfsii* followed by *T. viride* isolates. *G. deliquescens* restricted the growth considerably, while the gaseous metabolites of *T. hamatum* and *G. virens* caused less inhibition. The inhibitory effect of *G. roseum* was very low. Comparative efficacy of volatile fungitoxicants produced by test antagonists ageing 6 or more days was significantly higher than ageing less than 6 days. Ageing period of 9 days was optimum for maximum inhibitory effect. The effectivity of antagonists ageing 12 and 15 days marked a gradual decline except in *T. harzianum* which recorded complete growth inhibition of the pathogen at these enhanced ageing periods. Growth inhibition of the test pathogen was the least when inoculated simultaneously with antagonists.

The results herein indicated that *T. harzianum, T. viride* and *G. virens* were strongly antagonistic to *S. rolfsii* in dual culture. Hansen and Curl (1964) reported *T. viride* to be antagonistic to *S. rolfsii*. Chamswarng and Sangkaha (1988) recorded a marked inhibition of the fungus by the isolates of *Trichoderma* and *Gliocladium* under *in vitro* studies.

Table 4. Effect of volatile toxic substances produced by fungal antagonists on growth of *Sclerotium rolfsii*

Antagonist	Per cent growth inhibition by fungal antagonist of age (days)						Mean
	0	3	6	9	12	15	
Trichoderma viride –1	35.30 (36.46)	45.00 (42.13)	68.30 (55.75)	69.70 (56.59)	64.70 (53.54)	60.30 (50.97)	57.20 (49.24)[b]
T. viride –2	34.00 (35.67)	43.40 (41.18)	65.00 (53.73)	66.70 (54.75)	61.30 (51.55)	58.70 (50.00)	54.90 (47.81)[c]
T. harzianum	32.30 (34.65)	59.30 (50.38)	72.30 (58.27)	85.30 (67.49)	98.00 (82.40)	100.00 (87.13)	74.50 (63.38)[a]
T. hamatum	25.70 (30.44)	36.30 (37.06)	39.70 (39.04)	35.30 (36.46)	31.30 (34.03)	27.30 (31.51)	32.60 (34.76)[e]
Gliocladium virens	20.70 (27.03)	31.70 (34.25)	31.00 (33.82)	26.70 (31.07)	20.70 (27.03)	20.00 (26.57)	25.10 (29.96)[f]
G.deliquescens	21.70 (27.74)	41.00 (39.81)	60.00 (50.79)	57.00 (49.03)	51.30 (45.76)	47.30 (43.47)	46.40 (42.77)[d]
G. roseum	4.70 (12.41)	7.00 (15.29)	7.30 (15.63)	13.70 (21.69)	13.00 (21.12)	13.00 (21.12)	9.80 (17.88)[g]
Mean	24.90 (29.20)[t]	37.70 (37.16)[r]	49.10 (43.86)[q]	50.60 (45.30)[pq]	48.60 (45.06)[p]	46.70 (44.39)[q]	

$CD_{0.01}$

Antagonist = 0.89

Age = 0.84

Antagonist x Age = 2.17

The present investigations both under agar plate and well methods indicate that the non-volatile toxicants produced by *T. harzianum/G. virens* exerted significant inhibitory effect on *S. rolfsii* followed by the isolates of *T. viride*. The inhibitory action of the species/isolates of *Trichoderma* and *Gliocladium* may be postulated to the presence of antibiotic metabolites in their secretion as has also been suggested by Webster and Lomas (1964). These workers indicated that gliotoxin and viridin produced by *T. viride* were responsible for inhibiting the growth of *S. rolfsii*. The inhibitory effect of non-volatile substances of test antagonists evaluated under well method was considerably less than the corresponding inhibitions obtained in agar plate method. This may be attributed to the increased production of fungitoxic substances by antagonists on solid media than on liquid media as observed by Ribaldi (1954), and also to the rate of diffusion which is far less in solid medium resulting in increased concentration of fungitoxic substances per unit area.

The results emerging from the study also indicated that the gaseous-metabolites of *T. harzianum* were highly inhibitory against *S. rolfsii*, followed by *T. viride and G. deliquescens*. The volatile toxicants of *T. hamatum* and *G. virens* caused some inhibition only. Irrespective of the antagonist tried, the ageing period of 9 days was best suited for the production of volatile substances for affecting maximum inhibition of the test pathogen. Dennis and Webster (1971b) observed that *Trichoderma* species ageing 7-9 days were superior to others in producing volatile fungitoxic substances. Hutchinson and Cowan (1972) reported that differences in the amount of inhibition of assay fungi by gaseous metabolites from different species of *Trichoderma* could be accounted for by differences in the rate of production of carbon dioxide and concentration of acetaldehyde and ethanol.

REFERENCES

Chamswarng, C. and Sangkaha, K. (1988). *In vitro* screening for effective antagonists of *Sclerotium rolfsii* Sacc. a causal agent of tomato stem rot. *Kasetsart J. Natural Sci.* 22: 7-13.

Dennis, C. and Webster, J. (1971a). Antagonistic properties of species-groups of *Trichoderma*. I. Production of volatile antibiotics. *Trans. Br. Mycol. Soc.* 57: 25-39.

Dennis, C. and Webster, J. (1971b). Antagonistic properties of species –groups of *Trichoderma*. II. Production of volatile antibiotics. *Trans. Br. Mycol. Soc.* 57: 41-48.

Dennis, C. and Webster, J. (1971c). Antagonistic properties of species –groups of *Trichoderma*. III. Production of volatile antibiotics. *Trans. Br. Mycol. Soc.* 57: 363-369.

Gupta, R. (1987). Phytochemical industry in Himachal Pradesh. *Indian Hortic.* 32: 36-39.

Hansen, J.D. and Curl, E.A. (1964). Interaction of *Sclerotium rolfsii* and other micro-organims isolated from stolon tissue of *Trifolium repens*. *Phytopathology* 54: 1127-1132.

Hutchinson, S.A. and Cowan, M.E. (1972). Identification and biological effects of volatile metabolites from cultures of *Trichoderma harzianum*. *Trans. Br. Mycol. Soc.* 59: 71-77.

Ikoton, T. and AdeKunle, F. (1990). Inhibition of growth of some plant pathogenic fungi by some antagonistic micro-organisms isolated from soil. *J. Basic Microbiol.* 30: 95-98.

Johnson, L.F. (1957). Effect of antibiotics on the numbers of bacteria and fungi isolated from soil by dilution-plate method. *Phytopathology* 47: 630-631.

Mathur, S.B. and Sarbhoy, A.K. (1978). Biological control of *Sclerotium* root rot of sugarbeet. *Indian Phytopathol.* 31: 365-367.

Punja, Z.K. (1985). The biology, ecology and control of *Sclerotium rolfsii. Annu. Rev. Phytopathol.* 23: 97-127.

Ribaldi, M. (1954). Preliminary observations on the antibiotics activity of *Camarosporium* sp. *Ann. Sper. Agar. N.S.* 8: 1043-1053.

Singh, N. and Webster, J. (1973). Antagonism between *Stilbella erythrocephala* and other coprophilous fungi. *Trans. Br. Mycol. Soc.* 61: 487-493.

Vincent, J.H. (1947). Distortion of fungal hyphae in the presence of certain inhibitors. *Nature* 150: 850.

Webster, J. and Lomas, N. (1964). Does *Trichoderma viride* produce gliotoxin and viridin. *Trans. Br. Mycol. Soc.* 47: 537.

MUSHROOMS

Integrated Plant Disease Management (2005)
Eds. Sharma, R.C. and Sharma, J.N.
Scientific Publishers (India), Jodhpur, pp. 349 - 353.

53

Determination of thermal death point of different mushroom moulds

S.R. SHARMA and SATISH KUMAR

National Research Centre for Mushroom, Chambaghat, Solan – 173 213 (H.P.)

ABSTRACT

Mushrooms are attacked by a large number of parasitic and competitor moulds. These moulds take the heavy toll of the crop every year. In order to find the optimum temperature for eradication of these moulds, thermal death point was studied by dry heat and wet heat methods at different temperatures (30, 35, 40, 45, 50, 55, 60, 65 and 70°C) and time periods (10 min, 30 min, 2 h, 4 h and 6 h) . Under wet heat treatment, thermal death point after 10 minutes exposure was 60°C for *T.viride*, 55°C for *V. fungicola*, 50°C for *M. perniciosa* and *S. crysospermum* and 45°C for *C. dendroides*. Dry heat treatment for 10 min at 65°C in case of *T. viride*, 70°C for 1 h for *V. fungicola*, 55°C for 10 min for *M. perniciosa*, 65°C for 30 min for *S. crysospermum* and 40°C for 10 min for *C. dendroides* rendered the test fungi unviable.

INTRODUCTION

Mushroom cultivation involves several physical factors for the proper growth of mushrooms and their optimum yields. Physical factors greatly influence the crop right from Phase-I composting. Among the physical factors, temperature plays a vital role in every aspect of mushroom cultivation. Role of temperature in pasteurization of compost and casing mixture and cook out with live steam at the end of the cropping is well documented in the literature. Heating is required for the destruction of pests and pathogens in both Phase-II composting as well as in spent mushroom substrate. Temperature is the key factor which is easy to regulate in modern mushroom farms and by finding the difference between the optimum temperature used for the cultivation of mushrooms and optimum temperature for mushroom moulds, it might be possible to utilize this difference in order to reduce disease attack (Bech *et al.,* 1989). On the other hand, many pathogens have become resistant to methyl benzimidazole carbamate (MBC) fungicides following the extensive use (Gareth *et al.,* 1998). Moreover, mushrooms are very sensitive to pesticides and indoor nature of the crop and residue problem further limit the scope of liberal application of pesticides in mushrooms. Hence, physical factors especially heat can be used for the elimination of pests and pathogens right from beginning. In order to obtain detailed information required for killing the mycelium of the pathogenic fungi, *in vitro* studies were carried out and results are presented in this communication.

MATERIALS AND METHODS

Pure cultures of moulds viz., *Mycogone perniciosa, Verticillium fungicola, Cladobotryum dendroides, Trichoderma viride* and *Sepedonium crysospermum* were grown on malt extract agar medium (MEA) . After 7 days of growth these cultures were exposed to different temperatures (30, 35, 40, 45, 50, 55, 60, 65 and 70°C) for different durations of time periods (10 min, 30 min, 1h, 2 h, 4 h and 6 h) under dry heat (oven) and wet heat (water bath) conditions. For wet heat treatment scrapings of mycelial growth on MEA were taken in sterile distilled water and 10 ml suspension was exposed for different durations at a particular temperature in a serological water bath. Afterwards one ml of treated suspension was mixed in 100 ml MEA and poured in five Petri plates. These Petri plates were incubated at 25°C and observations on the viability were recorded at regular intervals of 24 h up to three days. For dry heat treatment uniform sterilized Whatman paper discs were dipped in fungal suspensions prepared in sterile water and these discs were exposed for different temperatures and duration in an oven. Afterwards these discs were inoculated aseptically in Petri plates containing MEA and observations on viability of test fungus were recorded after regular interval of 24 h up to three days.

RESULTS AND DISCUSSION

Perusal of the data (Tables 1,2) reveal the effect of the temperature time factor on the survival of the different moulds. Some information on the thermal sensitivity of the important soil-borne mushroom pathogens has been given by Wuest *et al.* (1970) and Wuest and Moore (1972). These authors treated test fungi in soil cultures for 15 and 30 min at various temperatures wherein all pathogens were eliminated. In the present investigations under wet heat method (Table 1), *V. fungicola* was found to survive when exposed to 50°C for 6 h. However, spores in aqueous suspension were reported to be killed at 38°C for 30 min exposure (Zaayen and Rutjens, 1981; Bech *et al.,* 1989). In case of *M. perniciosa* survival was recorded at 40°C for one hour exposure period. Beyond 45°C, no growth was recorded even at 10 min exposure period. Spores of *M. perniciosa* were found to survive at 42°C (Bech *et al.,* 1989). However, Zaayen and Rutjens (1981) reported that thermal death point of aqueous suspension of spores of *M. perniciosa* was 48°C at 30 min exposure. *C. dendroides* on the other hand proved to be more sensitive to temperature. Survival of this fungus was there at 40°C at 10 min exposure period. When exposed at 35°C for 30 min, no growth was recorded. However, thermal death point of aqueous suspension of spores was found to be 43°C for 30 min exposure period (Zaayen and Rutjens, 1981). According to Seth and Dar (1989), *C. dendroides* was killed at 50°C at a duration of 4 h. According to them pasteurization of compost at 60°C for 6 h renders it safe for economic mushroom growing. The pathogens *S. crysospermum* and *T. viride* were found to be quite heat tolerant as compared to other test moulds. *S. crysospermum* was found to survive at 50°C for one hour exposure period. At this temperature *T. viride* survived even when exposed for 4 h.

Table 1. Growth of different moulds after treatment at different temperatures and durations (wet method)

Mould	Time	Temperature (°C)								
		30	35	40	45	50	55	60	65	70
V. fungicola	10 min	+	+	+	+	+	-	-	-	-
	30 min	+	+	+	+	+	-	-	-	-
	1 h	+	+	+	+	+	-	-	-	-
	2 h	+	+	+	+	+	-	-	-	-
	4 h	+	+	+	+	+	-	-	-	-
	6 h	+	+	+	+	+	-	-	-	-
C. dendroides	10 min	+	+	+	-		-	-	-	-
	30 min	+	+	-	-	-	-	-	-	-
	1 h	+	+	-	-	-	-	-	-	-
	2 h	+	-	-	-	-	-	-	-	-
	4 h	+	-	-	-	-	-	-	-	-
	6 h	+	-	-	-	-	-	-	-	-
M. perniciosa	10 min	+	+	+	+	-	-	-	-	-
	30 min	+	+	+	-	-	-	-	-	-
	1 h	+	+	+	-	-	-	-	-	-
	2 h	+	+	-	-	-	-	-	-	-
	4 h	+	+	-	-	-	-	-	-	-
	6 h	+	+	-	-	-	-	-	-	-
S. crysospermum	10 min	+	+	+	+	+	-	-	-	-
	30 min	+	+	+	+	+	-	-	-	-
	1 h	+	+	+	+	+	-	-	-	-
	2 h	+	+	+	+	-	-	-	-	-
	4 h	+	+	+	-	-	-	-	-	-
	6 h	+	+	+	-	-	-	-	-	-
T. viride	10 min	+	+	+	+	+	+	-	-	-
	30 min	+	+	+	+	+	-	-	-	-
	1 h	+	+	+	+	+	-	-	-	-
	2 h	+	+	+	+	+	-	-	-	-
	4 h	+	+	+	+	+	-	-	-	-
	6 h	+	+	+	-	-	-	-	-	-

Table 2. Growth of different moulds after treatment at different temperatures and durations (dry method)

Mould	Time	Temperature (°C)								
		30	35	40	45	50	55	60	65	70
V. fungicola	10 min	+	+	+	+	+	+	+	+	+
	30 min	+	+	+	+	+	+	+	+	+
	1 h	+	+	+	+	+	+	+	+	-
	2 h	+	+	+	+	+	+	+	+	-
	4 h	+	+	+	+	+	+	+	-	-
	6 h	+	+	+	+	+	+	+	-	-
C. dendroides	10 min	+	+	-	-	-	-	-	-	-
	30 min	+	+	-	-	-	-	-	-	-
	1 h	+	-	-	-	-	-	-	-	-
	2 h	-	-	-	-	-	-	-	-	-
	4 h	-	-	-	-	-	-	-	-	-
	6 h	-	-	-	-	-	-	-	-	-
M. perniciosa	10 min	+	+	+	+	+	-	-	-	-
	30 min	+	+	+	+	-	-	-	-	-
	1 h	-	-	-	-	-	-	-	-	-
	2 h	-	-	-	-	-	-	-	-	-
	4 h	-	-	-	-	-	-	-	-	-
	6 h	-	-	-	-	-	-	-	-	-
S. crysospermum	10 min	+	+	+	+	+	+	+	+	+
	30 min	+	+	+	+	+	+	+	-	-
	1 h	+	+	+	+	+	-	-	-	-
	2 h	+	+	+	+	+	-	-	-	-
	4 h	+	+	+	+	+	-	-	-	-
	6 h	+	+	+	+	+	-	-	-	-
T. viride	10 min	+	+	+	+	+	+	+	-	-
	30 min	+	+	+	+	+	+	+	-	-
	1 h	+	+	+	+	+	-	-	-	-
	2 h	+	+	+	+	+	-	-	-	-
	4 h	+	+	+	+	+	-	-	-	-
	6 h	+	+	+	+	-	-	-	-	-

Under dry heat conditions (Table 2) *V. fungicola* survived at 70°C for exposure period of 30 min. Even the exposure at 65°C for 2 hours did not kill that pathogen. According to Bech *et al.* (1989) dry spores of *V. fungicola* survived at 125°C for 10 minutes exposure and 110°C for 30 min exposure time. In case of *M. perniciosa,* an exposure at 50°C for 10 min did not kill the fungus. Exposure beyond 50°C even for 10 min killed the test fungus. However, Bech *et al.* (1989) recorded survival of spores at 105°C for 10 min exposure period. Survival of *S. crysospermum* was recorded when exposed at 70°C for 10 min. Even exposure at 50°C for 6 h did not kill this pathogen. Similarly, *T. viride* survived at 60°C when exposed for 30 min. Both *S. crysospermum* and *T. viride* were found to be heat tolerant as compared to other test fungi.

Tolerance to heat may result in survival of the pathogen under Phase–II composting process which may create havoc in later stages of cropping. The lethal temperatures are different for mycelium and spores of the same fungus. Hot air treatment with no use of chemical cannot result in inactivation of the most of the pathogens in the growing rooms, however, heat treatment can be employed for treatment of the equipments in hot water for the elimination of the pathogens.

REFERENCES

Bech, K., Jacobsen, B.D. and Kovacs, G. (1989). Investigations on the influence of the temperature on growth and spore formation of *M. perniciosa* and *V. fungicola,* two pathogenic fungi of the cultivated mushroom. *Mush. Sci.* XII: 749-751.

Gareth, J.M., Damian, E., Elizabeth, M. and Averil, E.B. (1989). Identification of benzimidazole resistance in *Cladobotryum dendroides* using a PCR-based method. *Mycol. Res.* 102: 671-676.

Seth, P.K. and Dar, C.M. (1989). Studies on *Cladobotryum dendroides* (Bull: Merat) W. Gams. et. Hoozem, causing cobweb disease of *Agaricus bisporus* (Lange) Sing. and its control. *Mush. Sci.* XII : 711-723.

Wuest, P.J. and Moore, R.K. (1972). Additional data on the thermal death point for *Agaricus* sp. and for the spores of some major pathogens. *Mush. Sci.* XI: 393-402.

Wuest, P.J., Baker, K.F. and Conway, W.S. (1970). Sensitivity of selected mushroom pathogens to aerated steam. *Phytopathology* 60: 1274-1275.

Zaayen, A.V. and Rutjens, A.J. (1981). Thermal death point for two *Agaricus* spp. and for the spores of the some major pathogens. *Mush. Sci.* XI: 393-402.

Integrated Plant Disease Management (2005)
Eds. Sharma, R.C. and Sharma, J.N.
Scientific Publishers (India), Jodhpur, pp. 355 - 358.

54

Change in pH values of different casing substrates during cropping of *Agaricus bisporus* (Lange) Imbach

R.S. JARIAL and T.R. SHANDILYA

Department of Mycology and Plant Pathology, Dr.Y.S. Parmar University of Horticulture and Forestry, Nauni, Solan – 173 230 (H.P.)

ABSTACT

There was gradual fall in pH levels from the day of casing to the end of cropping in all the test combinations. In municipal waste fed to earthworms-leached + vermicompost - leached + farmyard manure it came down to 6.87, whereas in other two casing combinations i.e. municipal waste fed to earthworms - leached + vermicompost-leached and municipal waste fed to earthworms - leached + farmyard manure it lowered down to 6.65 and 6.88, respectively. On an average, pH lowered down from 7.64 on 0 day of casing to 6.87 at +56 days after casing.

INTRODUCTION

Mushroom (*Agaricus bisporus* (Lange) Imbach) production largely depends on covering colonized compost with a casing layer. The casing layer is a major source of variation in yield, quality and uniformity of commercial cropping (Nobel and Gaze, 1995). Casing involves covering the top of mushroom beds with a layer of appropriate soil mixture after mycelium has impregnated the compost. Casing is generally done to make a platform, where uniform fructification can take place and also to provide anchorage and essential reserves for developing sporophores of the mushroom. The pH of the casing material has direct impact on yield of mushrooms. Ivanovic (1993) compared different casing materials and recorded best yield with peat and river mud having pH 8.0. Singh *et al.* (2000) obtained good yields of mushrooms in the casing media having pH ranging between 6.8-7.16. A pH range between 7.0-8.0 has also been reported by several workers (Hawker, 1966; Atkins, 1972; Edwards, 1978; Shandilya, 1982).

MATERIALS AND METHODS

Ten grams sample from each casing combination i.e. municipal waste fed to earthworms-leached (MWEW-L) + vermicompost-leached (VC-L) + farmyard manure (FYM) (1:2:2 v/v), municipal waste fed to earthworms- leached + vermicompost-leached (1:2 v/v), municipal waste fed to earthworms- leached + farmyard manure (1:2 v/v) and farmyard manure + loam soil (1:1 v/v) was taken and weighed separately. To each sample 100 ml of distilled water was added and shaken well. Then the suspensions were kept for one hour at room temperature. After this, pH of each suspension was determined with the help of Elico digital (LI 127) pH meter. During cropping, pH was determined at 0 day (day of casing), +7, +14, +21, +28, +35, +42, +49 and +56 days after casing.

RESULTS AND DISCUSSION

It is clear from the data presented in Table 1 that pH dropped down from the date of casing to the end of cropping. In municipal waste fed to earthworms-leached + vermicompost-leached casing combination, the pH came down from 7.55 on the day of casing to 6.65 after +56 days of casing, whereas, in farmyard manure + loam soil it lowered from 7.85 on the day of casing to 7.08 after +56 days of casing. The pH in municipal waste fed to earthworms-leached + farmyard manure and municipal waste fed to earthworms-leached + vermicompost - leached + farmyard manure lowered down to 6.88 and 6.87 at +56 days of sampling from 7.63 and 7.53 on 0 day of casing, respectively. Irrespective of the casing combinations there was a general trend that the pH dropped down gradually as the cropping period progressed (Fig. 1). Similar observations have been made by Edwards and Flegg (1952), Allison and Kneebone (1963) and Shandilya (1989).This decrease in the pH of all the casing media with the progression of crop may be attributed to the production of oxalic acid by the mycelium of *Agaricus bisporus* (Vedder, 1978).

Table 1. Changes in pH value in different casing media with crop progression

Casing medium	pH value days after casing									Overall mean
	0	+7	+14	+21	+28	+35	+42	+49	+56	
Municipal waste fed to earthworms-leached + vermicompost-leached	7.55	7.50	7.42	7.39	7.21	7.00	6.90	6.77	6.65	7.15
Municipal waste fed to earthworms-leached + farmyard manure	7.63	7.53	7.43	7.40	7.30	7.22	7.10	6.99	6.88	7.27
Municipal waste fed to earthworms-leached + vermicompost-leached farmyard manure	7.53	7.44	7.37	7.29	7.25	7.19	7.09	7.01	6.87	7.22
Farmyard manure + loam soil (control)	7.85	7.75	7.57	7.40	7.32	7.25	7.18	7.11	7.08	7.39
Overall mean	7.64	7.56	7.45	7.37	7.30	7.16	7.01	6.97	6.87	

$CD_{0.05}$

Casing medium = 0.09

Interval = 0.14

Casing medium x Interval = 0.27

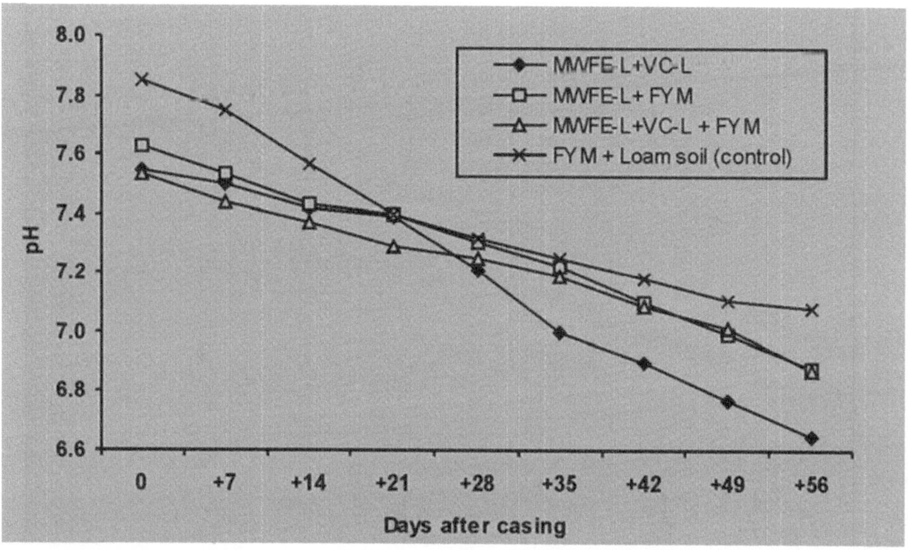

Fig. 1. Changes in pH with progression of crop in different casing media

REFERENCES

Allison, W.H. and Kneebone , L.R. (1963). Influence of compost pH and casing soil pH on mushroom production. *Mush. Sci.* 5: 81-89.

Atkins, F.C. (1972). *Mushroom Growing Today*. Faber and Faber, London, 182 p.

Edwards, R.L. (1978). Cultivation in Western countries : growing in houses. In: *The Biology and Cultivation of Edible Mushrooms* (Eds. Chang, S.T. and Hayes, W.A.), Academic Press, London, pp. 299-336.

Edwards, R.L. and Flegg, P.B. (1952). Casing soil. *Annu. Rep. Mush. Res. Stn.* Yaxely, 1951.pp. 21-43.

Hawker, L.E. (1966). Environment influence on reproduction. In: *The Fungi. An Advanced Treatise,* Vol. II. *The Fungal Organism.* (Eds. Ainsworth, G. and Sussman, A.S.), Academic Press, New York, pp. 435-469.

Ivanovic, B. (1993). The effect of physico-chemical properties of the casing mixture on the yield and quality of the cultivated mushroom. *Rev. Res. Work Fac. Agric. Belgrade* 38: 83-91.

Noble, R. and Gaze, R.H. (1995). Properties of casing peat types and additives and their influence on mushroom yield and quality. In: *Science and Cultivation of Edible Fungi* (Ed. Elliott), Balkema, Rotterdam, pp. 305-312.

Shandilya, T.R. (1982). Compost and casing research at Mushroom Research Centre, Solan during past seven years. *Indian J. Mush.* 8: 5-13.

Shandilya, T.R. (1989). Mushroom compost and casing research in India. *Mush. Sci.* 12: 743-752.

Singh, M., Singh, R.P. and Chaube, H.S. (2000). Impact of physico-chemical properties of casing on yield of *Agaricus bisporus* (Lange) Imbach. In: *Science and Cultivation of Edible Fungi* (Ed. Griensven, Van), Balkema, Rotterdam, pp. 441-446.

Vedder, P.J.C. (1978). *Modern Mushroom Growing.* Educabock B.& V., Industrieweg 1, Culemborg, The Netherlands.413 p.

Integrated Plant Disease Management (2005)
Eds. Sharma, R.C. and Sharma, J.N.
Scientific Publishers (India), Jodhpur, pp. 359 - 362.

55

Salt accumulation in different casing media with the crop progression of *Agaricus bisporus* (Lange) Imbach

R.S. JARIAL and T.R. SHANDILYA

Department of Mycology and Plant Pathology, Dr. Y.S. Parmar University of Horticulture and Forestry, Nauni, Solan – 173 230 (H.P.)

ABSTRACT

Changes in electrical conductivity in four test casing combinations viz., municipal waste fed to earthworms-leached + vermicompost- leached , municipal waste fed to earthworms -leached + farmyard manure, municipal waste fed to earthworms-leached + vermicompost- leached + farmyard manure and farmyard manure + loam soil was studied with the progression of crop. It was found that electrical conductivity increased significantly during the whole cropping span in all the four test casing substrates. The average electrical conductivity at the time of casing i.e. 0 day of casing was recorded to be 239.80 μ mhos/cm which increased to 742.80 μ mhos/cm after 56 days of casing.

INTRODUCTION

Casing involves covering the top of mushroom (*Agaricus bisporus* (Lange) Imbach) beds with a layer of appropriate soil mixture after mycelium has impregnated the compost. Casing is generally done to make a platform, where uniform fructification can take place and also to provide anchorage and essential reserves for developing sporophores of the mushroom. A good casing should have low availability of soluble inorganic ions (Hayes, 1981). The accumulation of ions in the casing layer ultimately affects the electrical conductivity of the casing soils (Yeo and Hayes, 1979) which also affects the fruiting of mushrooms. The electrical conductivity plays an important role in the production of *A. bisporus*, but it is not the sole controlling factor (Singh *et al.*, 2000). Keeping in view the importance of electrical conductivity of casing mixtures for mushroom production, the changes in electrical conductivity were recorded during the present investigations with the progression of crop.

MATERIALS AND METHODS

Ten grams sample from each casing combination i.e. municipal waste fed to earthworms-leached (MWFE-L) + vermicompost - leached (VC-L) + farmyard manure (FYM) (1:2:2 v/v), municipal waste fed to earthworms- leached + vermicompost-leached (1:2 v/v), municipal waste fed to earthworms- leached + farmyard manure (1:2 v/v) and farmyard manure + loam soil (1:1 v/v) was taken and weighed separately and to each sample 100 ml of distilled water was added. The mixture was allowed to stand in waterbath at 20°C for one hour and conductivity of collected samples were measured using a portable conductivity measuring bridge type MC3 (Electronic Instrument Limited). During cropping, electrical conductivity was measured at 0, 7, 14, 21, 28, 35, 42, 49 and 56 days after casing. Electrical conductivity was expressed as μ mhos/cm.

RESULTS AND DISCUSSION

A perusal of the data presented in Table 1 reveals that maximum accumulation of salts was recorded in farmyard manure + loam soil (765.60 μ mhos/cm) followed by municipal waste fed to earthworms-leached + vermicompost-leached + farmyard manure (748.20 μ mhos/cm) after 56 days of casing. Minimum accumulation of salts was recorded in municipal waste fed to earthworms - leached + vermicompost - leached (724.00 μ mhos/cm), casing combination amongst all the test media at the last day of sampling. The interaction of different combination treatments with the time elapsed after casing reveals that salt concentration was minimum in casing combination of municipal waste fed to earthworms-leached + vermicompost-leached + farmyard manure (225.60 μ mhos/cm) at 0 day (at the time of casing), whereas, maximum conductivity (765.60 μ mhos/cm) was recorded at 56 days after casing in the control treatment (Fig. 1). Flegg (1961) also reported that accumulation of soluble salts in the casing was a natural phenomenon. The same was further confirmed by Hayes *et al.* (1978). Shandliya (1989) also reported an increase in the concentration of calcium, potassium, sodium, chloride and sulphate ions in the casing layer during the cropping and were the main ions responsible for increase in salinity of casing media.

Table 1. Changes in electrical conductivity in different casing media with crop progression

Casing medium	Electrical conductivity (μ mhos/cm) days after casing									Overall mean
	0	+7	+14	+21	+28	+35	+42	+49	+56	
Municipal waste fed to earthworms-leached + vermicompost-leached	243.00	274.40	341.80	428.00	533.80	591.00	638.00	652.90	724.00	491.80
Municipal waste fed to earthworms-leached + farmyard manure	240.80	262.80	322.40	444.20	544.00	600.00	653.00	674.00	733.40	497.20
Municipal waste fed to earthworms-leached + vermicompost-leached + farmyard manure	225.60	297.80	344.60	462.00	564.20	613.60	682.60	700.80	748.20	515.52
Farmyard manure + loam soil (control)	249.80	314.80	385.40	489.60	579.80	619.60	694.80	720.20	765.60	535.50
Overall mean	239.80	287.40	348.50	455.90	555.40	606.00	667.10	686.85	742.80	

$CD_{0.05}$

Casing medium = 3.22

Interval = 4.83

Casing medium x Interval = 9.66

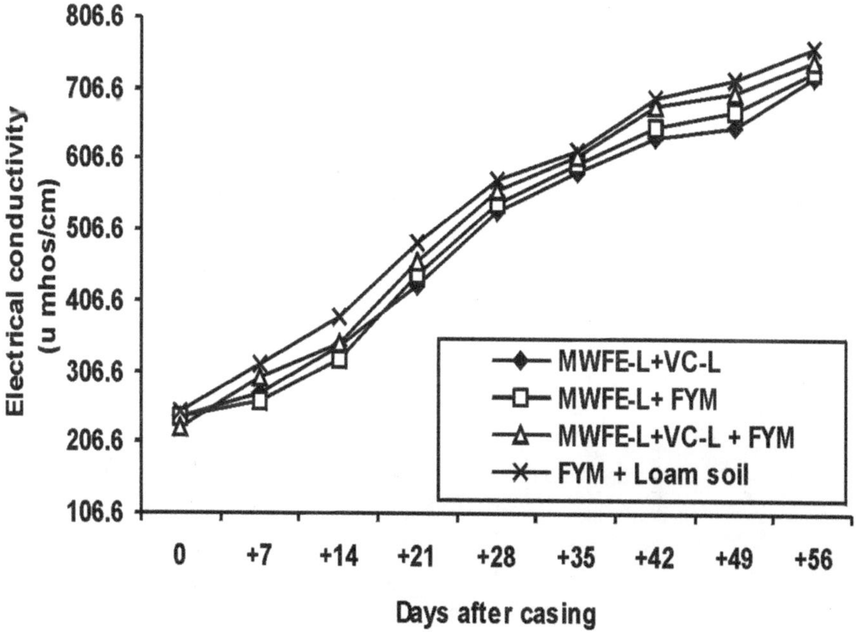

Fig. 1. Salt accumulation in different casing media with the progression of crop

REFERENCES

Flegg, P.B. (1961). Electrical conductivity of the casing layer in relation to fruiting of the cultivated mushroom. II Mushroom compost and composting: a review of literature. Annual Report 1960, Glasshouse Crops Research Institute, pp. 117-134.

Hayes, W.A. (1981). Inter related studies of physical, chemical and biological factors in casing soil and relationship with productivity in commercial culture of *A. bisporus* Lange (Pilat). *Mush. Sci.* 11: 103-129.

Hayes, W.A., Yeo, S.G., Cresswell, P.A. and Jakeman, K.J. (1978). Paper pulp mill byproducts as a casing medium for mushroom culture. *Mush. J.* 62: 38-44.

Shandilya, T.R. (1989). Mushroom compost and casing research in India. *Mush. Sci.* 12: 743-752.

Singh, M., Singh, R.P. and Chaube, H.S. (2000). Impact of physico-chemical properties of casing on yield of *Agaricus bisporus* (Lange) Imbach. In: *Science and Cultivation of Edible Fungi* (Ed. Griensven, Van.), Balkema, Rotterdam, pp. 441-446.

Yeo, S.G. and Hayes, W.A. (1979). A new medium for casing mushroom beds. *Mush. Sci.* 10: 217-229.